INTRODUCTORY BIOLOGY

Edited by
Roksana Khalid

ArclerPress

Introductory Biology

Edited by Roksana Khalid

Edition 2018

ArclerPress

Published by
Arcler Press LLC
708 3rd Avenue, 6th Floor
New York
NY 10017
United States of America
www.arclerpress.com

ISBN: 978-1-68094-539-3 (Hardback)

Printed in United States of America

Library of Congress Control Number: 2017933115

About the Editor

Roksana Khalid

Roksana obtained her MSc from McMaster University in 2016. Her interests span from microbiology to neuroscience and she is currently working on the gut-brain axis as a visiting researcher at the Karolinska Institutet.

Table of Contents

List of Abbreviations ix
Preface xi

1 HISTORY OF BIOLOGY

Introduction 1
Etymology of "Biology" 3
Ancient and Medieval Knowledge 3
Seventeenth and Eighteenth Centuries 7
19th Century: the Emergence of Biological Disciplines 9
Twentieth Century Biological Sciences 14
Twenty-First Century Biological Sciences 22
References 22

2 THEORY OF EVOLUTION

Introduction 27
History of Evolutionary Thought 29
Heredity 32
Variation 34
Mechanisms 37
Outcomes 43
Adaptation 44
Evolutionary History of Life 51
Applications 55
Social and Cultural Responses 56
References 57

3 ECOLOGY

Introduction 88

Levels, Scope, and Scale of Organization 91
Ecological Complexity 101
Relation to Evolution 102
Human Ecology 108
Relation to the Environment 109
References 119

4 PRINCIPLES OF BIOLOGICAL INHERITANCE
Introduction 140
Etymology 140
History 140
Features of Inheritance 144
Molecular Basis for Inheritance 147
Gene Expression 150
Genetic Change 154
References 159

5 GENETIC DIFFERENCES WITHIN AND BETWEEN POPULATIONS
Introduction 168
History 168
Four Processes 171
Applications 177
References 179

6 HUMAN NUTRITION
Introduction 185
Overview 186
Nutrients 187
Malnutrition 192
Global Nutrition Challenges 195
International Food Insecurity and Malnutrition 200
Nutrition Access Disparities 203
Nutrition Policy 204
Nutrition for Special Populations 206
References 210

7 HUMAN DISEASE AND THEIR CLASSIFICATION
Introduction 219
Terminology 220
Classifications 226
Causes 226
Prevention 228
Treatments 228
Epidemiology 228
Society and Culture 230
References 232

8 PLANT ANATOMY
Introduction 235
Structural Divisions 236
History 237
References 239

9 CLASSIFICATION SYSTEM IN BIOLOGY
Introduction 241
Definition and Associated Terms 242
Modern View 243
Historical Development 246
Summary 251
References 251

10 SYNTHETIC BIOLOGY
Introduction 257
History 258
Perspectives 259
Key Enabling Technologies 260
Research Examples 261
Applications 262
References 265

Index 273

List of Abbreviations

BSC	Biological Species Concept
DNA	deoxyribonucleic acid
RD	dietitians are registered
DALY	disability-adjusted life year
ID	intelligent design
LUCA	last universal common ancestor
LD	licensed
LTER	Long Term Ecological Network
NHMD	National hospital morbidity database
OPT	Occupied Palestinian Territory
PCR	polymerase chain reaction
PCR	polymerase chain reaction .
QALY	quality-adjusted life year
RDA	Recommended Dietary Allowances
RSI	repetitive stress injury
DSM	Statistical Manual of Mental Disor ders
SBOL	Synthetic Biology Open Language
TOBA	Taxonomic Outline of Bacteria and Archaea
YPLL	years of potential life lost

Preface

Biology is the study of the diverse forms, processes, and systems of life including microbes, fungi, plants, and animals. Biologists work in all natural environments, and agricultural and clinical venues as well. Biologists understand that life is connected across recent and geologic time, and across the molecular-to-biome spectrum. Biologists employ natural selection and evolution as mechanisms that explain both the diversity and connectedness of life. To understand this great biological complexity, biologists utilize integrative and comparative approaches for the resolution of the general processes, principles, and unifying themes that govern living systems. The field of biology is therefore very interdisciplinary, and biologists rely on knowledge from the physical sciences and mathematics to advance knowledge and create breakthroughs in applied and basic research problems. Aspects of biological science range from the study of molecular mechanisms in cells, to the classification and behaviour of organisms, how species evolve and interaction between ecosystems. Biology often overlaps with other sciences; for example, biochemistry and toxicology with biology, chemistry, and medicine; biophysics with biology and physics; stratigraphy with biology and geography; astrobiology with biology and astronomy. Social sciences such as geography, philosophy, psychology and sociology can also interact with biology, for example, in administration of biological resources, developmental biology, biogeography, evolutionary psychology and ethics. This text covers prospective biology majors and a grasp of the structure of evolutionary theory, the evidence for it, and the scope of its explanatory significance.

Content Coverage

Chapter One aims to present the history of biology traces the study of the living world from ancient to modern times.

Chapters Two and *Three* provide the theory of evolution and ecology. Ecology is focused on the examination of living organisms in the natural environment. It is the scientific analysis and study of interactions among organisms and their environment

Chapter Four covers the study of biological inheritance. This discipline is a major cornerstone of the larger field of biology. Trait inheritance and molecular inheritance mechanisms of genes are still primary principles of genetics in the 21st century, but modern genetics has expanded beyond inheritance to studying the function and behavior of genes.

Chapter Five and *Six* examine phenomena such as adaptation, speciation, population structure and essential nutrients necessary to support human life and health. Malnutrition and its consequences are immense contributors to deaths and disabilities worldwide. Promoting good nutrition helps children grow, promotes human development and advances economic growth and eradication of poverty.

Chapter Seven discusses about human disease and their classification. In humans, disease is often used more broadly to refer to any condition that causes pain, dysfunction, distress, social problems, or death to the person afflicted, or similar problems for those in contact with the person.

Chapter Eight presents an overview of plant anatomy or phytotomy i.e. the study of the internal structure of plants.

Chapter Nine is classification system in biology. In biology, kingdom is the second highest taxonomic rank below domain. Kingdoms are divided into smaller groups called phyla.

Chapter Ten elaborates an overview of synthetic biology as an emerging discipline that uses engineering principles to design and assemble biological component.

CHAPTER 1

History of Biology

Chapter Contents

- **Introduction**

Biology is the study of life on earth. The History of Biology however, focuses on the advent of life on earth, right from the ancient times. Biological discoveries have a remarkable impact on the human society. Traditionally, the history of biology is diversified into two wings – studies on medicine and theories of natural history. Medicines are not results of current biological discoveries.

INTRODUCTION

The history of biology traces the study of the living world from ancient to modern times. Although the concept of biology as a single coherent field arose in the 19th century, the biological sciences emerged from traditions of medicine and natural history reaching back to ayurveda, ancient Egyptian medicine and the works of Aristotle and Galen in the ancient Greco-Roman world. This ancient work was further developed in the Middle Ages by Muslim physicians and scholars such as Avicenna. During the European Renaissance and early modern period, biological thought was revolutionized in Europe by a renewed interest in empiricism and the discovery of many novel organisms. Prominent in this movement were Vesalius and Harvey, who used experimentation and careful observation in physiology, and naturalists such as Linnaeus and Buffon who began to classify the diversity of life and the fossil record, as well as the development and behavior of organisms. Microscopy revealed the previously unknown world of microorganisms, laying the groundwork for cell theory. The growing importance of natural theology, partly a response to the rise of mechanical philosophy, encouraged the growth of natural history (although it entrenched the argument from design).

Over the 18th and 19th centuries, biological sciences such as botany and zoology became increasingly professional scientific disciplines. Lavoisier and other physical scientists began to connect the animate and inanimate worlds through physics and chemistry. Explorer-naturalists such as Alexander von Humboldt investigated the interaction between organisms and their environment, and the ways this relationship depends on geography—laying the foundations for biogeography, ecology and ethology. Naturalists began to reject essentialism and consider the importance of extinction and the mutability of species. Cell theory provided a new perspective on the fundamental basis of life. These developments, as well as the results from embryology and paleontology, were synthesized in Charles Darwin's theory of evolution by natural selection. The end of the 19th century saw the fall of spontaneous generation and the rise of the germ theory of disease, though the mechanism of inheritance remained a mystery.

In the early 20th century, the rediscovery of Mendel's work led to the rapid development of genetics by Thomas Hunt Morgan and his students, and by the 1930s the combination of population genetics and natural selection in the "neo-Darwinian synthesis". New disciplines developed rapidly, especially after Watson and Crick proposed the structure of DNA. Following the establishment of the Central Dogma and the cracking of the genetic code, biology was largely split between organismal biology—the fields that deal with whole organisms and groups of organisms—and the fields related to cellular and molecular biology. By the late 20th century, new fields like genomics and proteomics were reversing this trend, with organismal biologists using molecular techniques, and molecular and cell biologists investigating the interplay between genes and the environment, as well as the genetics of natural populations of organisms.

The frontispiece to Erasmus Darwin's evolution-themed poem The Temple of Nature shows a goddess pulling back the veil from nature (in the person of Artemis). Allegory and metaphor have often played an important role in the history of biology.

Etymology of "Biology"

The word biology is formed by combining the Greek βίος (bios), meaning "life", and the suffix '-logy', meaning "science of", "knowledge of", "study of", based on the Greek verb λέγειν, 'legein' "to select", "to gather" (cf. the noun λόγος, 'logos' "word"). The term biology in its modern sense appears to have been introduced independently by Thomas Beddoes (in 1799), Karl Friedrich Burdach (in 1800), Gottfried Reinhold Treviranus (Biologie oder Philosophie der lebenden Natur, 1802) and Jean-Baptiste Lamarck (Hydrogéologie, 1802). The word itself appears in the title of Volume 3 of Michael Christoph Hanow's Philosophiae naturalis sive physicae dogmaticae: Geologia, biologia, phytologia generalis et dendrologia, published in 1766.

Before biology, there were several terms used for the study of animals and plants. Natural history referred to the descriptive aspects of biology, though it also included mineralogy and other non-biological fields; from the Middle Ages through the Renaissance, the unifying framework of natural history was the scala naturae or Great Chain of Being. Natural philosophy and natural theology encompassed the conceptual and metaphysical basis of plant and animal life, dealing with problems of why organisms exist and behave the way they do, though these subjects also included what is now geology, physics, chemistry, and astronomy. Physiology and (botanical) pharmacology were the province of medicine. Botany, zoology, and (in the case of fossils) geology replaced natural history and natural philosophy in the 18th and 19th centuries before biology was widely adopted. To this day, "botany" and "zoology" are widely used, although they have been joined by other sub-disciplines of biology, such as mycology and molecular biology.

Ancient and Medieval Knowledge

Early Cultures

The earliest humans must have had and passed on knowledge about plants and animals to increase their chances of survival. This may have included knowledge of human and animal anatomy and aspects of animal behavior (such as migration patterns). However, the first major turning point in biological knowledge came with the Neolithic Revolution about 10,000 years ago. Humans first domesticated plants for farming, then livestock animals to accompany the resulting sedentary societies.

The ancient cultures of Mesopotamia, Egypt, the Indian subcontinent, and China, among others, produced renowned surgeons and students of the natural sciences such as Susruta and Zhang Zhongjing, reflecting independent sophisticated systems of natural philosophy. However, the roots of modern biology are usually traced back to the secular tradition of ancient Greek philosophy.

Ancient Chinese Traditions

In ancient China, biological topics can be found dispersed across several different disciplines, including the work of herbologists, physicians, alchemists, and philosophers.

The Taoist tradition of Chinese alchemy, for example, can be considered part of the life sciences due to its emphasis on health (with the ultimate goal being the elixir of life). The system of classical Chinese medicine usually revolved around the theory of yin and yang, and the five phases. Taoist philosophers, such as Zhuangzi in the 4th century BCE, also expressed ideas related to evolution, such as denying the fixity of biological species and speculating that species had developed differing attributes in response to differing environments.

Ancient Indian Traditions

One of the oldest organised systems of medicine is known from the Indian subcontinent in the form of Ayurveda which originated around 1500 BCE from Atharvaveda (one of the four most ancient books of Indian knowledge, wisdom and culture).

The ancient Indian Ayurveda tradition independently developed the concept of three humours, resembling that of the four humours of ancient Greek medicine, though the Ayurvedic system included further complications, such as the body being composed of five elements and seven basic tissues. Ayurvedic writers also classified living things into four categories based on the method of birth (from the womb, eggs, heat & moisture, and seeds) and explained the conception of a fetus in detail. They also made considerable advances in the field of surgery, often without the use of human dissection or animal vivisection. One of the earliest Ayurvedic treatises was the Sushruta Samhita, attributed to Sushruta in the 6th century BCE. It was also an early materia medica, describing 700 medicinal plants, 64 preparations from mineral sources, and 57 preparations based on animal sources.

Ancient Mesopotamian Traditions

Ancient Mesopotamian medicine may be represented by Esagil-kin-apli, a prominent scholar of the 11th century BCE, who made a compilation of medical prescriptions and procedures, which he presented as exorcisms.

Ancient Egyptian Traditions

Over a dozen medical papyri have been preserved, most notably the Edwin Smith Papyrus (the oldest extant surgical handbook) and the Ebers Papyrus (a handbook of preparing and using materia medica for various diseases), both from the 16th century BCE.

Ancient Egypt is also known for developing embalming, which was used for mummification, in order to preserve human remains and forestall decomposition.

Ancient Greek and Roman Traditions

The pre-Socratic philosophers asked many questions about life but produced little systematic knowledge of specifically biological interest—though the attempts of the atomists to explain life in purely physical terms would recur periodically through the history of biology. However, the medical theories of Hippocrates and his followers, especially humorism, had a lasting impact.

The philosopher Aristotle was the most influential scholar of the living world from classical antiquity. Though his early work in natural philosophy was speculative, Aristotle's later biological writings were more empirical, focusing on biological causation and the diversity of life. He made countless observations of nature, especially the habits and attributes of plants and animals in the world around him, which he devoted considerable attention to categorizing. In all, Aristotle classified 540 animal species, and dissected at least 50. He believed that intellectual purposes, formal causes, guided all natural processes.

Frontispiece to a 1644 version of the expanded and illustrated edition of Historia Plantarum, originally written by Theophrastus around 300 BC

Aristotle, and nearly all Western scholars after him until the 18th century, believed that creatures were arranged in a graded scale of perfection rising from plants on up to humans: the scala naturae or Great Chain of Being. Aristotle's successor at the Lyceum, Theophrastus, wrote a series of books on botany—the History of Plants—which survived as the most important contribution of antiquity to botany, even into the Middle Ages. Many of Theophrastus' names survive into modern times, such as carpos for fruit, and pericarpion for seed vessel. Dioscorides wrote a pioneering and encyclopaedic pharmacopoeia, De Materia Medica, incorporating descriptions of some 600 plants and their uses in medicine. Pliny the Elder, in his Natural History, assembled a similarly encyclopaedic account of things in nature, including accounts of many plants and animals.

A few scholars in the Hellenistic period under the Ptolemies—particularly Herophilus of Chalcedon and Erasistratus of Chios—amended Aristotle's physiological work, even performing dissections and vivisections. Claudius Galen became the most important authority on medicine and anatomy. Though a few ancient atomists such as Lucretius challenged the teleological Aristotelian viewpoint that all aspects of life are the result of design or purpose, teleology (and after the rise of Christianity, natural theology) would remain central to biological thought essentially until the 18th and 19th centuries. Ernst W. Mayr argued that "Nothing of any real consequence happened in biology after Lucretius and Galen until the Renaissance." The ideas of the Greek traditions of natural history and medicine survived, but they were generally taken unquestioningly in medieval Europe.

Medieval and Islamic Knowledge

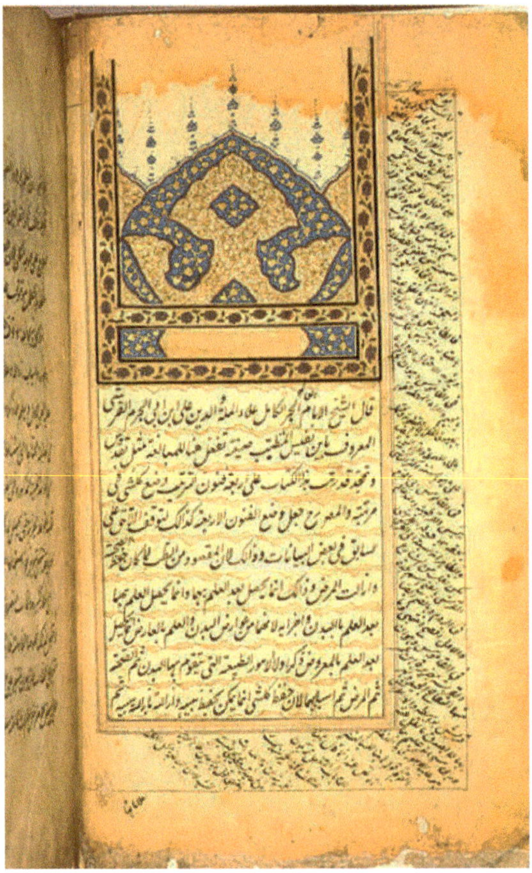

A biomedical work by Ibn al-Nafis, an early adherent of experimental dissection who discovered the pulmonary and coronary circulation

The decline of the Roman Empire led to the disappearance or destruction of much knowledge, though physicians still incorporated many aspects of the Greek tradition into training and practice. In Byzantium and the Islamic world, many of the Greek works were translated into Arabic and many of the works of Aristotle were preserved.

De arte venandi, by Frederick II, Holy Roman Emperor, was an influential medieval natural history text that explored bird morphology.

During the High Middle Ages, a few European scholars such as Hildegard of Bingen, Albertus Magnus and Frederick II expanded the natural history canon. The rise of European universities, though important for the development of physics and philosophy, had little impact on biological scholarship.

Seventeenth and Eighteenth Centuries

Systematizing, naming and classifying dominated natural history throughout much of the 17th and 18th centuries. Carl Linnaeus published a basic taxonomy for the natural world in 1735 (variations of which have been in use ever since), and in the 1750s introduced scientific names for all his species. While Linnaeus conceived of species as unchanging parts of a designed hierarchy, the other great naturalist of the 18th century, Georges-Louis Leclerc, Comte de Buffon, treated species as artificial categories and living forms as malleable—even suggesting the possibility of common descent. Though he was opposed to evolution, Buffon is a key figure in the history of evolutionary thought; his work would influence the evolutionary theories of both Lamarck and Darwin.

The discovery and description of new species and the collection of specimens became a passion of scientific gentlemen and a lucrative enterprise for entrepreneurs; many naturalists traveled the globe in search of scientific knowledge and adventure.

Cabinets of curiosities, such as that of Ole Worm, were centers of biological knowledge in the early modern period, bringing organisms from across the world together in one place. Before the Age of Exploration, naturalists had little idea of the sheer scale of biological diversity.

Extending the work of Vesalius into experiments on still living bodies (of both humans and animals), William Harvey and other natural philosophers investigated the roles of blood, veins and arteries. Harvey's De motu cordis in 1628 was the beginning of the end for Galenic theory, and alongside Santorio Santorio's studies of metabolism, it served as an influential model of quantitative approaches to physiology.

In the early 17th century, the micro-world of biology was just beginning to open up. A few lensmakers and natural philosophers had been creating crude microscopes since the late 16th century, and Robert Hooke published the seminal Micrographia based on observations with his own compound microscope in 1665. But it was not until Antony van Leeuwenhoek's dramatic improvements in lensmaking beginning in the 1670s—ultimately producing up to 200-fold magnification with a single lens—that scholars discovered spermatozoa, bacteria, infusoria and the sheer strangeness and diversity of microscopic life. Similar investigations by Jan Swammerdam led to new interest in entomology and built the basic techniques of microscopic dissection and staining.

As the microscopic world was expanding, the macroscopic world was shrinking. Botanists such as John Ray worked to incorporate the flood of newly discovered organisms shipped from across the globe into a coherent taxonomy, and a coherent theology (natural theology). Debate over another flood, the Noachian, catalyzed the development of paleontology; in 1669 Nicholas Steno published an essay on how the remains of living organisms could be trapped in layers of sediment and mineralized to produce fossils. Although Steno's ideas about fossilization were well known and much debated among natural philosophers, an organic origin for all fossils would not be accepted by all naturalists until the end of the 18th century due to philosophical and theological debate about issues such as the age of the earth and extinction.

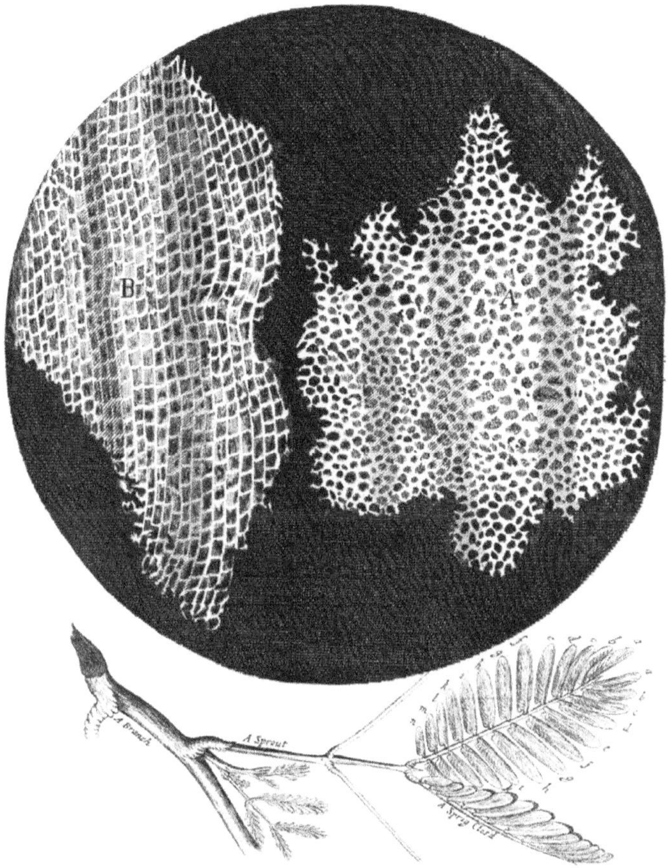

In Micrographia, Robert Hooke had applied the word cell to biological structures such as this piece of cork, but it was not until the 19th century that scientists considered cells the universal basis of life.

19th Century: the Emergence of Biological Disciplines

Up through the 19th century, the scope of biology was largely divided between medicine, which investigated questions of form and function (i.e., physiology), and natural history, which was concerned with the diversity of life and interactions among different forms of life and between life and non-life. By 1900, much of these domains overlapped, while natural history (and its counterpart natural philosophy) had largely given way to more specialized scientific disciplines—cytology, bacteriology, morphology, embryology, geography, and geology.

In the course of his travels, Alexander von Humboldt mapped the distribution of plants across landscapes and recorded a variety of physical conditions such as pressure and temperature.

Natural history and Natural Philosophy

Widespread travel by naturalists in the early-to-mid-19th century resulted in a wealth of new information about the diversity and distribution of living organisms. Of particular importance was the work of Alexander von Humboldt, which analyzed the relationship between organisms and their environment (i.e., the domain of natural history) using the quantitative approaches of natural philosophy (i.e., physics and chemistry). Humboldt's work laid the foundations of biogeography and inspired several generations of scientists.

Geology and Paleontology

The emerging discipline of geology also brought natural history and natural philosophy closer together; the establishment of the stratigraphic column linked the spatial distribution of organisms to their temporal distribution, a key precursor to concepts of evolution. Georges Cuvier and others made great strides in comparative anatomy and paleontology in the late 1790s and early 19th century. In a series of lectures and papers that made detailed comparisons between living mammals and fossil remains Cuvier was able to establish that the fossils were remains of species that had become extinct—rather than being remains of species still alive elsewhere in the world, as had been widely believed. Fossils discovered and described by Gideon Mantell, William Buckland, Mary Anning, and Richard Owen among others helped establish that there had been an 'age of reptiles' that had preceded even the prehistoric mammals. These discoveries captured the public imagination and focused attention on the history of life on earth. Most of these geologists held to catastrophism, but Charles Lyell's influential *Principles of Geology* (1830) popularised Hutton's uniformitarianism, a theory that explained the geological past and present on equal terms.

Evolution and Biogeography

The most significant evolutionary theory before Darwin's was that of Jean-Baptiste Lamarck; based on the inheritance of acquired characteristics (an inheritance mechanism that was widely accepted until the 20th century), it described a chain of development stretching from the lowliest microbe to humans. The British naturalist Charles Darwin, combining the biogeographical approach of Humboldt, the uniformitarian geology of Lyell, Thomas Malthus's writings on population growth, and his own morphological expertise, created a more successful evolutionary theory based on natural selection; similar evidence led Alfred Russel Wallace to independently reach the same conclusions.

The 1859 publication of Darwin's theory in On the Origin of Species by Means of Natural Selection, or the Preservation of Favoured Races in the Struggle for Life is often considered the central event in the history of modern biology. Darwin's established credibility as a naturalist, the sober tone of the work, and most of all the sheer strength and volume of evidence presented, allowed Origin to succeed where previous evolutionary works such as the anonymous Vestiges of Creation had failed. Most scientists were convinced of evolution and common descent by the end of the 19th century. However, natural selection would not be accepted as the primary mechanism of evolution until well into the 20th century, as most contemporary theories of heredity seemed incompatible with the inheritance of random variation.

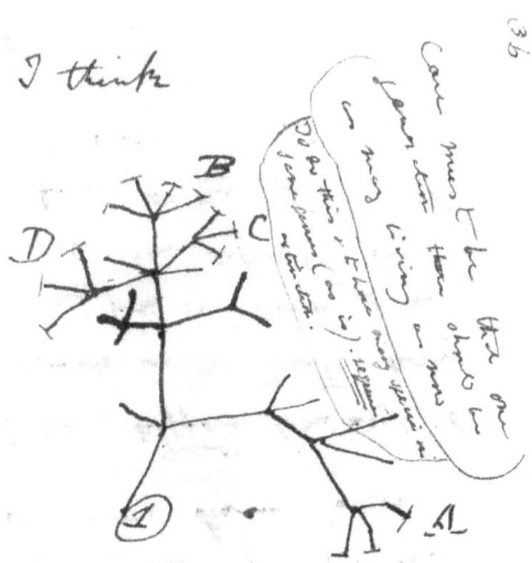

Charles Darwin's first sketch of an evolutionary tree from his *First Notebook on Transmutation of Species* (1837)

Wallace, following on earlier work by de Candolle, Humboldt and Darwin, made major contributions to zoogeography. Because of his interest in the transmutation hypothesis, he paid particular attention to the geographical distribution of closely allied species during his field work first in South America and then in the Malay archipelago. While in the archipelago he identified the Wallace line, which runs through the Spice Islands dividing the fauna of the archipelago between an Asian zone and a New Guinea/Australian zone.

His key question, as to why the fauna of islands with such similar climates should be so different, could only be answered by considering their origin. In 1876 he wrote *The Geographical Distribution of Animals*, which was the standard reference work for over half a century, and a sequel, *Island Life*, in 1880 that focused on island biogeography. He extended the six-zone system developed by Philip Sclater for describing the geographical distribution of birds to animals of all kinds. His method of tabulating data on animal groups in geographic zones highlighted the discontinuities; and his appreciation of evolution allowed him to propose rational explanations, which had not been done before.

The scientific study of heredity grew rapidly in the wake of Darwin's *Origin of Species* with the work of Francis Galton and the biometricians. The origin of genetics is usually traced to the 1866 work of the monkGregor Mendel, who would later be credited with the laws of inheritance. However, his work was not recognized as significant until 35 years afterward. In the meantime, a variety of theories of inheritance (based on pangenesis, orthogenesis, or other mechanisms) were debated and investigated vigorously. Embryology and ecology also became central biological fields, especially as linked to evolution and popularized in the work of Ernst Haeckel. Most of the 19th century work on heredity, however, was not in the realm of natural history, but that of experimental physiology.

Physiology

Over the course of the 19th century, the scope of physiology expanded greatly, from a primarily medically oriented field to a wide-ranging investigation of the physical and chemical processes of life—including plants, animals, and even microorganisms in addition to man. *Living things as machines* became a dominant metaphor in biological (and social) thinking.

Innovative laboratory glassware and experimental methods developed by Louis Pasteur and other biologists contributed to the young field of bacteriology in the late 19th century.

Cell theory, Embryology and Germ Theory

Advances in microscopy also had a profound impact on biological thinking. In the early 19th century, a number of biologists pointed to the central importance of the cell. In 1838 and 1839, Schleiden and Schwann began promoting the ideas that (1) the basic unit of organisms is the cell and (2) that individual cells have all the characteristics of life, though they opposed the idea that (3) all cells come from the division of other cells. Thanks to the work of Robert Remak and Rudolf Virchow, however, by the 1860s most biologists accepted all three tenets of what came to be known as cell theory.

Cell theory led biologists to re-envision individual organisms as interdependent assemblages of individual cells. Scientists in the rising field of cytology, armed with increasingly powerful microscopes and new staining methods, soon found that even single cells were far more complex than the homogeneous fluid-filled chambers described by earlier microscopists. Robert Brown had described the nucleus in 1831, and by the end of the 19th century cytologists identified many of the key cell components: chromosomes, centrosomes mitochondria, chloroplasts, and other structures made visible through staining. Between 1874 and 1884 Walther Flemming described the discrete stages of mitosis, showing that they were not artifacts of staining but occurred in living cells, and moreover, that chromosomes doubled in number just before the cell divided and a daughter cell was produced. Much of the research on cell reproduction came together in August Weismann's theory of heredity: he identified the nucleus (in particular chromosomes) as the hereditary material, proposed the distinction between somatic cells and germ cells (arguing that chromosome number must be halved for germ cells, a precursor to the concept of meiosis), and adopted Hugo de Vries's theory of pangenes. Weismannism was extremely influential, especially in the new field of experimental embryology.

By the mid-1850s the miasma theory of disease was largely superseded by the germ theory of disease, creating extensive interest in microorganisms and their interactions with other forms of life. By the 1880s, bacteriology was becoming a coherent discipline, especially through the work of Robert Koch, who introduced methods for growing pure cultures on agar gels containing specific nutrients in Petri dishes. The long-held idea that living organisms could easily originate from nonliving matter (spontaneous generation) was attacked in a series of experiments carried out by Louis Pasteur, while debates over vitalism vs. mechanism (a perennial issue since the time of Aristotle and the Greek atomists) continued apace.

Rise of Organic Chemistry and Experimental Physiology

In chemistry, one central issue was the distinction between organic and inorganic substances, especially in the context of organic transformations such as fermentation and putrefaction. Since Aristotle these had been considered essentially biological (*vital*) processes. However, Friedrich Wöhler, Justus Liebig and other pioneers of the rising field of organic chemistry—building on the work of Lavoisier—showed that the organic world could often be analyzed by physical and chemical methods. In 1828 Wöhler showed that the organic substance urea could be created by chemical means that do not involve life, providing a powerful challenge to vitalism. Cell extracts ("ferments") that could effect chemical transformations were discovered, beginning with diastase in 1833. By the end of the 19th century the concept of enzymes was well established, though equations of chemical kinetics would not be applied to enzymatic reactions until the early 20th century.

Physiologists such as Claude Bernard explored (through vivisection and other experimental methods) the chemical and physical functions of living bodies to an unprecedented degree, laying the groundwork for endocrinology (a field that developed quickly after the discovery of the first hormone, secretin, in 1902), biomechanics, and the study of nutrition and digestion. The importance and diversity of experimental physiology methods, within both medicine and biology, grew dramatically over the second half of the 19th century. The control and manipulation of life processes became a central concern, and experiment was placed at the center of biological education.

Twentieth Century Biological Sciences

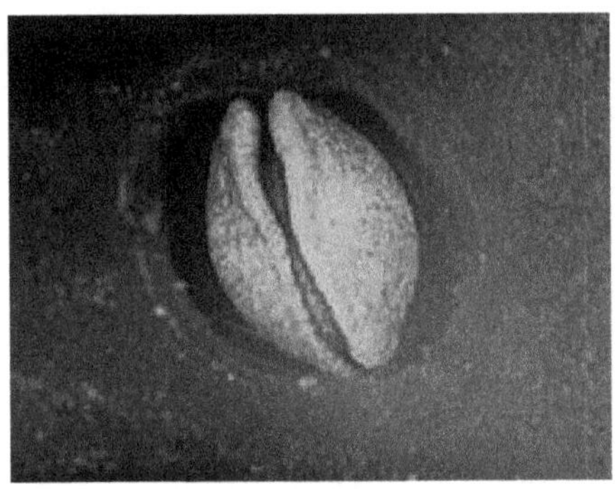

Embryonic development of a salamander, filmed in the 1920s

At the beginning of the 20th century, biological research was largely a professional endeavour. Most work was still done in the natural history mode, which emphasized morphological and phylogenetic analysis over experiment-based causal explanations. However, anti-vitalist experimental physiologists and embryologists, especially in Europe, were increasingly influential. The tremendous success of experimental approaches to development, heredity, and metabolism in the 1900s and 1910s demonstrated the power of experimentation in biology. In the following decades, experimental work replaced natural history as the dominant mode of research.

Ecology and Environmental Science

In the early 20th century, naturalists were faced with increasing pressure to add rigor and preferably experimentation to their methods, as the newly prominent laboratory-based biological disciplines had done. Ecology had emerged as a combination of biogeography with the biogeochemical cycle concept pioneered by chemists; field biologists developed quantitative methods such as the quadrat and adapted laboratory instruments and cameras for the field to further set their work apart from traditional natural history. Zoologists and botanists did what they could to mitigate the unpredictability of the living world, performing laboratory experiments and studying semi-controlled natural environments such as gardens; new institutions like the Carnegie Station for Experimental Evolution and the Marine Biological Laboratory provided more controlled environments for studying organisms through their entire life cycles.

The ecological succession concept, pioneered in the 1900s and 1910s by Henry Chandler Cowles and Frederic Clements, was important in early plant ecology. Alfred Lotka's predator-prey equations, G. Evelyn Hutchinson's studies of the biogeography and biogeochemical structure of lakes and rivers (limnology) and Charles Elton's studies of animal food chains were pioneers among the succession of quantitative methods that colonized the developing ecological specialties. Ecology became an independent discipline in the 1940s and 1950s after Eugene P. Odum synthesized many of the concepts of ecosystem ecology, placing relationships between groups of organisms (especially material and energy relationships) at the center of the field.

In the 1960s, as evolutionary theorists explored the possibility of multiple units of selection, ecologists turned to evolutionary approaches. In population ecology, debate over group selection was brief but vigorous; by 1970, most biologists agreed that natural selection was rarely effective above the level of individual organisms. The evolution of ecosystems, however, became a lasting research focus. Ecology expanded rapidly with the rise of the environmental movement; the International Biological Program attempted to apply the methods of big science (which had been so successful in the physical sciences) to ecosystem ecology and pressing environmental issues, while smaller-scale independent efforts such as island biogeography and the Hubbard Brook Experimental Forest helped redefine the scope of an increasingly diverse discipline.

Classical Genetics, the Modern Synthesis, and Evolutionary Theory

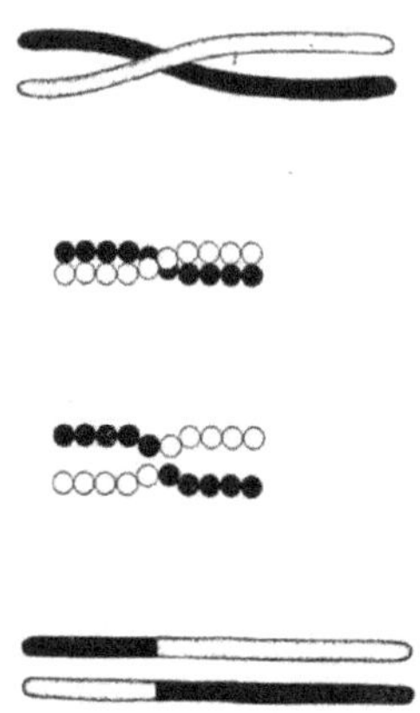

Thomas Hunt Morgan's illustration of crossing over, part of the Mendelian-chromosome theory of heredity

1900 marked the so-called *rediscovery of Mendel*: Hugo de Vries, Carl Correns, and Erich von Tschermak independently arrived at Mendel's laws (which were not actually present in Mendel's work). Soon after, cytologists (cell biologists) proposed that chromosomes were the hereditary material. Between 1910 and 1915, Thomas Hunt Morgan and the "Drosophilists" in his fly lab forged these two ideas—both controversial—into the "Mendelian-chromosome theory" of heredity. They quantified the phenomenon of genetic linkage and postulated that genes reside on chromosomes like beads on string; they hypothesized crossing over to explain linkage and constructed genetic maps of the fruit fly *Drosophila melanogaster*, which became a widely used model organism.

Hugo de Vries tried to link the new genetics with evolution; building on his work with heredity and hybridization, he proposed a theory of mutationism, which was widely accepted in the early 20th century. Lamarckism also had many adherents. Darwinism was seen as incompatible with the continuously variable traits studied by biometricians, which seemed only partially heritable. In the 1920s and 1930s—following the acceptance of the Mendelian-chromosome theory— the emergence of the discipline of population genetics, with the work of R.A. Fisher, J.B.S. Haldane and Sewall Wright, unified the idea of evolution by natural selection with Mendelian genetics, producing the modern synthesis. The inheritance of acquired characters was rejected, while mutationism gave way as genetic theories matured.

In the second half of the century the ideas of population genetics began to be applied in the new discipline of the genetics of behavior, sociobiology, and, especially in humans, evolutionary psychology. In the 1960s W.D. Hamilton and others developed game theory approaches to explain altruism from an evolutionary perspective through kin selection. The possible origin of higher organisms through endosymbiosis, and contrasting approaches to molecular evolution in the gene-centered view (which held selection as the predominant cause of evolution) and the neutral theory (which made genetic drift a key factor) spawned perennial debates over the proper balance of adaptationism and contingency in evolutionary theory.

In the 1970s Stephen Jay Gould and Niles Eldredge proposed the theory of punctuated equilibrium which holds that stasis is the most prominent feature of the fossil record, and that most evolutionary changes occur rapidly over relatively short periods of time. In 1980 Luis Alvarez and Walter Alvarez proposed the hypothesis that an impact event was responsible for the Cretaceous–Paleogene extinction event. Also in the early 1980s, statistical analysis of the fossil record of marine organisms published by Jack Sepkoski and David M. Raup led to a better appreciation of the importance of mass extinction events to the history of life on earth.

Biochemistry, Microbiology, and Molecular Biology

By the end of the 19th century all of the major pathways of drug metabolism had been discovered, along with the outlines of protein and fatty acid metabolism and urea synthesis. In the early decades of the 20th century, the minor components of foods in human nutrition, the vitamins, began to be isolated and synthesized. Improved laboratory techniques such as chromatography and electrophoresis led to rapid advances in physiological chemistry, which—as *biochemistry*—began to achieve independence from its medical origins. In the 1920s and 1930s, biochemists—led by Hans Krebs and Carl and Gerty Cori—began to work out many of the central metabolic pathways of life: the citric acid cycle, glycogenesis and glycolysis, and the synthesis of steroids and porphyrins. Between the 1930s and 1950s, Fritz Lipmann and others established the role of ATP as the universal carrier of energy in the cell, and mitochondria as the powerhouse of the cell. Such traditionally biochemical work continued to be very actively pursued throughout the 20th century and into the 21st.

Origins of Molecular Biology

Following the rise of classical genetics, many biologists—including a new wave of physical scientists in biology—pursued the question of the gene and its physical nature. Warren Weaver—head of the science division of the Rockefeller Foundation—issued grants to promote research that applied the methods of physics and chemistry to basic biological problems, coining the term *molecular biology* for this approach in 1938; many of the significant biological breakthroughs of the 1930s and 1940s were funded by the Rockefeller Foundation.

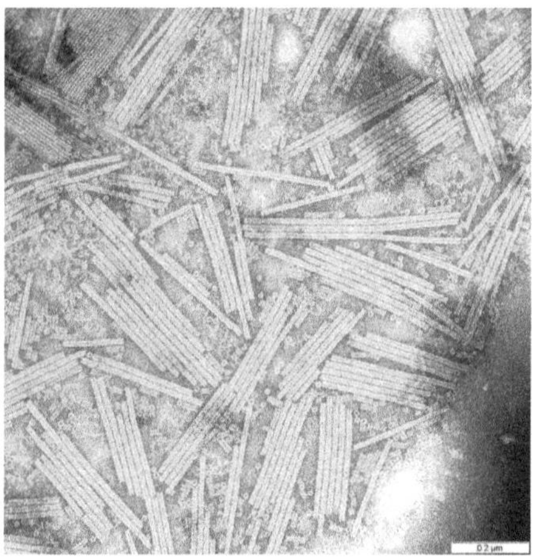

Wendell Stanley's crystallization of tobacco mosaic virus as a pure nucleoprotein in 1935 convinced many scientists that heredity might be explained purely through physics and chemistry.

Like biochemistry, the overlapping disciplines of bacteriology and virology (later combined as *microbiology*), situated between science and medicine, developed rapidly in the early 20th century. Félix d'Herelle's isolation of bacteriophage during World War I initiated a long line of research focused on phage viruses and the bacteria they infect.

The development of standard, genetically uniform organisms that could produce repeatable experimental results was essential for the development of molecular genetics. After early work with *Drosophila* and maize, the adoption of simpler model systems like the bread mold *Neurospora crassa* made it possible to connect genetics to biochemistry, most importantly with Beadle and Tatum's one gene-one enzyme hypothesis in 1941. Genetics experiments on even simpler systems like tobacco mosaic virus and bacteriophage, aided by the new technologies of electron microscopy and ultracentrifugation, forced scientists to re-evaluate the literal meaning of *life*; virus heredity and reproducing nucleoprotein cell structures outside the nucleus ("plasmagenes") complicated the accepted Mendelian-chromosome theory.

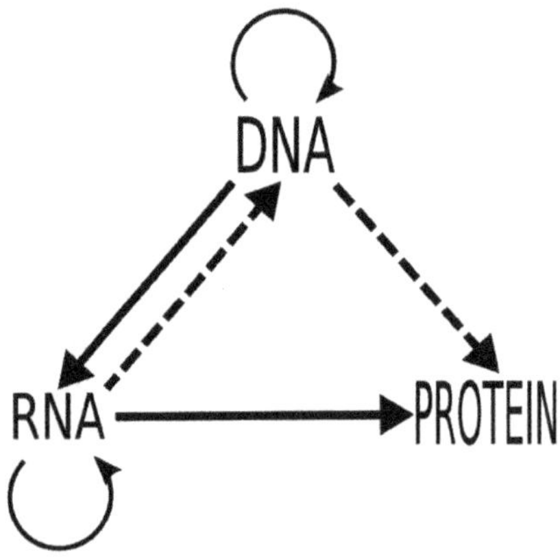

The "central dogma of molecular biology" (originally a "dogma" only in jest) was proposed by Francis Crick in 1958. This is Crick's reconstruction of how he conceived of the central dogma at the time. The solid lines represent (as it seemed in 1958) known modes of information transfer, and the dashed lines represent postulated ones.

Oswald Avery showed in 1943 that DNA was likely the genetic material of the chromosome, not its protein; the issue was settled decisively with the 1952 Hershey–Chase experiment—one of many contributions from the so-called phage group centered around physicist-turned-biologist Max Delbrück. In 1953 James Watson and Francis Crick, building on the work of Maurice Wilkins and Rosalind Franklin, suggested that the structure of DNA was a double helix. In their famous paper "Molecular structure of Nucleic Acids", Watson and Crick noted coyly, "It has not escaped our notice that the specific pairing we have postulated immediately suggests a possible copying mechanism for the genetic material." After the 1958 Meselson–Stahl experiment confirmed the semiconservative replication of DNA, it was clear to most biologists that nucleic acid sequence must somehow determine amino acid sequence in proteins; physicist George Gamow proposed that a fixed genetic code connected proteins and DNA. Between 1953 and 1961, there were few known biological sequences—either DNA or protein—but an abundance of proposed code systems, a situation made even more complicated by expanding knowledge of the intermediate role of RNA. To actually decipher the code, it took an extensive series of experiments in biochemistry and bacterial genetics, between 1961 and 1966—most importantly the work of Nirenberg and Khorana.

Expansion of Molecular Biology

In addition to the Division of Biology at Caltech, the Laboratory of Molecular Biology (and its precursors) at Cambridge, and a handful of other institutions, the Pasteur Institute became a major center for molecular biology research in the late 1950s. Scientists at Cambridge, led by Max Perutz and John Kendrew, focused on the rapidly developing field of structural biology, combining X-ray crystallography with Molecular modelling and the new computational possibilities of digital computing (benefiting both directly and indirectly from the military funding of science). A number of biochemists led by Frederick Sanger later joined the Cambridge lab, bringing together the study of macromolecular structure and function. At the Pasteur Institute, François Jacob and Jacques Monod followed the 1959 PaJaMo experiment with a series of publications regarding the *lac* operon that established the concept of gene regulation and identified what came to be known as messenger RNA. By the mid-1960s, the intellectual core of molecular biology—a model for the molecular basis of metabolism and reproduction— was largely complete.

The late 1950s to the early 1970s was a period of intense research and institutional expansion for molecular biology, which had only recently become a somewhat coherent discipline. In what organismic biologist E. O. Wilson called «The Molecular Wars», the methods and practitioners of molecular biology spread rapidly, often coming to dominate departments and even entire disciplines. Molecularization was particularly important in genetics, immunology, embryology, and neurobiology, while the idea that life is controlled by a «genetic program»—a metaphor Jacob and Monod introduced from the emerging fields of cybernetics and computer science—became an influential perspective throughout biology.Immunology in particular became linked with molecular biology, with innovation flowing both ways: the clonal selection theory developed by Niels Jerne and Frank Macfarlane Burnet in the mid-1950s helped shed light on the general mechanisms of protein synthesis.

Resistance to the growing influence of molecular biology was especially evident in evolutionary biology. Protein sequencing had great potential for the quantitative study of evolution (through the molecular clock hypothesis), but leading evolutionary biologists questioned the relevance of molecular biology for answering the big questions of evolutionary causation. Departments and disciplines fractured as organismic biologists asserted their importance and independence: Theodosius Dobzhansky made the famous statement that «nothing in biology makes sense except in the light of evolution» as a response to the molecular challenge. The issue became even more critical after 1968; Motoo Kimura›s neutral theory of molecular evolution suggested that natural selection was not the ubiquitous cause of evolution, at least at the molecular level, and that molecular evolution might be a fundamentally different process from morphological evolution. (Resolving this «molecular/morphological paradox» has been a central focus of molecular evolution research since the 1960s.)

Biotechnology, Genetic Engineering, and Genomics

Biotechnology in the general sense has been an important part of biology since the late 19th century. With the industrialization of brewing and agriculture, chemists and biologists became aware of the great potential of human-controlled biological processes. In particular, fermentation proved a great boon to chemical industries. By the early 1970s, a wide range of biotechnologies were being developed, from drugs like penicillin and steroids to foods like *Chlorella* and single-cell protein to gasohol—as well as a wide range of hybrid high-yield crops and agricultural technologies, the basis for the Green Revolution.

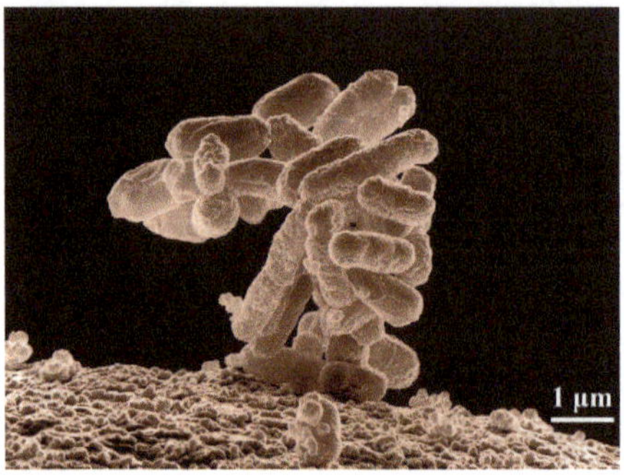

Carefully engineered strains of the bacterium *Escherichia coli* are crucial tools in biotechnology as well as many other biological fields.

Recombinant DNA

Biotechnology in the modern sense of genetic engineering began in the 1970s, with the invention of recombinant DNA techniques. Restriction enzymes were discovered and characterized in the late 1960s, following on the heels of the isolation, then duplication, then synthesis of viral genes. Beginning with the lab of Paul Berg in 1972 (aided by *EcoRI* from Herbert Boyer's lab, building on work with ligase by Arthur Kornberg's lab), molecular biologists put these pieces together to produce the first transgenic organisms. Soon after, others began using plasmid vectors and adding genes for antibiotic resistance, greatly increasing the reach of the recombinant techniques.

Wary of the potential dangers (particularly the possibility of a prolific bacteria with a viral cancer-causing gene), the scientific community as well as a wide range of scientific outsiders reacted to these developments with both enthusiasm and fearful restraint. Prominent molecular biologists led by Berg suggested a temporary moratorium on recombinant DNA research until the dangers could be assessed and policies could be created. This moratorium was largely respected, until the participants in the 1975 Asilomar Conference on Recombinant DNA created policy recommendations and concluded that the technology could be used safely.

Following Asilomar, new genetic engineering techniques and applications developed rapidly. DNA sequencing methods improved greatly (pioneered by Frederick Sanger and Walter Gilbert), as did oligonucleotide synthesis and transfection techniques. Researchers learned to control the expression of transgenes, and were soon racing—in both academic and industrial contexts—to create organisms capable of expressing human genes for the production of human hormones. However, this was a more daunting task than molecular biologists had expected; developments between 1977 and 1980 showed that, due to the phenomena of split genes and splicing, higher organisms had a much more complex system of gene expression than the bacteria models of earlier studies. The first such race, for synthesizing human insulin, was won by Genentech. This marked the beginning of the biotech boom (and with it, the era of gene patents), with an unprecedented level of overlap between biology, industry, and law.

Molecular Systematics and Genomics

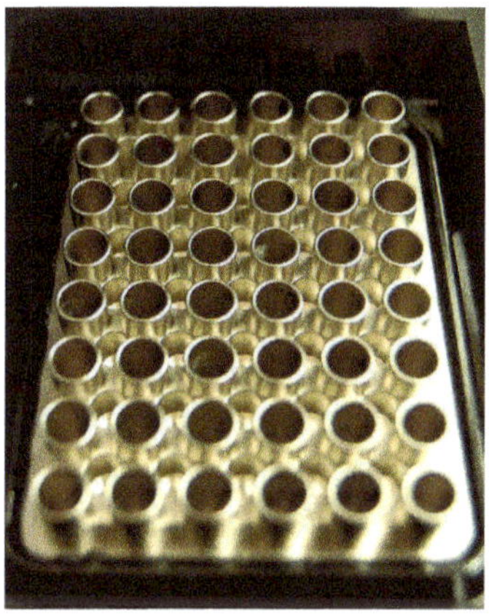

Inside of a 48-well thermal cycler, a device used to perform polymerase chain reaction on many samples at once

By the 1980s, protein sequencing had already transformed methods of scientific classification of organisms (especially cladistics) but biologists soon began to use RNA and DNA sequences as characters; this expanded the significance of molecular evolution within evolutionary biology, as the results of molecular systematics could be compared with traditional evolutionary trees based on morphology. Following the pioneering ideas of Lynn Margulis on endosymbiotic theory, which holds that some of the organelles of eukaryotic cells originated from free living prokaryotic organisms through symbiotic relationships, even the overall division of the tree of life was revised. Into the 1990s, the five domains (Plants, Animals, Fungi, Protists, and Monerans) became three (the Archaea, the Bacteria, and the Eukarya) based on Carl Woese's pioneering molecular systematics work with 16S rRNA sequencing.

The development and popularization of the polymerase chain reaction (PCR) in mid-1980s (by Kary Mullis and others at Cetus Corp.) marked another watershed in the history of modern biotechnology, greatly increasing the ease and speed of genetic analysis. Coupled with the use of expressed sequence tags, PCR led to the discovery of many more genes than could be found through traditional biochemical or genetic methods and opened the possibility of sequencing entire genomes.

The unity of much of the morphogenesis of organisms from fertilized egg to adult began to be unraveled after the discovery of the homeobox genes, first in fruit flies, then in other insects and animals, including humans. These developments led to advances in the field of evolutionary developmental biology towards understanding how the various body plans of the animal phyla have evolved and how they are related to one another.

The Human Genome Project—the largest, most costly single biological study ever undertaken—began in 1988 under the leadership of James D. Watson, after preliminary work with genetically simpler model organisms such as *E. coli*, *S. cerevisiae* and *C. elegans*. Shotgun sequencing and gene discovery methods pioneered by Craig Venter—and fueled by the financial promise of gene patents with Celera Genomics— led to a public–private sequencing competition that ended in compromise with the first draft of the human DNA sequence announced in 2000.

Twenty-First Century Biological Sciences

At the beginning of the 21st century, biological sciences converged with previously differentiated new and classic disciplines like Physics into research fields like Biophysics. Advances were made in analytical chemistry and physics instrumentation including improved sensors, optics, tracers, instrumentation, signal processing, networks, robots, satellites, and compute power for data collection, storage, analysis, modeling, visualization, and simulations. These technology advances allowed theoretical and experimental research including internet publication of molecular biochemistry, biological systems, and ecosystems science. This enabled worldwide access to better measurements, theoretical models, complex simulations, theory predictive model experimentation, analysis, worldwide internet observational data reporting, open peer-review, collaboration, and internet publication. New fields of biological sciences research emerged including Bioinformatics, Neuroscience, Theoretical biology, Computational genomics, Astrobiology and Synthetic Biology.

REFERENCES

1. Agar, Jon. Science in the Twentieth Century and Beyond. Polity Press: Cambridge, 2012. ISBN 978-0-7456-3469-2
2. Allen, Garland E. Thomas Hunt Morgan: The Man and His Science. Princeton University Press: Princeton, 1978. ISBN 0-691-08200-6
3. Allen, Garland E. Life Science in the Twentieth Century. Cambridge University Press, 1975.
4. Annas, Julia Classical Greek Philosophy. In Boardman, John; Griffin, Jasper; Murray, Oswyn (ed.) The Oxford History of the Classical World. Oxford University Press: New York, 1986. ISBN 0-19-872112-9
5. Barnes, Jonathan Hellenistic Philosophy and Science. In Boardman, John; Griffin, Jasper; Murray, Oswyn (ed.) The Oxford History of the Classical World. Oxford University Press: New York, 1986. ISBN 0-19-872112-9
6. Bowler, Peter J. The Earth Encompassed: A History of the Environmental Sciences. W. W. Norton & Company: New York, 1992. ISBN 0-393-32080-4
7. Bowler, Peter J. The Eclipse of Darwinism: Anti-Darwinian Evolution Theories in the Decades around 1900. The Johns Hopkins University Press: Baltimore, 1983. ISBN 0-8018-2932-1
8. Bowler, Peter J. Evolution: The History of an Idea. University of California Press, 2003. ISBN 0-520-23693-9.

9. Browne, Janet. The Secular Ark: Studies in the History of Biogeography. Yale University Press: New Haven, 1983. ISBN 0-300-02460-6
10. Bud, Robert. The Uses of Life: A History of Biotechnology. Cambridge University Press: London, 1993. ISBN 0-521-38240-8
11. Caldwell, John. "Drug metabolism and pharmacogenetics: the British contribution to fields of international significance." British Journal of Pharmacology, Vol. 147, Issue S1 (January 2006), pp S89–S99.
12. Coleman, William Biology in the Nineteenth Century: Problems of Form, Function, and Transformation. Cambridge University Press: New York, 1977. ISBN 0-521-29293-X
13. Creager, Angela N. H. The Life of a Virus: Tobacco Mosaic Virus as an Experimental Model, 1930–1965. University of Chicago Press: Chicago, 2002. ISBN 0-226-12025-2
14. Creager, Angela N. H. "Building Biology across the Atlantic," essay review in Journal of the History of Biology, Vol. 36, No. 3 (September 2003), pp. 579–589.
15. de Chadarevian, Soraya. Designs for Life: Molecular Biology after World War II. Cambridge University Press: Cambridge, 2002. ISBN 0-521-57078-6
16. Dietrich, Michael R. "Paradox and Persuasion: Negotiating the Place of Molecular Evolution within Evolutionary Biology," in Journal of the History of Biology, Vol. 31 (1998), pp. 85–111.
17. Davies, Kevin. Cracking the Genome: Inside the Race to Unlock Human DNA. The Free Press: New York, 2001. ISBN 0-7432-0479-4
18. Fruton, Joseph S. Proteins, Enzymes, Genes: The Interplay of Chemistry and Biology. Yale University Press: New Haven, 1999. ISBN 0-300-07608-8
19. Gottweis, Herbert. Governing Molecules: The Discursive Politics of Genetic Engineering in Europe and the United States. MIT Press: Cambridge, MA, 1998. ISBN 0-262-07189-4
20. Gould, Stephen Jay. The Structure of Evolutionary Theory. The Belknap Press of Harvard University Press: Cambridge, 2002. ISBN 0-674-00613-5
21. Hagen, Joel B. An Entangled Bank: The Origins of Ecosystem Ecology. Rutgers University Press: New Brunswick, 1992. ISBN 0-8135-1824-5
22. Hall, Stephen S. Invisible Frontiers: The Race to Synthesize a Human Gene. Atlantic Monthly Press: New York, 1987. ISBN 0-87113-147-1
23. Holmes, Frederic Lawrence. Meselson, Stahl, and the Replication of DNA: A History of "The Most Beautiful Experiment in Biology". Yale University Press: New Haven, 2001. ISBN 0-300-08540-0
24. Junker, Thomas. Geschichte der Biologie. C. H. Beck: München, 2004.
25. Kay, Lily E. The Molecular Vision of Life: Caltech, The Rockefeller Foundation, and the Rise of the New Biology. Oxford University Press: New York, 1993. ISBN 0-19-511143-5

26. Kohler, Robert E. Lords of the Fly: Drosophila Genetics and the Experimental Life. Chicago University Press: Chicago, 1994. ISBN 0-226-45063-5
27. Kohler, Robert E. Landscapes and Labscapes: Exploring the Lab-Field Border in Biology. University of Chicago Press: Chicago, 2002. ISBN 0-226-45009-0
28. Krimsky, Sheldon. Biotechnics and Society: The Rise of Industrial Genetics. Praeger Publishers: New York, 1991. ISBN 0-275-93860-3
29. Larson, Edward J. Evolution: The Remarkable History of a Scientific Theory. The Modern Library: New York, 2004. ISBN 0-679-64288-9
30. Lennox, James (2006-02-15). "Aristotle's Biology". Stanford Encyclopedia of Philosophy. Retrieved October 28, 2006.
31. Lovejoy, Arthur O. The Great Chain of Being: A Study of the History of an Idea. Harvard University Press, 1936. Reprinted by Harper & Row, ISBN 0-674-36150-4, 2005 paperback: ISBN 0-674-36153-9.
32. Magner, Lois N. A History of the Life Sciences, third edition. Marcel Dekker, Inc.: New York, 2002. ISBN 0-8247-0824-5
33. Mason, Stephen F. A History of the Sciences. Collier Books: New York, 1956.
34. Mayr, Ernst. The Growth of Biological Thought: Diversity, Evolution, and Inheritance. The Belknap Press of Harvard University Press: Cambridge, Massachusetts, 1982. ISBN 0-674-36445-7
35. Mayr, Ernst and William B. Provine, eds. The Evolutionary Synthesis: Perspectives on the Unification of Biology. Harvard University Press: Cambridge, 1998. ISBN 0-674-27226-9
36. Morange, Michel. A History of Molecular Biology, translated by Matthew Cobb. Harvard University Press: Cambridge, 1998. ISBN 0-674-39855-6
37. Rabinbach, Anson. The Human Motor: Energy, Fatigue, and the Origins of Modernity. University of California Press, 1992. ISBN 0-520-07827-6
38. Rabinow, Paul. Making PCR: A Story of Biotechnology. University of Chicago Press: Chicago, 1996. ISBN 0-226-70146-8
39. Rudwick, Martin J.S. The Meaning of Fossils. The University of Chicago Press: Chicago, 1972. ISBN 0-226-73103-0
40. Raby, Peter. Bright Paradise: Victorian Scientific Travellers. Princeton University Press: Princeton, 1997. ISBN 0-691-04843-6
41. Rothman, Sheila M. and David J. Rothman. The Pursuit of Perfection: The Promise and Perils of Medical Enhancement. Vintage Books: New York, 2003. ISBN 0-679-75835-6
42. Sapp, Jan. Genesis: The Evolution of Biology. Oxford University Press: New York, 2003. ISBN 0-19-515618-8
43. Secord, James A. Victorian Sensation: The Extraordinary Publication, Reception, and Secret Authorship of Vestiges of the Natural History of Creation. University of Chicago Press: Chicago, 2000. ISBN 0-226-74410-8
44. Serafini, Anthony The Epic History of Biology, Perseus Publishing, 1993.
45. Sulston, John. The Common Thread: A Story of Science, Politics, Ethics and the Human Genome. National Academy Press, 2002. ISBN 0-309-08409-1
46. Smocovitis, Vassiliki Betty. Unifying Biology: The Evolutionary Synthesis and Evolutionary Biology. Princeton University Press: Princeton, 1996. ISBN 0-691-03343-9

47. Summers, William C. Félix d'Herelle and the Origins of Molecular Biology, Yale University Press: New Haven, 1999. ISBN 0-300-07127-2

48. Sturtevant, A. H. A History of Genetics. Cold Spring Harbor Laboratory Press: Cold Spring Harbor, 2001. ISBN 0-87969-607-9

49. Thackray, Arnold, ed. Private Science: Biotechnology and the Rise of the Molecular Sciences. University of Pennsylvania Press: Philadelphia, 1998. ISBN 0-8122-3428-6

50. Wilson, Edward O. Naturalist. Island Press, 1994.

51. Zimmer, Carl. Evolution: the triumph of an idea. HarperCollins: New York, 2001. ISBN 0-06-113840-1

CHAPTER 2

Theory of Evolution

Chapter Contents

- **Introduction**

Evolutionary biology is an integral part of biology in general -- the study and theory of evolution in organisms. More than just a subfield, one might view this field as the lens through which all of biology should be viewed, creationists notwithstanding. Evolutionary biology concerns itself with the origin of species through genetic variation and natural selection as well as the shared descent of species from common ancestors.

INTRODUCTION

Evolution is change in the heritable characteristics of biological populations over successive generations. Evolutionary processes give rise to biodiversity at every level of biological organisation, including the levels of species, individual organisms, and molecules.

All life on Earth shares a common ancestor known as the last universal common ancestor (LUCA), which lived approximately 3.5–3.8 billion years ago, although a study in 2015 found "remains of biotic life" from 4.1 billion years ago in ancient rocks in Western Australia. In July 2016, scientists reported identifying a set of 355 genes from the LUCA of all organisms living on Earth.

Repeated formation of new species (speciation), change within species (anagenesis), and loss of species (extinction) throughout the evolutionary history of life on Earth are demonstrated by shared sets of morphological and biochemical traits, including shared DNA sequences. These shared traits are more similar among species that share a more recent common ancestor, and can be used to reconstruct a biological "tree of life" based on evolutionary relationships (phylogenetics), using both existing species and fossils. The fossil record includes a progression from early biogenic graphite, to microbial mat fossils, to fossilized multicellular organisms. Existing patterns of biodiversity have been shaped both by speciation and by extinction. More than 99 percent of all species that ever lived on Earth are estimated to be extinct. Estimates of Earth's current species range from 10 to 14 million, of which about 1.9 million are estimated to have been named and 1.6 million documented in a central database to date. More recently, in May 2016, scientists reported that 1 trillion species are estimated to be on Earth currently with only one-thousandth of one percent described.

In the mid-19th century, Charles Darwin formulated the scientific theory of evolution by natural selection, published in his book On the Origin of Species (1859). Evolution by natural selection is a process demonstrated by the observation that more offspring are produced than can possibly survive, along with three facts about populations: 1) traits vary among individuals with respect to morphology, physiology, and behaviour (phenotypic variation), 2) different traits confer different rates of survival and reproduction (differential fitness), and 3) traits can be passed from generation to generation (heritability of fitness). Thus, in successive generations members of a population are replaced by progeny of parents better adapted to survive and reproduce in the biophysical environment in which natural selection takes place. This teleonomy is the quality whereby the process of natural selection creates and preserves traits that are seemingly fitted for the functional roles they perform. Natural selection, including sexual selection, is the only known cause of adaptation but not the only known cause of evolution. Other, nonadaptive evolutionary processes include mutation, genetic drift and gene migration.

In the early 20th century the modern evolutionary synthesis integrated classical genetics with Darwin's theory of evolution by natural selection through the discipline of population genetics. The importance of natural selection as a cause of evolution was accepted into other branches of biology. Moreover, previously held notions about evolution, such as orthogenesis, evolutionism, and other beliefs about innate "progress" within the largest-scale trends in evolution, became obsolete scientific theories. Scientists continue to study various aspects of evolutionary biology by forming and testing hypotheses, constructing mathematical models of theoretical biology and biological theories, using observational data, and performing experiments in both the field and the laboratory.

In terms of practical application, an understanding of evolution has been instrumental to developments in numerous scientific and industrial fields, including agriculture, human and veterinary medicine, and the life sciences in general. Discoveries in evolutionary biology have made a significant impact not just in the traditional branches of biology but also in other academic disciplines, including biological anthropology, and evolutionary psychology. Evolutionary computation, a sub-field of artificial intelligence, involves the application of Darwinian principles to problems in computer science.

History of Evolutionary Thought

Lucretius

Alfred Russel Wallace

Thomas Robert Malthus

Statistician Ronald Fisher

In 1842, Charles Darwin penned his first sketch of *On the Origin of Species*.

The proposal that one type of organism could descend from another type goes back to some of the first pre-Socratic Greek philosophers, such as Anaximander and Empedocles. Such proposals survived into Roman times. The poet and philosopher Lucretius followed Empedocles in his masterwork *De rerum natura* (*On the Nature of Things*). In contrast to these materialistic views, Aristotelianism considered all natural things as actualisations of fixed natural possibilities, known as forms. This was part of a medieval teleological understanding of nature in which all things have an intended role to play in a divine cosmic order. Variations of this idea became the standard understanding of the Middle Ages and were integrated into Christian learning, but Aristotle did not demand that real types of organisms always correspond one-for-one with exact metaphysical forms and specifically gave examples of how new types of living things could come to be.

In the 17th century, the new method of modern science rejected the Aristotelian approach. It sought explanations of natural phenomena in terms of physical laws that were the same for all visible things and that did not require the existence of any fixed natural categories or divine cosmic order. However, this new approach was slow to take root in the biological sciences, the last bastion of the concept of fixed natural types. John

Ray applied one of the previously more general terms for fixed natural types, species, to plant and animal types, but he strictly identified each type of living thing as a species and proposed that each species could be defined by the features that perpetuated themselves generation after generation. The biological classification introduced by Carl Linnaeus in 1735 explicitly recognized the hierarchical nature of species relationships, but still viewed species as fixed according to a divine plan.

Other naturalists of this time speculated on the evolutionary change of species over time according to natural laws. In 1751, Pierre Louis Maupertuis wrote of natural modifications occurring during reproduction and accumulating over many generations to produce new species. Georges-Louis Leclerc, Comte de Buffon suggested that species could degenerate into different organisms, and Erasmus Darwin proposed that all warm-blooded animals could have descended from a single microorganism (or "filament"). The first full-fledged evolutionary scheme was Jean-Baptiste Lamarck's "transmutation" theory of 1809, which envisaged spontaneous generation continually producing simple forms of life that developed greater complexity in parallel lineages with an inherent progressive tendency, and postulated that on a local level these lineages adapted to the environment by inheriting changes caused by their use or disuse in parents. (The latter process was later called Lamarckism.) These ideas were condemned by established naturalists as speculation lacking empirical support. In particular, Georges Cuvier insisted that species were unrelated and fixed, their similarities reflecting divine design for functional needs. In the meantime, Ray's ideas of benevolent design had been developed by William Paley into the *Natural Theology or Evidences of the Existence and Attributes of the Deity* (1802), which proposed complex adaptations as evidence of divine design and which was admired by Charles Darwin.

The crucial break from the concept of constant typological classes or types in biology came with the theory of evolution through natural selection, which was formulated by Charles Darwin in terms of variable populations. Partly influenced by *An Essay on the Principle of Population* (1798) by Thomas Robert Malthus, Darwin noted that population growth would lead to a "struggle for existence" in which favorable variations prevailed as others perished. In each generation, many offspring fail to survive to an age of reproduction because of limited resources. This could explain the diversity of plants and animals from a common ancestry through the working of natural laws in the same way for all types of organism. Darwin developed his theory of "natural selection" from 1838 onwards and was writing up his "big book" on the subject when Alfred Russel Wallace sent him a version of virtually the same theory in 1858. Their separate papers were presented together at a 1858 meeting of the Linnean Society of London. At the end of 1859, Darwin›s publication of his abstract as *On the Origin of Species* explained natural selection in detail and in a way that led to an increasingly wide acceptance of concepts of evolution. Thomas Henry Huxley applied Darwin's ideas to humans, using paleontology and comparative anatomy to provide strong evidence that humans and apes shared a common ancestry. Some were disturbed by this since it implied that humans did not have a special place in the universe.

Precise mechanisms of reproductive heritability and the origin of new traits remained a mystery. Towards this end, Darwin developed his provisional theory of pangenesis. In 1865, Gregor Mendel reported that traits were inherited in a predictable manner through the independent assortment and segregation of elements (later known as genes). Mendel›s laws of inheritance eventually supplanted most of Darwin›s pangenesis theory. August Weismann made the important distinction between germ cells that give rise to gametes (such as sperm and egg cells) and the somatic cells of the body, demonstrating that heredity passes through the germ line only. Hugo de Vries connected Darwin's pangenesis theory to Weismann's germ/soma cell distinction and proposed that Darwin's pangenes were concentrated in the cell nucleus and when expressed they could move into the cytoplasm to change the cells structure. De Vries was also one of the researchers who made Mendel›s work well-known, believing that Mendelian traits corresponded to the transfer of heritable variations along the germline. To explain how new variants originate, de Vries developed a mutation theory that led to a temporary rift between those who accepted Darwinian evolution and biometricians who allied with de Vries. In the 1930s, pioneers in the field of population genetics, such as Ronald Fisher, Sewall Wright and J. B. S. Haldane set the foundations of evolution onto a robust statistical philosophy. The false contradiction between Darwin›s theory, genetic mutations, and Mendelian inheritance was thus reconciled.

In the 1920s and 1930s a modern evolutionary synthesis connected natural selection, mutation theory, and Mendelian inheritance into a unified theory that applied generally to any branch of biology. The modern synthesis was able to explain patterns observed across species in populations, through fossil transitions in palaeontology, and even complex cellular mechanisms in developmental biology. The publication of the structure of DNA by James Watson and Francis Crick in 1953 demonstrated a physical mechanism for inheritance. Molecular biology improved our understanding of the relationship between genotype and phenotype. Advancements were also made in phylogenetic systematics, mapping the transition of traits into a comparative and testable framework through the publication and use of evolutionary trees. In 1973, evolutionary biologist Theodosius Dobzhansky penned that "nothing in biology makes sense except in the light of evolution," because it has brought to light the relations of what first seemed disjointed facts in natural history into a coherent explanatory body of knowledge that describes and predicts many observable facts about life on this planet.

Since then, the modern synthesis has been further extended to explain biological phenomena across the full and integrative scale of the biological hierarchy, from genes to species. This extension, known as evolutionary developmental biology and informally called evo-devo, emphasises how changes between generations (evolution) acts on patterns of change within individual organisms (development).

Heredity

Evolution in organisms occurs through changes in heritable traits—the inherited characteristics of an organism. In humans, for example, eye colour is an inherited characteristic and an individual might inherit the "brown-eye trait" from one of their parents. Inherited traits are controlled by genes and the complete set of genes within an organism's genome (genetic material) is called its genotype.

The complete set of observable traits that make up the structure and behaviour of an organism is called its phenotype. These traits come from the interaction of its genotype with the environment. As a result, many aspects of an organism›s phenotype are not inherited. For example, suntanned skin comes from the interaction between a person›s genotype and sunlight; thus, suntans are not passed on to people›s children. However, some people tan more easily than others, due to differences in genotypic variation; a striking example are people with the inherited trait of albinism, who do not tan at all and are very sensitive to sunburn.

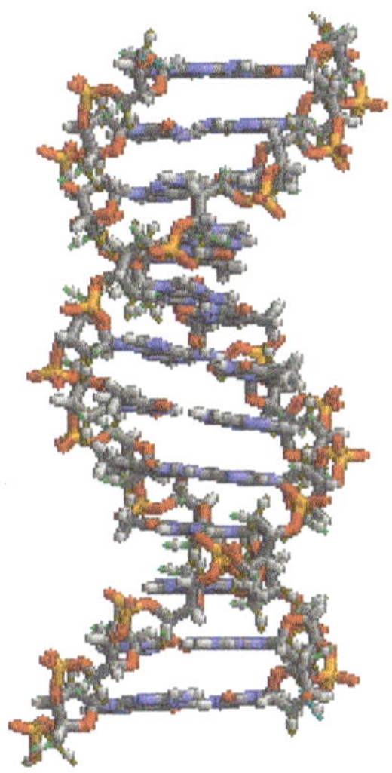

DNA structure. Bases are in the centre, surrounded by phosphate–sugar chains in a double helix.

Heritable traits are passed from one generation to the next via DNA, a molecule that encodes genetic information. DNA is a long biopolymer composed of four types of bases. The sequence of bases along a particular DNA molecule specify the genetic information, in a manner similar to a sequence of letters spelling out a sentence. Before a cell divides, the DNA is copied, so that each of the resulting two cells will inherit the DNA sequence. Portions of a DNA molecule that specify a single functional unit are called genes; different genes have different sequences of bases. Within cells, the long strands of DNA form condensed structures called chromosomes. The specific location of a DNA sequence within a chromosome is known as a locus. If the DNA sequence at a locus varies between individuals, the different forms of this sequence are called alleles. DNA sequences can change through mutations, producing new alleles. If a mutation occurs within a gene, the new allele may affect the trait that the gene controls, altering the phenotype of the organism. However, while this simple correspondence between an allele and a trait works in some cases, most traits are more complex and are controlled by quantitative trait loci (multiple interacting genes).

Recent findings have confirmed important examples of heritable changes that cannot be explained by changes to the sequence of nucleotides in the DNA. These phenomena are classed as epigenetic inheritance systems. DNA methylation marking chromatin, self-sustaining metabolic loops, gene silencing by RNA interference and the three-dimensional conformation of proteins (such as prions) are areas where epigenetic inheritance systems have been discovered at the organismic level. Developmental biologists suggest that complex interactions in genetic networks and communication among cells can lead to heritable variations that may underlay some of the mechanics in developmental plasticity and canalisation. Heritability may also occur at even larger scales. For example, ecological inheritance through the process of niche construction is defined by the regular and repeated activities of organisms in their environment. This generates a legacy of effects that modify and feed back into the selection regime of subsequent generations. Descendants inherit genes plus environmental characteristics generated by the ecological actions of ancestors. Other examples of heritability in evolution that are not under the direct control of genes include the inheritance of cultural traits and symbiogenesis.

Variation

White peppered moth

Black morph in peppered moth evolution

An individual organism's phenotype results from both its genotype and the influence from the environment it has lived in. A substantial part of the phenotypic variation in a population is caused by genotypic variation.The modern evolutionary synthesis defines evolution as the change over time in this genetic variation. The frequency of one particular allele will become more or less prevalent relative to other forms of that gene. Variation disappears when a new allele reaches the point of fixation—when it either disappears from the population or replaces the ancestral allele entirely.

Natural selection will only cause evolution if there is enough genetic variation in a population. Before the discovery of Mendelian genetics, one common hypothesis was blending inheritance. But with blending inheritance, genetic variance would be rapidly lost, making evolution by natural selection implausible. The Hardy–Weinberg principle provides the solution to how variation is maintained in a population with Mendelian inheritance. The frequencies of alleles (variations in a gene) will remain constant in the absence of selection, mutation, migration and genetic drift.

Variation comes from mutations in the genome, reshuffling of genes through sexual reproduction and migration between populations (gene flow). Despite the constant introduction of new variation through mutation and gene flow, most of the genome of a species is identical in all individuals of that species.However, even relatively small differences in genotype can lead to dramatic differences in phenotype: for example, chimpanzees and humans differ in only about 5% of their genomes.

Mutation

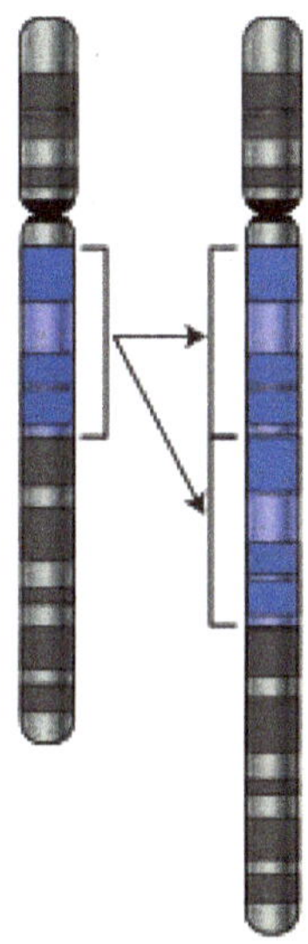

Duplication of part of a chromosome

Mutations are changes in the DNA sequence of a cell's genome. When mutations occur, they may alter the product of a gene, or prevent the gene from functioning, or have no effect. Based on studies in the fly *Drosophila melanogaster*, it has been suggested that if a mutation changes a protein produced by a gene, this will probably be harmful, with about 70% of these mutations having damaging effects, and the remainder being either neutral or weakly beneficial.

Mutations can involve large sections of a chromosome becoming duplicated (usually by genetic recombination), which can introduce extra copies of a gene into a genome. Extra copies of genes are a major source of the raw material needed for new genes to evolve. This is important because most new genes evolve within gene families from pre-existing genes that share common ancestors. For example, the human eye uses four genes to make structures that sense light: three for colour vision and one for night vision; all four are descended from a single ancestral gene.

New genes can be generated from an ancestral gene when a duplicate copy mutates and acquires a new function. This process is easier once a gene has been duplicated because it increases the redundancy of the system; one gene in the pair can acquire a new function while the other copy continues to perform its original function. Other types of mutations can even generate entirely new genes from previously noncoding DNA.

The generation of new genes can also involve small parts of several genes being duplicated, with these fragments then recombining to form new combinations with new functions. When new genes are assembled from shuffling pre-existing parts, domains act as modules with simple independent functions, which can be mixed together to produce new combinations with new and complex functions. For example, polyketide synthases are large enzymes that make antibiotics; they contain up to one hundred independent domains that each catalyse one step in the overall process, like a step in an assembly line.

Sex and Recombination

In asexual organisms, genes are inherited together, or *linked*, as they cannot mix with genes of other organisms during reproduction. In contrast, the offspring of sexual organisms contain random mixtures of their parents' chromosomes that are produced through independent assortment. In a related process called homologous recombination, sexual organisms exchange DNA between two matching chromosomes. Recombination and reassortment do not alter allele frequencies, but instead change which alleles are associated with each other, producing offspring with new combinations of alleles. Sex usually increases genetic variation and may increase the rate of evolution.

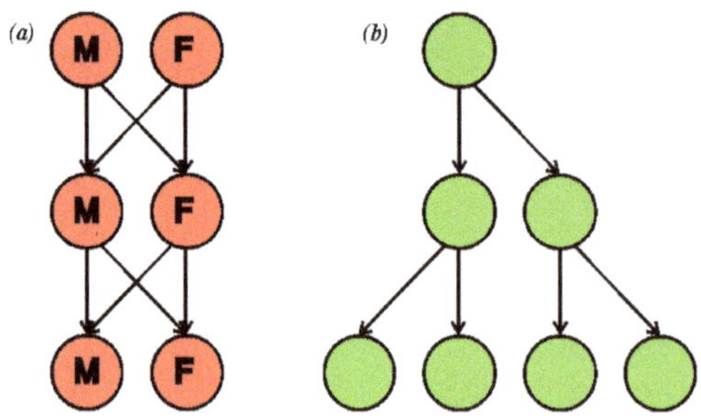

This diagram illustrates the *twofold cost of sex*. If each individual were to contribute to the same number of offspring (two), *(a)* the sexual population remains the same size each generation, where the *(b)*Asexual reproduction population doubles in size each generation.

The two-fold cost of sex was first described by John Maynard Smith. The first cost is that in sexually dimorphic species only one of the two sexes can bear young. (This cost does not apply to hermaphroditic species, like most plants and many invertebrates.) The second cost is that any individual who reproduces sexually can only pass on 50% of its genes to any individual offspring, with even less passed on as each new generation passes. Yet sexual reproduction is the more common means of reproduction among eukaryotes and multicellular organisms. The Red Queen hypothesis has been used to explain the significance of sexual reproduction as a means to enable continual evolution and adaptation in response to coevolution with other species in an ever-changing environment.

Gene Flow

Gene flow is the exchange of genes between populations and between species. It can therefore be a source of variation that is new to a population or to a species. Gene flow can be caused by the movement of individuals between separate populations of organisms, as might be caused by the movement of mice between inland and coastal populations, or the movement of pollen between heavy metal tolerant and heavy metal sensitive populations of grasses.

Gene transfer between species includes the formation of hybrid organisms and horizontal gene transfer. Horizontal gene transfer is the transfer of genetic material from one organism to another organism that is not its offspring; this is most common among bacteria. In medicine, this contributes to the spread of antibiotic resistance, as when one bacteria acquires resistance genes it can rapidly transfer them to other species.Horizontal transfer of genes from bacteria to eukaryotes such as the yeast *Saccharomyces cerevisiae* and the adzuki bean weevil *Callosobruchus chinensis* has occurred. An example of larger-scale transfers are the eukaryotic bdelloid rotifers, which have received a range of genes from bacteria, fungi and plants. Viruses can also carry DNA between organisms, allowing transfer of genes even across biological domains.

Large-scale gene transfer has also occurred between the ancestors of eukaryotic cells and bacteria, during the acquisition of chloroplasts and mitochondria. It is possible that eukaryotes themselves originated from horizontal gene transfers between bacteria and archaea.

Mechanisms

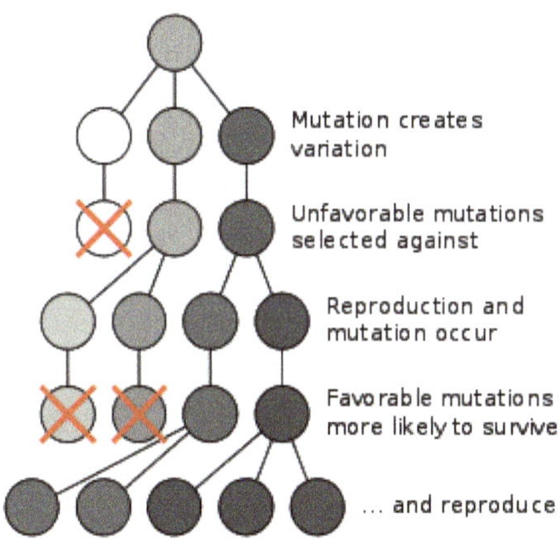

Mutation followed by natural selection results in a population with darker colouration.

From a Neo-Darwinian perspective, evolution occurs when there are changes in the frequencies of alleles within a population of interbreeding organisms. For example, the allele for black colour in a population of moths becoming more common. Mechanisms that can lead to changes in allele frequencies include natural selection, genetic drift, genetic hitchhiking, mutation and gene flow.

Natural Selection

Evolution by means of natural selection is the process by which traits that enhance survival and reproduction become more common in successive generations of a population. It has often been called a "self-evident" mechanism because it necessarily follows from three simple facts:

- Variation exists within populations of organisms with respect to morphology, physiology, and behaviour (phenotypic variation).
- Different traits confer different rates of survival and reproduction (differential fitness).
- These traits can be passed from generation to generation (heritability of fitness).

More offspring are produced than can possibly survive, and these conditions produce competition between organisms for survival and reproduction. Consequently, organisms with traits that give them an advantage over their competitors are more likely to pass on their traits to the next generation than those with traits that do not confer an advantage.

The central concept of natural selection is the evolutionary fitness of an organism. Fitness is measured by an organism's ability to survive and reproduce, which determines the size of its genetic contribution to the next generation. However, fitness is not the same as the total number of offspring: instead fitness is indicated by the proportion of subsequent generations that carry an organism's genes. For example, if an organism could survive well and reproduce rapidly, but its offspring were all too small and weak to survive, this organism would make little genetic contribution to future generations and would thus have low fitness.

If an allele increases fitness more than the other alleles of that gene, then with each generation this allele will become more common within the population. These traits are said to be selected *for*." Examples of traits that can increase fitness are enhanced survival and increased fecundity. Conversely, the lower fitness caused by having a less beneficial or deleterious allele results in this allele becoming rarer—they are "selected *against*." Importantly, the fitness of an allele is not a fixed characteristic; if the environment changes, previously neutral or harmful traits may become beneficial and previously beneficial traits become harmful. However, even if the direction of selection does reverse in this way, traits that were lost in the past may not re-evolve in an identical form.

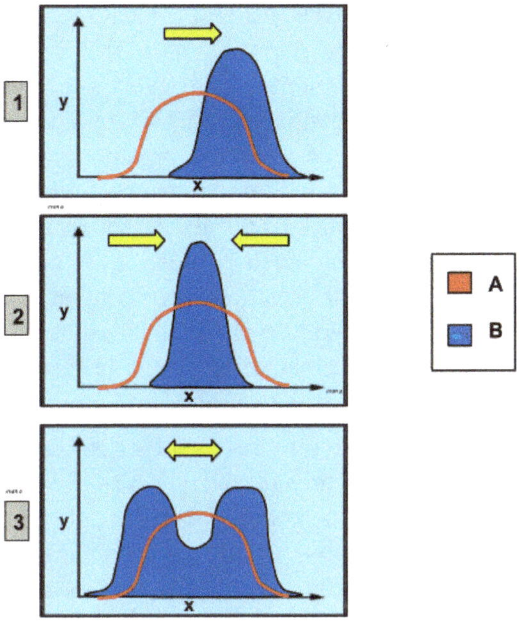

These charts depict the different types of genetic selection. On each graph, the x-axis variable is the type of phenotypic trait and the y-axis variable is the number of organisms. Group A is the original population and Group B is the population after selection.

- Graph 1 shows directional selection, in which a single extreme phenotype is favored.
- Graph 2 depicts stabilizing selection, where the intermediate phenotype is favored over the extreme traits.• Graph 3 shows disruptive selection, in which the extreme phenotypes are favored over the intermediate.

Natural selection within a population for a trait that can vary across a range of values, such as height, can be categorised into three different types. The first is directional selection, which is a shift in the average value of a trait over time—for example, organisms slowly getting taller. Secondly, disruptive selection is selection for extreme trait values and often results in two different values becoming most common, with selection against the average value. This would be when either short or tall organisms had an advantage, but not those of medium height. Finally, in stabilising selection there is selection against extreme trait values on both ends, which causes a decrease in variance around the average value and less diversity. This would, for example, cause organisms to slowly become all the same height.

A special case of natural selection is sexual selection, which is selection for any trait that increases mating success by increasing the attractiveness of an organism to potential mates. Traits that evolved through sexual selection are particularly prominent among males of several animal species. Although sexually favoured, traits such as cumbersome antlers, mating calls, large body size and bright colours often attract predation, which compromises the survival of individual males. This survival disadvantage is balanced by higher reproductive success in males that show these hard-to-fake, sexually selected traits.

Natural selection most generally makes nature the measure against which individuals and individual traits, are more or less likely to survive. Nature in this sense refers to an ecosystem, that is, a system in which organisms interact with every other element, physical as well as biological, in their local environment. Eugene Odum, a founder of ecology, defined an ecosystem as: Any unit that includes all of the organisms...in a given area interacting with the physical environment so that a flow of energy leads to clearly defined trophic structure, biotic diversity and material cycles (ie: exchange of materials between living and nonliving parts) within the system. Each population within an ecosystem occupies a distinct niche, or position, with distinct relationships to other parts of the system. These relationships involve the life history of the organism, its position in the food chain and its geographic range. This broad understanding of nature enables scientists to delineate specific forces which, together, comprise natural selection.

Natural selection can act at different levels of organisation, such as genes, cells, individual organisms, groups of organisms and species. Selection can act at multiple levels simultaneously. An example of selection occurring below the level of the individual organism are genes called transposons, which can replicate and spread throughout a genome. Selection at a level above the individual, such as group selection, may allow the evolution of cooperation, as discussed below.

Biased Mutation

In addition to being a major source of variation, mutation may also function as a mechanism of evolution when there are different probabilities at the molecular level for different mutations to occur, a process known as mutation bias. If two genotypes, for example one with the nucleotide G and another with the nucleotide A in the same position, have the same fitness, but mutation from G to A happens more often than mutation from A to G, then genotypes with A will tend to evolve. Different insertion vs. deletion mutation biases in different taxa can lead to the evolution of different genome sizes. Developmental or mutational biases have also been observed in morphological evolution. For example, according to the phenotype-first theory of evolution, mutations can eventually cause the genetic assimilation of traits that were previously induced by the environment.

Mutation bias effects are superimposed on other processes. If selection would favor either one out of two mutations, but there is no extra advantage to having both, then the mutation that occurs the most frequently is the one that is most likely to become fixed in a population.Mutations leading to the loss of function of a gene are much more common than mutations that produce a new, fully functional gene. Most loss of function mutations are selected against. But when selection is weak, mutation bias towards loss of function can affect evolution. For example, pigments are no longer useful when animals live in the darkness of caves, and tend to be lost. This kind of loss of function can occur because of mutation bias, and/or because the function had a cost, and once the benefit of the function disappeared, natural selection leads to the loss. Loss of sporulation ability in *Bacillus subtilis* during laboratory evolution appears to have been caused by mutation bias, rather than natural selection against the cost of maintaining sporulation ability. When there is no selection for loss of function, the speed at which loss evolves depends more on the mutation rate than it does on the effective population size, indicating that it is driven more by mutation bias than by genetic drift. In parasitic organisms, mutation bias leads to selection pressures as seen in Ehrlichia. Mutations are biased towards antigenic variants in outer-membrane proteins.

Genetic Drift

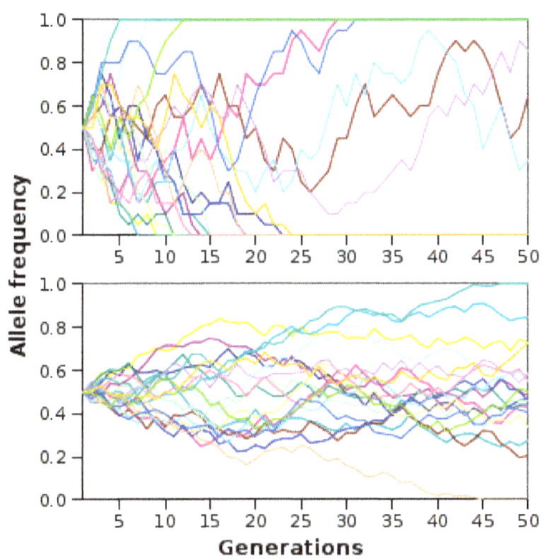

Simulation of genetic drift of 20 unlinked alleles in populations of 10 (top) and 100 (bottom). Drift to fixation is more rapid in the smaller population.

Genetic drift is the change in allele frequency from one generation to the next that occurs because alleles are subject to sampling error. As a result, when selective forces are absent or relatively weak, allele frequencies tend to drift upward or downward randomly (in a random walk). This drift halts when an allele eventually becomes fixed, either by disappearing from the population, or replacing the other alleles entirely. Genetic drift may therefore eliminate some alleles from a population due to chance alone. Even in the absence of selective forces, genetic drift can cause two separate populations that began with the same genetic structure to drift apart into two divergent populations with different sets of alleles.

It is usually difficult to measure the relative importance of selection and neutral processes, including drift. The comparative importance of adaptive and non-adaptive forces in driving evolutionary change is an area of current research.

The neutral theory of molecular evolution proposed that most evolutionary changes are the result of the fixation of neutral mutations by genetic drift. Hence, in this model, most genetic changes in a population are the result of constant mutation pressure and genetic drift. This form of the neutral theory is now largely abandoned, since it does not seem to fit the genetic variation seen in nature. However, a more recent and better-supported version of this model is the nearly neutral theory, where a mutation that would be effectively neutral in a small population is not necessarily neutral in a large population. Other alternative theories propose that genetic drift is dwarfed by other stochastic forces in evolution, such as genetic hitchhiking, also known as genetic draft.

The time for a neutral allele to become fixed by genetic drift depends on population size, with fixation occurring more rapidly in smaller populations. The number of individuals in a population is not critical, but instead a measure known as the effective population size. The effective population is usually smaller than the total population since it takes into account factors such as the level of inbreeding and the stage of the lifecycle in which the population is the smallest. The effective population size may not be the same for every gene in the same population.

Genetic Hitchhiking

Recombination allows alleles on the same strand of DNA to become separated. However, the rate of recombination is low (approximately two events per chromosome per generation). As a result, genes close together on a chromosome may not always be shuffled away from each other and genes that are close together tend to be inherited together, a phenomenon known as linkage. This tendency is measured by finding how often two alleles occur together on a single chromosome compared to expectations, which is called their linkage disequilibrium. A set of alleles that is usually inherited in a group is called a haplotype. This can be important when one allele in a particular haplotype is strongly beneficial: natural selection can drive a selective sweep that will also cause the other alleles in the haplotype to become more common in the population; this effect is called genetic hitchhiking or genetic draft. Genetic draft caused by the fact that some neutral genes are genetically linked to others that are under selection can be partially captured by an appropriate effective population size.

Gene Flow

Gene flow involves the exchange of genes between populations and between species. The presence or absence of gene flow fundamentally changes the course of evolution. Due to the complexity of organisms, any two completely isolated populations will eventually evolve genetic incompatibilities through neutral processes, as in the Bateson-Dobzhansky-Muller model, even if both populations remain essentially identical in terms of their adaptation to the environment.

If genetic differentiation between populations develops, gene flow between populations can introduce traits or alleles which are disadvantageous in the local population and this may lead to organisms within these populations evolving mechanisms that prevent mating with genetically distant populations, eventually resulting in the appearance of new species. Thus, exchange of genetic information between individuals is fundamentally important for the development of the biological species concept.

During the development of the modern synthesis, Sewall Wright developed his shifting balance theory, which regarded gene flow between partially isolated populations as an important aspect of adaptive evolution. However, recently there has been substantial criticism of the importance of the shifting balance theory.

Outcomes

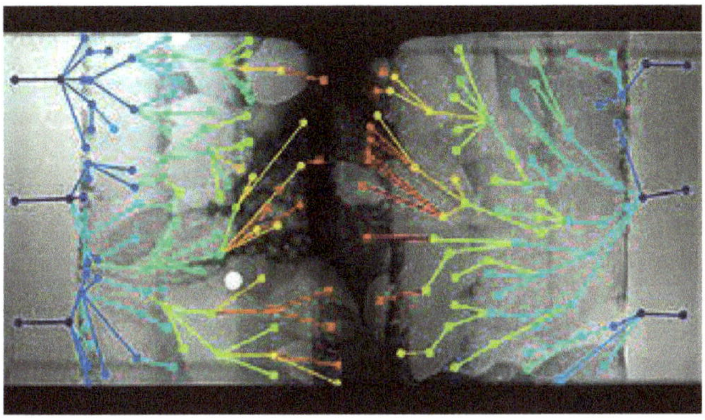

A visual demonstration of rapid antibiotic resistance evolution by *E. coli* growing across a plate with increasing concentrations of trimethoprim.

Evolution influences every aspect of the form and behaviour of organisms. Most prominent are the specific behavioural and physical adaptations that are the outcome of natural selection. These adaptations increase fitness by aiding activities such as finding food, avoiding predators or attracting mates. Organisms can also respond to selection by cooperating with each other, usually by aiding their relatives or engaging in mutually beneficial symbiosis. In the longer term, evolution produces new species through splitting ancestral populations of organisms into new groups that cannot or will not interbreed.

These outcomes of evolution are distinguished based on time scale as macroevolution versus microevolution. Macroevolution refers to evolution that occurs at or above the level of species, in particular speciation and extinction; whereas microevolution refers to smaller evolutionary changes within a species or population, in particular shifts in gene frequency and adaptation. In general, macroevolution is regarded as the outcome of long periods of microevolution. Thus, the distinction between micro- and macroevolution is not a fundamental one—the difference is simply the time involved. However, in macroevolution, the traits of the entire species may be important. For instance, a large amount of variation among individuals allows a species to rapidly adapt to new habitats, lessening the chance of it going extinct, while a wide geographic range increases the chance of speciation, by making it more likely that part of the population will become isolated. In this sense, microevolution and macroevolution might involve selection at different levels—with microevolution acting on genes and organisms, versus macroevolutionary processes such as species selection acting on entire species and affecting their rates of speciation and extinction.

A common misconception is that evolution has goals, long-term plans, or an innate tendency for progress, as expressed in beliefs such as orthogenesis and evolutionism; realistically however, evolution has no long-term goal and does not necessarily produce greater complexity. Although complex species have evolved, they occur as a side effect of the overall number of organisms increasing and simple forms of life still remain more common in the biosphere. For example, the overwhelming majority of species are microscopic prokaryotes, which form about half the world›s biomass despite their small size, and constitute the vast majority of Earth's biodiversity.Simple organisms have therefore been the dominant form of life on Earth throughout its history and continue to be the main form of life up to the present day, with complex life only appearing more diverse because it is more noticeable. Indeed, the evolution of microorganisms is particularly important to modern evolutionary research, since their rapid reproduction allows the study of experimental evolution and the observation of evolution and adaptation in real time.

Adaptation

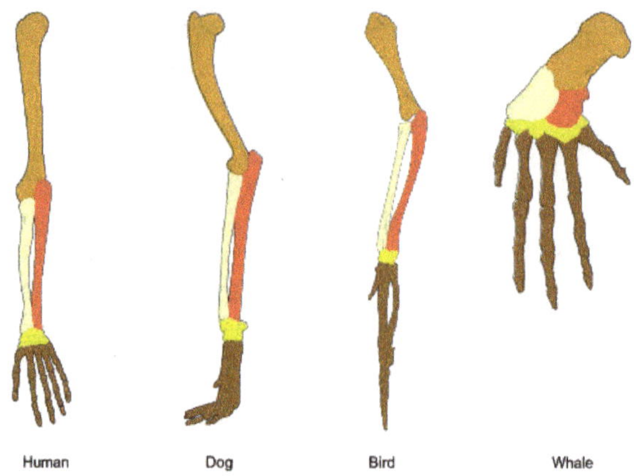

Homologous bones in the limbs of tetrapods. The bones of these animals have the same basic structure, but have been adapted for specific uses.

Adaptation is the process that makes organisms better suited to their habitat. Also, the term adaptation may refer to a trait that is important for an organism›s survival. For example, the adaptation of horses› teeth to the grinding of grass. By using the term *adaptation* for the evolutionary process and *adaptive trait* for the product (the bodily part or function), the two senses of the word may be distinguished. Adaptations are produced by natural selection.The following definitions are due to Theodosius Dobzhansky:

- *Adaptation* is the evolutionary process whereby an organism becomes better able to live in its habitat or habitats.
- *Adaptedness* is the state of being adapted: the degree to which an organism is able to live and reproduce in a given set of habitats.
- An *adaptive trait* is an aspect of the developmental pattern of the organism which enables or enhances the probability of that organism surviving and reproducing.

Adaptation may cause either the gain of a new feature, or the loss of an ancestral feature. An example that shows both types of change is bacterial adaptation to antibiotic selection, with genetic changes causing antibiotic resistance by both modifying the target of the drug, or increasing the activity of transporters that pump the drug out of the cell. Other striking examples are the bacteria *Escherichia coli* evolving the ability to use citric acid as a nutrient in a long-term laboratory experiment, *Flavobacterium* evolving a novel enzyme that allows these bacteria to grow on the by-products of nylon manufacturing, and the soil bacterium *Sphingobium* evolving an entirely new metabolic pathway that degrades the synthetic pesticidepentachlorophenol. An interesting but still controversial idea is that some adaptations might increase the ability of organisms to generate genetic diversity and adapt by natural selection (increasing organisms› evolvability).

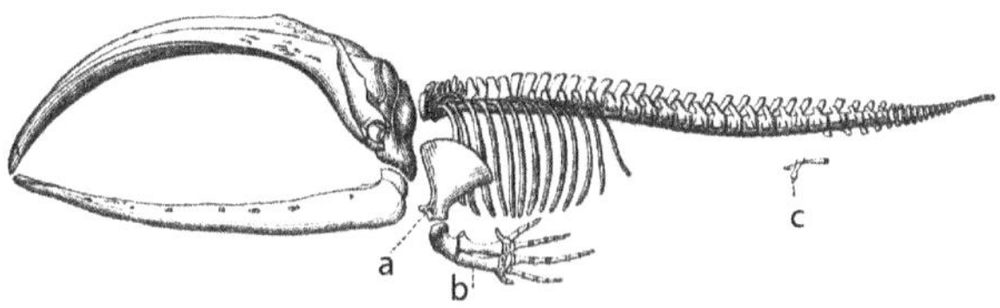

A baleen whale skeleton, *a* and *b* label flipper bones, which were adapted from front leg bones: while *c* indicates vestigial leg bones, suggesting an adaptation from land to sea.

Adaptation occurs through the gradual modification of existing structures. Consequently, structures with similar internal organisation may have different functions in related organisms. This is the result of a single ancestral structure being adapted to function in different ways. The bones within bat wings, for example, are very similar to those in mice feet and primate hands, due to the descent of all these structures from a common mammalian ancestor. However, since all living organisms are related to some extent, even organs that appear to have little or no structural similarity, such as arthropod, squid and vertebrate eyes, or the limbs and wings of arthropods and vertebrates, can depend on a common set of homologous genes that control their assembly and function; this is called deep homology.

During evolution, some structures may lose their original function and become vestigial structures. Such structures may have little or no function in a current species, yet have a clear function in ancestral species, or other closely related species. Examples include pseudogenes,the non-functional remains of eyes in blind cave-dwelling fish, wings in flightless birds, the presence of hip bones in whales and snakes, and sexual traits in organisms that reproduce via asexual reproduction. Examples of vestigial structures in humans include wisdom teeth, the coccyx, the vermiform appendix, and other behavioural vestiges such as goose bumps and primitive reflexes.

However, many traits that appear to be simple adaptations are in fact exaptations: structures originally adapted for one function, but which coincidentally became somewhat useful for some other function in the process. One example is the African lizard *Holaspis guentheri*, which developed an extremely flat head for hiding in crevices, as can be seen by looking at its near relatives. However, in this species, the head has become so flattened that it assists in gliding from tree to tree—an exaptation. Within cells, molecular machines such as the bacterial flagellaand protein sorting machinery evolved by the recruitment of several pre-existing proteins that previously had different functions. Another example is the recruitment of enzymes from glycolysis and xenobiotic metabolism to serve as structural proteins called crystallins within the lenses of organisms' eyes.

An area of current investigation in evolutionary developmental biology is the developmental basis of adaptations and exaptations. This research addresses the origin and evolution of embryonic development and how modifications of development and developmental processes produce novel features. These studies have shown that evolution can alter development to produce new structures, such as embryonic bone structures that develop into the jaw in other animals instead forming part of the middle ear in mammals. It is also possible for structures that have been lost in evolution to reappear due to changes in developmental genes, such as a mutation in chickens causing embryos to grow teeth similar to those of crocodiles. It is now becoming clear that most alterations in the form of organisms are due to changes in a small set of conserved genes.

Coevolution

Common garter snake (*Thamnophis sirtalis sirtalis*) has evolved resistance to the defensive substance tetrodotoxin in its amphibian prey.

Interactions between organisms can produce both conflict and cooperation. When the interaction is between pairs of species, such as a pathogen and a host, or a predator and its prey, these species can develop matched sets of adaptations. Here, the evolution of one species causes adaptations in a second species. These changes in the second species then, in turn, cause new adaptations in the first species. This cycle of selection and response is called coevolution. An example is the production of tetrodotoxin in the rough-skinned newt and the evolution of tetrodotoxin resistance in its predator, the common garter snake. In this predator-prey pair, an evolutionary arms race has produced high levels of toxin in the newt and correspondingly high levels of toxin resistance in the snake.

Cooperation

Not all co-evolved interactions between species involve conflict. Many cases of mutually beneficial interactions have evolved. For instance, an extreme cooperation exists between plants and the mycorrhizal fungi that grow on their roots and aid the plant in absorbing nutrients from the soil. This is a reciprocal relationship as the plants provide the fungi with sugars from photosynthesis. Here, the fungi actually grow inside plant cells, allowing them to exchange nutrients with their hosts, while sending signals that suppress the plant immune system.

Coalitions between organisms of the same species have also evolved. An extreme case is the eusociality found in social insects, such as bees, termites and ants, where sterile insects feed and guard the small number of organisms in a colony that are able to reproduce. On an even smaller scale, the somatic cells that make up the body of an animal limit their reproduction so they can maintain a stable organism, which then supports a small number of the animal's germ cells to produce offspring. Here, somatic cells respond to specific signals that instruct them whether to grow, remain as they are, or die. If cells ignore these signals and multiply inappropriately, their uncontrolled growth causes cancer.

Such cooperation within species may have evolved through the process of kin selection, which is where one organism acts to help raise a relative›s offspring. This activity is selected for because if the *helping* individual contains alleles which promote the helping activity, it is likely that its kin will *also* contain these alleles and thus those alleles will be passed on. Other processes that may promote cooperation include group selection, where cooperation provides benefits to a group of organisms.

Speciation

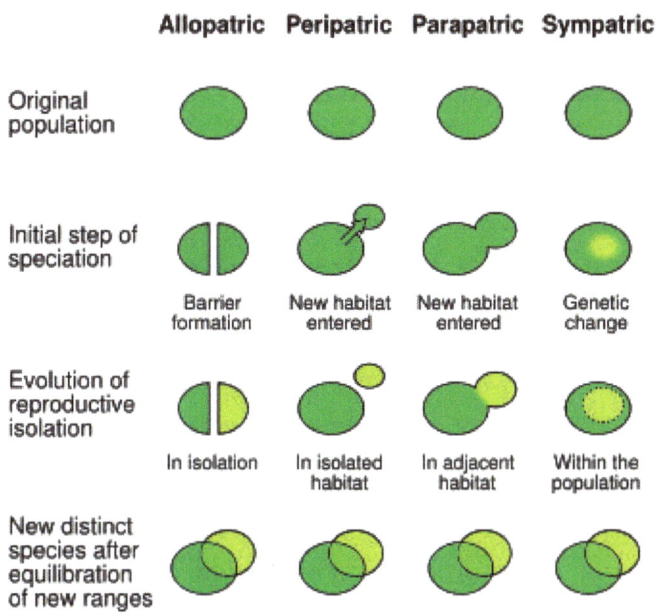

The four mechanisms of speciation

Speciation is the process where a species diverges into two or more descendant species.

There are multiple ways to define the concept of "species." The choice of definition is dependent on the particularities of the species concerned. For example, some species concepts apply more readily toward sexually reproducing organisms while others lend themselves better toward asexual organisms. Despite the diversity of various species concepts, these various concepts can be placed into one of three broad philosophical approaches: interbreeding, ecological and phylogenetic. The Biological Species Concept (BSC) is a classic example of the interbreeding approach. Defined by Ernst Mayr in 1942, the BSC states that "species are groups of actually or potentially interbreeding natural populations, which are reproductively isolated from other such groups." Despite its wide and long-term use, the BSC like others is not without controversy, for example because these concepts cannot be applied to prokaryotes, and this is called the species problem. Some researchers have attempted a unifying monistic definition of species, while others adopt a pluralistic approach and suggest that there may be different ways to logically interpret the definition of a species.

Barriers to reproduction between two diverging sexual populations are required for the populations to become new species. Gene flow may slow this process by spreading the new genetic variants also to the other populations. Depending on how far two species have diverged since their most recent common ancestor, it may still be possible for them to produce offspring, as with horses and donkeys mating to produce mules.Such hybrids are generally infertile. In this case, closely related species may regularly interbreed, but hybrids will be selected against and the species will remain distinct. However, viable hybrids are occasionally formed and these new species can either have properties intermediate between their parent species, or possess a totally new phenotype. The importance of hybridisation in producing new species of animals is unclear, although cases have been seen in many types of animals, with the gray tree frog being a particularly well-studied example.

Speciation has been observed multiple times under both controlled laboratory conditions and in nature. In sexually reproducing organisms, speciation results from reproductive isolation followed by genealogical divergence. There are four mechanisms for speciation. The most common in animals is allopatric speciation, which occurs in populations initially isolated geographically, such as by habitat fragmentation or migration. Selection under these conditions can produce very rapid changes in the appearance and behaviour of organisms. As selection and drift act independently on populations isolated from the rest of their species, separation may eventually produce organisms that cannot interbreed.

The second mechanism of speciation is peripatric speciation, which occurs when small populations of organisms become isolated in a new environment. This differs from allopatric speciation in that the isolated populations are numerically much smaller than the parental population. Here, the founder effect causes rapid speciation after an increase in inbreeding increases selection on homozygotes, leading to rapid genetic change.

The third mechanism of speciation is parapatric speciation. This is similar to peripatric speciation in that a small population enters a new habitat, but differs in that there is no physical separation between these two populations. Instead, speciation results from the evolution of mechanisms that reduce gene flow between the two populations. Generally this occurs when there has been a drastic change in the environment within the parental species› habitat. One example is the grass *Anthoxanthum odoratum*, which can undergo parapatric speciation in response to localised metal pollution from mines. Here, plants evolve that have resistance to high levels of metals in the soil. Selection against interbreeding with the metal-sensitive parental population produced a gradual change in the flowering time of the metal-resistant plants, which eventually produced complete reproductive isolation. Selection against hybrids between the two populations may cause *reinforcement*, which is the evolution of traits that promote mating within a species, as well as character displacement, which is when two species become more distinct in appearance.

Geographical isolation of finches on the Galápagos Islands produced over a dozen new species.

Finally, in sympatric speciation species diverge without geographic isolation or changes in habitat. This form is rare since even a small amount of gene flow may remove genetic differences between parts of a population. Generally, sympatric speciation in animals requires the evolution of both genetic differences and non-random mating, to allow reproductive isolation to evolve.

One type of sympatric speciation involves crossbreeding of two related species to produce a new hybrid species. This is not common in animals as animal hybrids are usually sterile. This is because during meiosis the homologous chromosomes from each parent are from different species and cannot successfully pair. However, it is more common in plants because plants often double their number of chromosomes, to form polyploids. This allows the chromosomes from each parental species to form matching pairs during meiosis, since each parent›s chromosomes are represented by a pair already. An example of such a speciation event is when the plant species *Arabidopsis thaliana* and *Arabidopsis arenosa* crossbred to give the new species *Arabidopsis suecica*. This happened about 20,000 years ago, and the speciation process has been repeated in the laboratory, which allows the study of the genetic mechanisms involved in this process. Indeed, chromosome doubling within a species may be a common cause of reproductive isolation, as half the doubled chromosomes will be unmatched when breeding with undoubled organisms.

Speciation events are important in the theory of punctuated equilibrium, which accounts for the pattern in the fossil record of short "bursts" of evolution interspersed with relatively long periods of stasis, where species remain relatively unchanged. In this theory, speciation and rapid evolution are linked, with natural selection and genetic drift acting most strongly on organisms undergoing speciation in novel habitats or small populations. As a result, the periods of stasis in the fossil record correspond to the parental population and the organisms undergoing speciation and rapid evolution are found in small populations or geographically restricted habitats and therefore rarely being preserved as fossils.

Extinction

Tyrannosaurus rex. Non-aviandinosaurs died out in the Cretaceous–Paleogene extinction event at the end of the Cretaceous period.

Extinction is the disappearance of an entire species. Extinction is not an unusual event, as species regularly appear through speciation and disappear through extinction. Nearly all animal and plant species that have lived on Earth are now extinct, and extinction appears to be the ultimate fate of all species. These extinctions have happened continuously throughout the history of life, although the rate of extinction spikes in occasional mass extinction events. The Cretaceous–Paleogene extinction event, during which the non-avian dinosaurs became extinct, is the most well-known, but the earlier Permian–Triassic extinction event was even more severe, with approximately 96% of all marine species driven to extinction. The Holocene extinction event is an ongoing mass extinction associated with humanity's expansion across the globe over the past few thousand years. Present-day extinction rates are 100–1000 times greater than the background rate and up to 30% of current species may be extinct by the mid 21st century. Human activities are now the primary cause of the ongoing extinction event; global warming may further accelerate it in the future.

The role of extinction in evolution is not very well understood and may depend on which type of extinction is considered. The causes of the continuous low-level extinction events, which form the majority of extinctions, may be the result of competition between species for limited resources (the competitive exclusion principle). If one species can out-compete another, this could produce species selection, with the fitter species surviving and the other species being driven to extinction. The intermittent mass extinctions are also important, but instead of acting as a selective force, they drastically reduce diversity in a nonspecific manner and promote bursts of rapid evolution and speciation in survivors.

Evolutionary History of Life

Origin of Life

The Earth is about 4.54 billion years old. The earliest undisputed evidence of life on Earth dates from at least 3.5 billion years ago, during the Eoarchean Era after a geological crust started to solidify following the earlier molten Hadean Eon. Microbial mat fossils have been found in 3.48 billion-year-old sandstone in Western Australia. Other early physical evidence of a biogenic substance is graphite in 3.7 billion-year-old metasedimentary rocks discovered in Western Greenland as well as "remains of biotic life" found in 4.1 billion-year-old rocks in Western Australia. According to one of the researchers, "If life arose relatively quickly on Earth … then it could be common in the universe."

More than 99 percent of all species, amounting to over five billion species, that ever lived on Earth are estimated to be extinct. Estimates on the number of Earth's current species range from 10 million to 14 million, of which about 1.9 million are estimated to have been named and 1.6 million documented in a central database to date, leaving at least 80 percent not yet described.

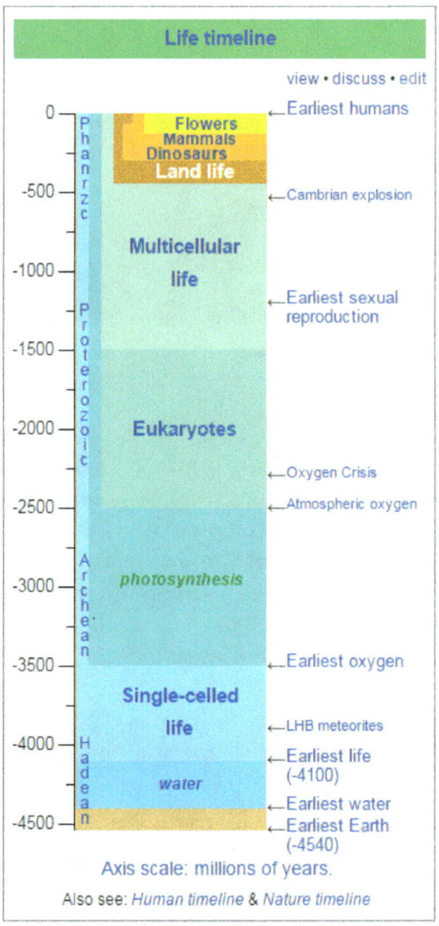

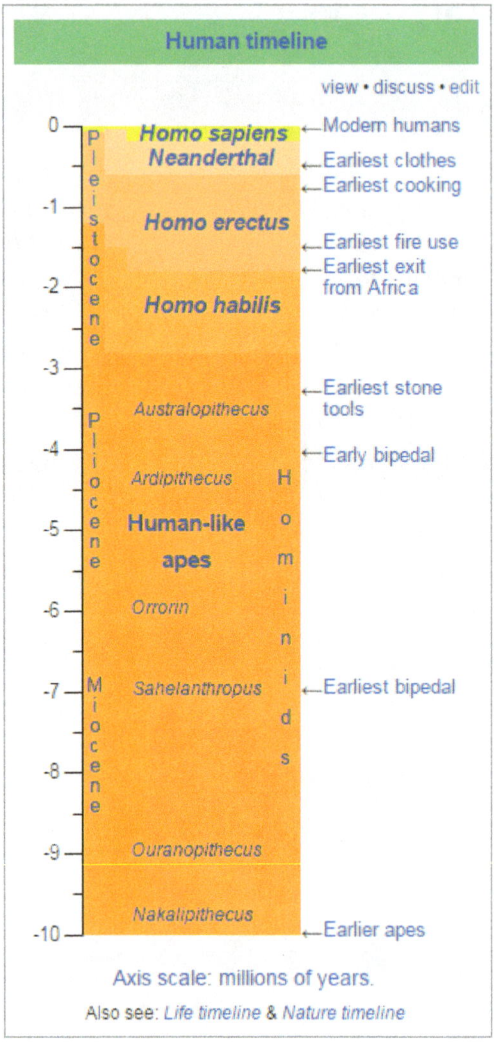

Highly energetic chemistry is thought to have produced a self-replicating molecule around 4 billion years ago, and half a billion years later the last common ancestor of all life existed. The current scientific consensus is that the complex biochemistry that makes up life came from simpler chemical reactions. The beginning of life may have included self-replicating molecules such as RNA and the assembly of simple cells.

Common Descent

All organisms on Earth are descended from a common ancestor or ancestral gene pool. Current species are a stage in the process of evolution, with their diversity the product of a long series of speciation and extinction events. The common descent of organisms was first deduced from four simple facts about organisms: First, they have geographic distributions that cannot be explained by local adaptation. Second, the diversity of life is not a set of completely unique organisms, but organisms that share morphological similarities. Third, vestigial traits with no clear purpose resemble functional ancestral traits and finally, that organisms can be classified using these similarities into a hierarchy of nested groups—similar to a family tree. However, modern research has suggested that, due to horizontal gene transfer, this "tree of life" may be more complicated than a simple branching tree since some genes have spread independently between distantly related species.

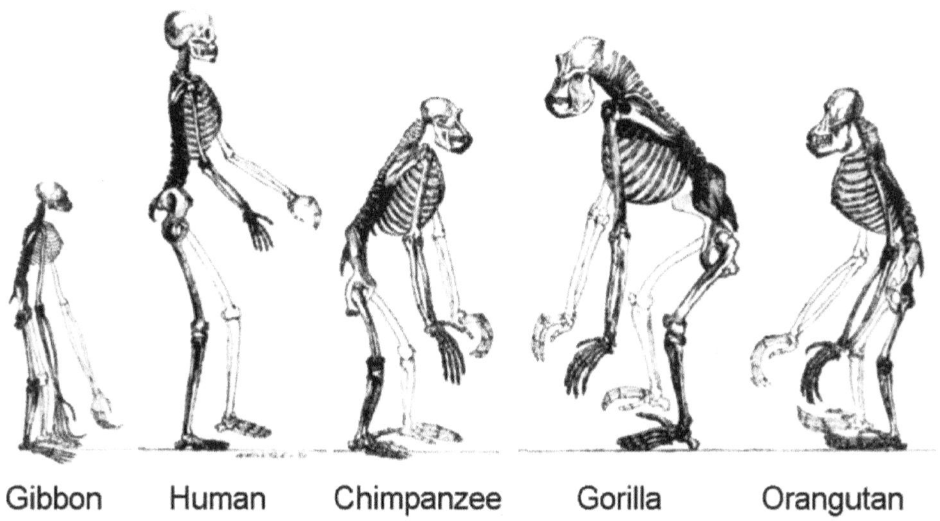

The hominoids are descendants of a common ancestor.

Past species have also left records of their evolutionary history. Fossils, along with the comparative anatomy of present-day organisms, constitute the morphological, or anatomical, record. By comparing the anatomies of both modern and extinct species, paleontologists can infer the lineages of those species. However, this approach is most successful for organisms that had hard body parts, such as shells, bones or teeth. Further, as prokaryotes such as bacteria and archaea share a limited set of common morphologies, their fossils do not provide information on their ancestry.

More recently, evidence for common descent has come from the study of biochemical similarities between organisms. For example, all living cells use the same basic set of nucleotides and amino acids. The development of molecular genetics has revealed the record of evolution left in organisms' genomes: dating when species diverged through the molecular clock produced by mutations. For example, these DNA sequence comparisons have revealed that humans and chimpanzees share 98% of their genomes and analysing the few areas where they differ helps shed light on when the common ancestor of these species existed.

Evolution of Life

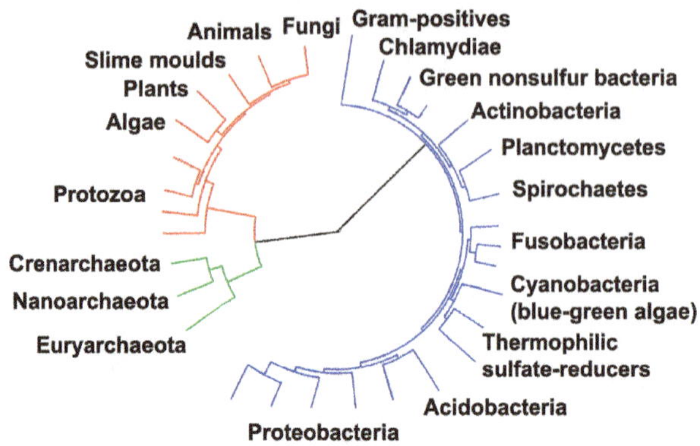

Evolutionary tree showing the divergence of modern species from their common ancestor in the centre. The three domains are coloured, with bacteria blue, archaea green and eukaryotes red.

Prokaryotes inhabited the Earth from approximately 3–4 billion years ago. No obvious changes in morphology or cellular organisation occurred in these organisms over the next few billion years. The eukaryotic cells emerged between 1.6–2.7 billion years ago. The next major change in cell structure came when bacteria were engulfed by eukaryotic cells, in a cooperative association called endosymbiosis. The engulfed bacteria and the host cell then underwent coevolution, with the bacteria evolving into either mitochondria or hydrogenosomes. Another engulfment of cyanobacterial-like organisms led to the formation of chloroplasts in algae and plants.

The history of life was that of the unicellular eukaryotes, prokaryotes and archaea until about 610 million years ago when multicellular organisms began to appear in the oceans in the Ediacaran period. The evolution of multicellularity occurred in multiple independent events, in organisms as diverse as sponges, brown algae, cyanobacteria, slime moulds and myxobacteria. In January 2016, scientists reported that, about 800 million years ago, a minor genetic change in a single molecule called GK-PID may have allowed organisms to go from a single cell organism to one of many cells.

Soon after the emergence of these first multicellular organisms, a remarkable amount of biological diversity appeared over approximately 10 million years, in an event called the Cambrian explosion. Here, the majority of types of modern animals appeared in the fossil record, as well as unique lineages that subsequently became extinct. Various triggers for the Cambrian explosion have been proposed, including the accumulation of oxygen in the atmosphere from photosynthesis.

About 500 million years ago, plants and fungi colonised the land and were soon followed by arthropods and other animals. Insects were particularly successful and even today make up the majority of animal species. Amphibians first appeared around 364 million years ago, followed by early amniotes and birds around 155 million years ago (both from "reptile-like lineages), mammals around 129 million years ago, homininae around 10 million years ago and modern humans around 250,000 years ago. However, despite the evolution of these large animals, smaller organisms similar to the types that evolved early in this process continue to be highly successful and dominate the Earth, with the majority of both biomass and species being prokaryotes.

Applications

Concepts and models used in evolutionary biology, such as natural selection, have many applications.

Artificial selection is the intentional selection of traits in a population of organisms. This has been used for thousands of years in the domestication of plants and animals. More recently, such selection has become a vital part of genetic engineering, with selectable markers such as antibiotic resistance genes being used to manipulate DNA. Proteins with valuable properties have evolved by repeated rounds of mutation and selection (for example modified enzymes and new antibodies) in a process called directed evolution.

Understanding the changes that have occurred during an organism›s evolution can reveal the genes needed to construct parts of the body, genes which may be involved in human genetic disorders. For example, the Mexican tetra is an albino cavefish that lost its eyesight during evolution. Breeding together different populations of this blind fish produced some offspring with functional eyes, since different mutations had occurred in the isolated populations that had evolved in different caves. This helped identify genes required for vision and pigmentation.

Many human diseases are not static phenomena, but capable of evolution. Viruses, bacteria, fungi and cancers evolve to be resistant to host immune defences, as well as pharmaceutical drugs. These same problems occur in agriculture with pesticide and herbicideresistance. It is possible that we are facing the end of the effective life of most of available antibiotics and predicting the evolution and evolvability of our pathogens and devising strategies to slow or circumvent it is requiring deeper knowledge of the complex forces driving evolution at the molecular level.

In computer science, simulations of evolution using evolutionary algorithms and artificial life started in the 1960s and were extended with simulation of artificial selection. Artificial evolution became a widely recognised optimisation method as a result of the work of Ingo Rechenberg in the 1960s. He used evolution strategies to solve complex engineering problems. Genetic algorithms in particular became popular through the writing of John Henry Holland. Practical applications also include automatic evolution of computer programmes. Evolutionary algorithms are now used to solve multi-dimensional problems more efficiently than software produced by human designers and also to optimise the design of systems.

Social and Cultural Responses

As evolution became widely accepted in the 1870s, caricatures of Charles Darwin with an ape or monkey body symbolised evolution.

In the 19th century, particularly after the publication of *On the Origin of Species* in 1859, the idea that life had evolved was an active source of academic debate centred on the philosophical, social and religious implications of evolution. Today, the modern evolutionary synthesis is accepted by a vast majority of scientists. However, evolution remains a contentious concept for some theists.

While various religions and denominations have reconciled their beliefs with evolution through concepts such as theistic evolution, there are creationists who believe that evolution is contradicted by the creation myths found in their religions and who raise various objections to evolution. As had been demonstrated by responses to the publication of *Vestiges of the Natural History of Creation* in 1844, the most controversial aspect of evolutionary biology is the implication of human evolution that humans share common ancestry with apes and that the mental and moral faculties of humanity have the same types of natural causes as other inherited traits in animals. In some countries, notably the United States, these tensions between science and religion have fuelled the current creation–evolution controversy, a religious conflict focusing on politics and public education. While other scientific fields such as cosmology and Earth science also conflict with literal interpretations of many religious texts, evolutionary biology experiences significantly more opposition from religious literalists.

The teaching of evolution in American secondary school biology classes was uncommon in most of the first half of the 20th century. The Scopes Trial decision of 1925 caused the subject to become very rare in American secondary biology textbooks for a generation, but it was gradually re-introduced later and became legally protected with the 1968 *Epperson v. Arkansas* decision. Since then, the competing religious belief of creationism was legally disallowed in secondary school curricula in various decisions in the 1970s and 1980s, but it returned in pseudoscientific form as intelligent design (ID), to be excluded once again in the 2005 *Kitzmiller v. Dover Area School District* case.

REFERENCES

1. Hall & Hallgrímsson 2008, pp. 4–6
2. "Evolution Resources". National Academies of Sciences, Engineering, and Medicine. 2016.
3. Hall & Hallgrímsson 2008, pp. 3–5
4. Kampourakis 2014, pp. 127–129
5. Doolittle, W. Ford (February 2000). "Uprooting the Tree of Life"(PDF). Scientific American. Stuttgart: Georg von Holtzbrinck Publishing Group. 282 (2): 90–95. doi:10.1038/scientificamerican0200-90. ISSN 0036-8733. PMID 10710791. Archived from the original(PDF) on 2006-09-07. Retrieved 2015-04-05.
6. Glansdorff, Nicolas; Ying Xu; Labedan, Bernard (July 9, 2008). "The Last Universal Common Ancestor: emergence, constitution and genetic legacy of an elusive forerunner". Biology Direct. London: BioMed Central. 3: 29. doi:10.1186/1745-6150-3-29. ISSN 1745-6150. PMC 2478661 . PMID 18613974.
7. Schopf, J. William; Kudryavtsev, Anatoliy B.; Czaja, Andrew D.; Tripathi, Abhishek B. (October 5, 2007). "Evidence of Archean life: Stromatolites and microfossils". Precambrian Research. Amsterdam, the Netherlands: Elsevier. 158 (3–4): 141–155. doi:10.1016/j.precamres.2007.04.009. ISSN 0301-9268.
8. Borenstein, Seth (October 19, 2015). "Hints of life on what was thought to be desolate early Earth". Excite. Yonkers, NY: Mindspark Interactive Network. Associated Press. Retrieved 2015-10-20.
9. Bell, Elizabeth A.; Boehnike, Patrick; Harrison, T. Mark; et al. (November 24, 2015). "Potentially biogenic carbon preserved in a 4.1 billion-year-old zircon" (PDF). Proc. Natl. Acad. Sci. U.S.A. Washington, D.C.: National Academy of Sciences. 112 (47): 14518–14521. doi:10.1073/pnas.1517557112. ISSN 0027-8424. PMC 4664351 . PMID 26483481. Retrieved 2015-12-30.
10. Wade, Nicholas (July 25, 2016). "Meet Luca, the Ancestor of All Living Things". New York Times. Retrieved July 25, 2016.
11. Panno 2005, pp. xv-16
12. Ohtomo, Yoko; Kakegawa, Takeshi; Ishida, Akizumi; et al. (January 2014). "Evidence for biogenic graphite in early Archaean Isua metasedimentary rocks". Nature Geoscience. London: Nature Publishing Group. 7 (1): 25–28. Bibcode:2014NatGe...7...25O. doi:10.1038/ngeo2025. ISSN 1752-0894.

13. Borenstein, Seth (November 13, 2013). "Oldest fossil found: Meet your microbial mom". Excite. Yonkers, NY: Mindspark Interactive Network. Associated Press. Retrieved 2015-05-31.
14. Pearlman, Jonathan (November 13, 2013). "'Oldest signs of life on Earth found'". The Daily Telegraph. London: Telegraph Media Group. Retrieved 2014-12-15.
15. Noffke, Nora; Christian, Daniel; Wacey, David; Hazen, Robert M. (November 16, 2013). "Microbially Induced Sedimentary Structures Recording an Ancient Ecosystem in the ca. 3.48 Billion-Year-Old Dresser Formation, Pilbara, Western Australia". Astrobiology. New Rochelle, NY: Mary Ann Liebert, Inc. 13 (12): 1103–1124. Bibcode:2013AsBio..13.1103N. doi:10.1089/ast.2013.1030. ISSN 1531-1074. PMC 3870916 . PMID 24205812.
16. Futuyma 2004, p. 33
17. Stearns & Stearns 1999, p. x
18. Novacek, Michael J. (November 8, 2014). "Prehistory's Brilliant Future". The New York Times. New York: The New York Times Company. ISSN 0362-4331. Retrieved 2014-12-25.
19. Mora, Camilo; Tittensor, Derek P.; Adl, Sina; et al. (August 23, 2011). "How Many Species Are There on Earth and in the Ocean?". PLOS Biology. San Francisco, CA: Public Library of Science. 9 (8): e1001127. doi:10.1371/journal.pbio.1001127. ISSN 1545-7885. PMC 3160336 . PMID 21886479.
20. Miller & Spoolman 2012, p. 62
21. Chapman, Arthur D. (2009). Numbers of Living Species in Australia and the World. 2nd edition. Canberra: Australian Biological Resources Study. ISBN 978-0-642-56860-1. Retrieved 2016-11-06.
22. "Catalogue of Life: 2016 Annual Checklist". 2016. Retrieved 2016-11-06.
23. Staff (2 May 2016). "Researchers find that Earth may be home to 1 trillion species". National Science Foundation. Retrieved 6 May 2016.
24. Lewontin, R. C. (November 1970). "The Units of Selection"(PDF). Annual Review of Ecology and Systematics. Palo Alto, CA: Annual Reviews. 1: 1–18. doi:10.1146/annurev.es.01.110170.000245. ISSN 1545-2069. JSTOR 2096764.
25. Darwin 1859, Chapter XIV
26. Scott-Phillips, T. C.; Laland, K. N.; Shuker, D. M.; Dickins, T. E.; West, S. A. (2014). "The Niche Construction Perspective: A Critical Appraisal". Evolution. 68: 1231–1243. doi:10.1111/evo.12332.
27. Provine 1988, pp. 49–79
28. NAS 2008, pp. R11–R12
29. Ayala & Avise 2014[page needed]
30. NAS 2008, p. 17
31. Moore, Decker & Cotner 2010, p. 454
32. Futuyma, Douglas J., ed. (1999). "Evolution, Science, and Society: Evolutionary Biology and the National Research Agenda" (PDF)(Executive summary). New

Brunswick, NJ: Office of University Publications, Rutgers, The State University of New Jersey. OCLC 43422991. Archived from the original (PDF) on 2012-01-31. Retrieved 2014-11-24.

33. Darwin 1909, p. 53
34. Kirk, Raven & Schofield 1983, pp. 100–142, 280–321
35. Lucretius. "Book V, lines 855–877". De Rerum Natura. Perseus Digital Library. Edited and translated by William Ellery Leonard (1916). Medford/Somerville, MA: Tufts University. OCLC 33233743. Retrieved 2014-11-25.
36. Sedley, David (2003). "Lucretius and the New Empedocles" (PDF). Leeds International Classical Studies. Leeds, West Yorkshire, England: Leeds International Classics Seminar. 2 (4). ISSN 1477-3643. Retrieved 2014-11-25.
37. Torrey, Harry Beal; Felin, Frances (March 1937). "Was Aristotle an Evolutionist?". The Quarterly Review of Biology. 12 (1): 1–18. doi:10.1086/394520. ISSN 0033-5770. JSTOR 2808399.
38. Hull, David L. (December 1967). "The Metaphysics of Evolution". The British Journal for the History of Science. Cambridge: Cambridge University Press. 3 (4): 309–337. doi:10.1017/S0007087400002892. JSTOR 4024958.
39. Mason 1962, pp. 43–44
40. Mayr 1982, pp. 256–257
41. Ray 1686
42. Waggoner, Ben (July 7, 2000). "Carl Linnaeus (1707-1778)". Evolution (Online exhibit). Berkeley, CA: University of California Museum of Paleontology. Retrieved 2012-02-11.
43. Bowler 2003, pp. 73–75
44. "Erasmus Darwin (1731-1802)". Evolution (Online exhibit). Berkeley, CA: University of California Museum of Paleontology. October 4, 1995. Retrieved February 11, 2012.
45. Lamarck 1809
46. Nardon & Grenier 1991, p. 162
47. Gould 2002[page needed]
48. Ghiselin, Michael T. (September–October 1994). "The Imaginary Lamarck: A Look at Bogus 'History' in Schoolbooks". The Textbook Letter. Sausalito, CA: The Textbook League. OCLC 23228649. Retrieved 2008-01-23.
49. Magner 2002[page needed]
50. Jablonka, Eva; Lamb, Marion J. (August 2007). "Précis of Evolution in Four Dimensions". Behavioural and Brain Sciences. Cambridge: Cambridge University Press. 30 (4): 353–365. doi:10.1017/S0140525X07002221. ISSN 0140-525X.
51. Burkhardt & Smith 1991
52. "Darwin, C. R. to Lubbock, John". Darwin Correspondence Project. Cambridge, UK: University of Cambridge. Retrieved 2014-12-01. Letter 2532, November

22, 1859.

53. Sulloway, Frank J. (June 2009). "Why Darwin rejected intelligent design". Journal of Biosciences. Bangalore: Indian Academy of Sciences. 34 (2): 173–183. doi:10.1007/s12038-009-0020-8. ISSN 0250-5991. PMID 19550032.
54. Dawkins 1990[page needed]
55. Sober, Elliott (June 16, 2009). "Did Darwin write the Origin backwards?". Proc. Natl. Acad. Sci. U.S.A. Washington, D.C.: National Academy of Sciences. 106 (Suppl. 1): 10048–10055. Bibcode:2009PNAS..10610048S. doi:10.1073/pnas.0901109106. ISSN 0027-8424.
56. Mayr 2002, p. 165
57. Bowler 2003, pp. 145–146
58. Sokal, Robert R.; Crovello, Theodore J. (March–April 1970). "The Biological Species Concept: A Critical Evaluation". The American Naturalist. Chicago, IL: University of Chicago Press on behalf of the American Society of Naturalists. 104 (936): 127–153. doi:10.1086/282646. ISSN 0003-0147. JSTOR 2459191.
59. Darwin, Charles; Wallace, Alfred (August 20, 1858). "On the Tendency of Species to form Varieties; and on the Perpetuation of Varieties and Species by Natural Means of Selection". Journal of the Proceedings of the Linnean Society of London. Zoology. 3 (9): 45–62. doi:10.1111/j.1096-3642.1858.tb02500.x. ISSN 1096-3642. Retrieved 2007-05-13.
60. Desmond, Adrian J. (July 17, 2014). "Thomas Henry Huxley". Encyclopædia Britannica Online. Chicago, IL: Encyclopædia Britannica, Inc. Retrieved 2014-12-02.
61. Liu, Y. S.; Zhou, X. M.; Zhi, M. X.; Li, X. J.; Wang, Q. L. (September 2009). "Darwin's contributions to genetics". Journal of Applied Genetics. Poznań: Institute of Plant Genetics, Polish Academy of Sciences. 50 (3): 177–184. doi:10.1007/BF03195671. ISSN 1234-1983. PMID 19638672.
62. Weiling, Franz (July 1991). "Historical study: Johann Gregor Mendel 1822–1884". American Journal of Medical Genetics. 40 (1): 1–25; discussion 26. doi:10.1002/ajmg.1320400103. PMID 1887835.
63. Wright 1984, p. 480
64. Provine 1971
65. Stamhuis, Ida H.; Meijer, Onno G.; Zevenhuizen, Erik J. A. (June 1999). "Hugo de Vries on Heredity, 1889-1903: Statistics, Mendelian Laws, Pangenes, Mutations". Isis. Chicago, IL: University of Chicago Press. 90 (2): 238–267. doi:10.1086/384323. ISSN 0021-1753. JSTOR 237050. PMID 10439561.
66. Quammen 2006[page needed]
67. Bowler 1989[page needed]
68. Watson, J. D.; Crick, F. H. C. (April 25, 1953). "Molecular Structure of Nucleic Acids: A Structure for Deoxyribose Nucleic Acid" (PDF). Nature. London: Nature Publishing Group. 171 (4356): 737–738. Bibcode:1953Natur.171..737W. doi:10.1038/171737a0. ISSN 0028-0836. PMID 13054692. Retrieved 2014-12-

04. It has not escaped our notice that the specific pairing we have postulated immediately suggests a possible copying mechanism for the genetic material.

69. Hennig 1999, p. 280
70. Wiley & Lieberman 2011[page needed]
71. Dobzhansky, Theodosius (March 1973). "Nothing in Biology Makes Sense Except in the Light of Evolution". The American Biology Teacher. McLean, VA: National Association of Biology Teachers. 35 (3): 125–129. doi:10.2307/4444260.
72. Kutschera, Ulrich; Niklas, Karl J. (June 2004). "The modern theory of biological evolution: an expanded synthesis". Naturwissenschaften. Heidelberg: Springer-Verlag Heidelberg. 91 (6): 255–276. Bibcode:2004NW.....91..255K. doi:10.1007/s00114-004-0515-y. ISSN 1432-1904. PMID 15241603.
73. Cracraft & Bybee 2005[page needed]
74. Avise, John C.; Ayala, Francisco J. (May 11, 2010). "In the light of evolution IV: The human condition" (PDF). Proc. Natl. Acad. Sci. U.S.A. Washington, D.C.: National Academy of Sciences. 107 (Suppl. 2): 8897–8901. doi:10.1073/pnas.100321410. ISSN 0027-8424. Retrieved 2014-12-29.
75. Sturm, Richard A.; Frudakis, Tony N. (August 2004). "Eye colour: portals into pigmentation genes and ancestry". Trends in Genetics. Cambridge, MA: Cell Press. 20 (8): 327–332. doi:10.1016/j.tig.2004.06.010. ISSN 0168-9525. PMID 15262401.
76. Pearson, Helen (May 25, 2006). "Genetics: What is a gene?". Nature. London: Nature Publishing Group. 441 (7092): 398–401. Bibcode:2006Natur.441..398P. doi:10.1038/441398a. ISSN 0028-0836. PMID 16724031.
77. Visscher, Peter M.; Hill, William G.; Wray, Naomi R. (April 2008). "Heritability in the genomics era — concepts and misconceptions". Nature Reviews Genetics. London: Nature Publishing Group. 9 (4): 255–266. doi:10.1038/nrg2322. ISSN 1471-0056. PMID 18319743.
78. Oetting, William S.; Brilliant, Murray H.; King, Richard A. (August 1996). "The clinical spectrum of albinism in humans". Molecular Medicine Today. Cambridge, MA: Cell Press. 2 (8): 330–335. doi:10.1016/1357-4310(96)81798-9. ISSN 1357-4310. PMID 8796918.
79. Futuyma 2005[page needed]
80. Phillips, Patrick C. (November 2008). "Epistasis—the essential role of gene interactions in the structure and evolution of genetic systems". Nature Reviews Genetics. London: Nature Publishing Group. 9 (11): 855–867. doi:10.1038/nrg2452. ISSN 1471-0056. PMC 2689140 . PMID 18852697.
81. Rongling Wu; Min Lin (March 2006). "Functional mapping — how to map and study the genetic architecture of dynamic complex traits". Nature Reviews Genetics. London: Nature Publishing Group. 7 (3): 229–237. doi:10.1038/nrg1804. ISSN 1471-0056. PMID 16485021.
82. Jablonka, Eva; Raz, Gal (June 2009). "Transgenerational Epigenetic Inheritance: Prevalence, Mechanisms, and Implications for the Study of Heredity and

Evolution". The Quarterly Review of Biology. Chicago, IL: University of Chicago Press. 84 (2): 131–176. doi:10.1086/598822. ISSN 0033-5770. PMID 19606595.

83. Bossdorf, Oliver; Arcuri, Davide; Richards, Christina L.; Pigliucci, Massimo (May 2010). "Experimental alteration of DNA methylation affects the phenotypic plasticity of ecologically relevant traits in Arabidopsis thaliana". Evolutionary Ecology. Dordrecht, the Netherlands: Springer Netherlands. 24 (3): 541–553. doi:10.1007/s10682-010-9372-7. ISSN 0269-7653.
84. Jablonka & Lamb 2005[page needed]
85. Jablonka, Eva; Lamb, Marion J. (December 2002). "The Changing Concept of Epigenetics". Annals of the New York Academy of Sciences. Hoboken, NJ: Wiley-Blackwell on behalf of the New York Academy of Sciences. 981 (1): 82–96. Bibcode:2002NYASA.981...82J. doi:10.1111/j.1749-6632.2002.tb04913.x. ISSN 0077-8923. PMID 12547675.
86. Laland, Kevin N.; Sterelny, Kim (September 2006). "Perspective: Seven Reasons (Not) to Neglect Niche Construction". Evolution. Hoboken, NJ: John Wiley & Sons on behalf of the Society for the Study of Evolution. 60 (9): 1751–1762. doi:10.1111/j.0014-3820.2006.tb00520.x. ISSN 0014-3820.
87. Chapman, Michael J.; Margulis, Lynn (December 1998). "Morphogenesis by symbiogenesis" (PDF). International Microbiology. Barcelona: Spanish Society for Microbiology. 1 (4): 319–326. ISSN 1139-6709. PMID 10943381. Retrieved 2014-12-09.
88. Wilson, David Sloan; Wilson, Edward O. (December 2007). "Rethinking the Theoretical Foundation of Sociobiology" (PDF). The Quarterly Review of Biology. Chicago, IL: University of Chicago Press. 82 (4): 327–348. doi:10.1086/522809. ISSN 0033-5770. PMID 18217526.
89. Amos, William; Harwood, John (February 28, 1998). "Factors affecting levels of genetic diversity in natural populations". Philosophical Transactions of the Royal Society B: Biological Sciences. London: Royal Society. 353 (1366): 177–186. doi:10.1098/rstb.1998.0200. ISSN 0962-8436. PMC 1692205 . PMID 9533122.
90. Ewens 2004[page needed]
91. Butlin, Roger K.; Tregenza, Tom (February 28, 1998). "Levels of genetic polymorphism: marker loci versus quantitative traits". Philosophical Transactions of the Royal Society B: Biological Sciences. London: Royal Society. 353 (1366): 187–198. doi:10.1098/rstb.1998.0201. ISSN 0962-8436. PMC 1692210 . PMID 9533123.
92. Butlin, Roger K.; Tregenza, Tom (December 29, 2000). "Correction for Butlin and Tregenza, Levels of genetic polymorphism: marker loci versus quantitative traits". Philosophical Transactions of the Royal Society B: Biological Sciences. London: Royal Society. 355 (1404): 1865. doi:10.1098/rstb.2000.2000. ISSN 0962-8436. Some of the values in table 1 on p. 193 were given incorrectly. The errors do not affect the conclusions drawn in the paper. The corrected table is reproduced below.
93. Wetterbom, Anna; Sevov, Marie; Cavelier, Lucia; Bergström, Tomas F. (November

2006). "Comparative Genomic Analysis of Human and Chimpanzee Indicates a Key Role for Indels in Primate Evolution". Journal of Molecular Evolution. New York: Springer-Verlag New York. 63 (5): 682–690. doi:10.1007/s00239-006-0045-7. ISSN 0022-2844. PMID 17075697.

94. Sawyer, Stanley A.; Parsch, John; Zhang Zhi; Hartl, Daniel L. (Apr 17, 2007). "Prevalence of positive selection among nearly neutral amino acid replacements in Drosophila". Proc. Natl. Acad. Sci. U.S.A. Washington, D.C.: National Academy of Sciences. 104 (16): 6504–6510. Bibcode:2007PNAS..104.6504S. doi:10.1073/pnas.0701572104. ISSN 0027-8424. PMC 1871816 . PMID 17409186.

95. Hastings, P. J.; Lupski, James R.; Rosenberg, Susan M.; Ira, Grzegorz (August 2009). "Mechanisms of change in gene copy number". Nature Reviews Genetics. London: Nature Publishing Group. 10 (8): 551–564. doi:10.1038/nrg2593. ISSN 1471-0056. PMC 2864001 . PMID 19597530.

96. Carroll, Grenier & Weatherbee 2005[page needed]

97. Harrison, Paul M.; Gerstein, Mark (May 17, 2002). "Studying Genomes Through the Aeons: Protein Families, Pseudogenes and Proteome Evolution". Journal of Molecular Biology. Amsterdam, the Netherlands: Elsevier. 318 (5): 1155–1174. doi:10.1016/S0022-2836(02)00109-2. ISSN 0022-2836. PMID 12083509.

98. Bowmaker, James K. (May 1998). "Evolution of colour vision in vertebrates". Eye. London: Nature Publishing Group on behalf of the Royal College of Ophthalmologists. 12 (3b): 541–547. doi:10.1038/eye.1998.143. ISSN 0950-222X. PMID 9775215.

99. Gregory, T. Ryan; Hebert, Paul D. N. (April 1999). "The Modulation of DNA Content: Proximate Causes and Ultimate Consequences". Genome Research. Cold Spring Harbor, NY: Cold Spring Harbor Laboratory Press. 9 (4): 317–324. doi:10.1101/gr.9.4.317. ISSN 1088-9051. PMID 10207154. Retrieved 2014-12-11.

100. Hurles, Matthew (July 13, 2004). "Gene Duplication: The Genomic Trade in Spare Parts". PLOS Biology. San Francisco, CA: Public Library of Science. 2 (7): e206. doi:10.1371/journal.pbio.0020206. ISSN 1545-7885. PMC 449868 . PMID 15252449.

101. Liu, Na; Okamura, Katsutomo; Tyler, David M.; et al. (October 2008). "The evolution and functional diversification of animal microRNA genes". Cell Research. London: Nature Publishing Group on behalf of the Shanghai Institutes for Biological Sciences. 18 (10): 985–996. doi:10.1038/cr.2008.278. ISSN 1001-0602. PMC 2712117 . PMID 18711447. Retrieved 2014-12-11.

102. Siepel, Adam (October 2009). "Darwinian alchemy: Human genes from noncoding DNA". Genome Research. Cold Spring Harbor, NY: Cold Spring Harbor Laboratory Press. 19 (10): 1693–1695. doi:10.1101/gr.098376.109. ISSN 1088-9051. PMC 2765273 . PMID 19797681. Retrieved 2014-12-11.

103. Orengo, Christine A.; Thornton, Janet M. (July 2005). "Protein families and their evolution—a structural perspective". Annual Review of Biochemistry. Palo Alto, CA: Annual Reviews. 74: 867–900. doi:10.1146/annurev.biochem.74.082803.133029. ISSN 0066-4154. PMID 15954844.

104. Long, Manyuan; Betrán, Esther; Thornton, Kevin; Wang, Wen (November 2003). "The origin of new genes: glimpses from the young and old". Nature Reviews Genetics. London: Nature Publishing Group. 4 (11): 865–875. doi:10.1038/nrg1204. ISSN 1471-0056. PMID 14634634.

105. Wang, Minglei; Caetano-Anollés, Gustavo (January 14, 2009). "The Evolutionary Mechanics of Domain Organization in Proteomes and the Rise of Modularity in the Protein World". Structure. Cambridge, MA: Cell Press. 17 (1): 66–78. doi:10.1016/j.str.2008.11.008. ISSN 1357-4310. PMID 19141283.

106. Weissman, Kira J.; Müller, Rolf (April 14, 2008). "Protein–Protein Interactions in Multienzyme Megasynthetases". ChemBioChem. Weinheim, Germany: Wiley-VCH. 9 (6): 826–848. doi:10.1002/cbic.200700751. ISSN 1439-4227. PMID 18357594.

107. Radding, Charles M. (December 1982). "Homologous Pairing and Strand Exchange in Genetic Recombination". Annual Review of Genetics. Palo Alto, CA: Annual Reviews. 16: 405–437. doi:10.1146/annurev.ge.16.120182.002201. ISSN 0066-4197. PMID 6297377.

108. Agrawal, Aneil F. (September 5, 2006). "Evolution of Sex: Why Do Organisms Shuffle Their Genotypes?". Current Biology. Cambridge, MA: Cell Press. 16 (17): R696–R704. doi:10.1016/j.cub.2006.07.063. ISSN 0960-9822. PMID 16950096.

109. Peters, Andrew D.; Otto, Sarah P. (June 2003). "Liberating genetic variance through sex". BioEssays. Hoboken, NJ: John Wiley & Sons. 25 (6): 533–537. doi:10.1002/bies.10291. ISSN 0265-9247. PMID 12766942.

110. Goddard, Matthew R.; Godfray, H. Charles J.; Burt, Austin (March 31, 2005). "Sex increases the efficacy of natural selection in experimental yeast populations". Nature. London: Nature Publishing Group. 434 (7033): 636–640. Bibcode:2005Natur.434..636G. doi:10.1038/nature03405. ISSN 0028-0836. PMID 15800622.

111. Maynard Smith 1978[page needed]

112. Ridley 1993[page needed]

113. Van Valen, Leigh (1973). "A New Evolutionary Law" (PDF). Evolutionary Theory. Chicago, IL: University of Chicago. 1: 1–30. ISSN 0093-4755. Retrieved 2014-12-24.

114. Hamilton, W. D.; Axelrod, Robert; Tanese, Reiko (May 1, 1990). "Sexual reproduction as an adaptation to resist parasites (a review)". Proc. Natl. Acad. Sci. U.S.A. Washington, D.C.: National Academy of Sciences. 87 (9): 3566–3573. Bibcode:1990PNAS...87.3566H. doi:10.1073/pnas.87.9.3566. ISSN 0027-8424. PMID 2185476.

115. Birdsell & Wills 2003, pp. 113–117

116. Morjan, Carrie L.; Rieseberg, Loren H. (June 2004). "How species evolve collectively: implications of gene flow and selection for the spread of advantageous alleles". Molecular Ecology. Hoboken, NJ: Wiley-Blackwell. 13 (6): 1341–1356. doi:10.1111/j.1365-294X.2004.02164.x. ISSN 0962-1083. PMC 2600545 . PMID 15140081.

117. Boucher, Yan; Douady, Christophe J.; Papke, R. Thane; et al. (December 2003). "Lateral gene transfer and the origins of prokaryotic groups". Annual Review of Genetics. Palo Alto, CA: Annual Reviews. 37: 283–328. doi:10.1146/annurev.genet.37.050503.084247. ISSN 0066-4197. PMID 14616063.

118. Walsh, Timothy R. (October 2006). "Combinatorial genetic evolution of multiresistance". Current Opinion in Microbiology. Amsterdam, the Netherlands: Elsevier. 9 (5): 476–482. doi:10.1016/j.mib.2006.08.009. ISSN 1369-5274. PMID 16942901.

119. Kondo, Natsuko; Nikoh, Naruo; Ijichi, Nobuyuki; et al. (October 29, 2002). "Genome fragment of Wolbachia endosymbiont transferred to X chromosome of host insect". Proc. Natl. Acad. Sci. U.S.A. Washington, D.C.: National Academy of Sciences. 99 (22): 14280–14285. Bibcode:2002PNAS...9914280K. doi:10.1073/pnas.222228199. ISSN 0027-8424. PMC 137875 . PMID 12386340.

120. Sprague, George F., Jr. (December 1991). "Genetic exchange between kingdoms". Current Opinion in Genetics & Development. Amsterdam, the Netherlands: Elsevier. 1 (4): 530–533. doi:10.1016/S0959-437X(05)80203-5. ISSN 0959-437X. PMID 1822285.

121. Gladyshev, Eugene A.; Meselson, Matthew; Arkhipova, Irina R. (May 30, 2008). "Massive Horizontal Gene Transfer in Bdelloid Rotifers". Science. Washington, D.C.: American Association for the Advancement of Science. 320 (5880): 1210–1213. Bibcode:2008Sci...320.1210G. doi:10.1126/science.1156407. ISSN 0036-8075. PMID 18511688.

122. Baldo, Angela M.; McClure, Marcella A. (September 1999). "Evolution and Horizontal Transfer of dUTPase-Encoding Genes in Viruses and Their Hosts". Journal of Virology. Washington, D.C.: American Society for Microbiology. 73 (9): 7710–7721. ISSN 0022-538X. PMC 104298 . PMID 10438861.

123. Rivera, Maria C.; Lake, James A. (September 9, 2004). "The ring of life provides evidence for a genome fusion origin of eukaryotes". Nature. London: Nature Publishing Group. 431 (7005): 152–155. Bibcode:2004Natur.431..152R. doi:10.1038/nature02848. ISSN 0028-0836. PMID 15356622.

124. Hurst, Laurence D. (February 2009). "Fundamental concepts in genetics: genetics and the understanding of selection". Nature Reviews Genetics. London: Nature Publishing Group. 10 (2): 83–93. doi:10.1038/nrg2506. ISSN 1471-0056. PMID 19119264.

125. Orr, H. Allen (August 2009). "Fitness and its role in evolutionary genetics". Nature Reviews Genetics. London: Nature Publishing Group. 10 (8): 531–539. doi:10.1038/nrg2603. ISSN 1471-0056. PMC 2753274 . PMID 19546856.

126. Haldane, J. B. S. (March 14, 1959). "The Theory of Natural Selection To-Day". Nature. London: Nature Publishing Group. 183 (4663): 710–713. Bibcode:1959Natur.183..710H. doi:10.1038/183710a0. ISSN 0028-0836. PMID 13644170.

127. Lande, Russell; Arnold, Stevan J. (November 1983). "The Measurement of Selection on Correlated Characters". Evolution. Hoboken, NJ: John Wiley &

Sons on behalf of the Society for the Study of Evolution. 37 (6): 1210–1226. doi:10.2307/2408842. ISSN 0014-3820. JSTOR 2408842.

128. Goldberg, Emma E.; Igić, Boris (November 2008). "On phylogenetic tests of irreversible evolution". Evolution. Hoboken, NJ: John Wiley & Sons on behalf of the Society for the Study of Evolution. 62 (11): 2727–2741. doi:10.1111/j.1558-5646.2008.00505.x. ISSN 0014-3820. PMID 18764918.
129. Collin, Rachel; Miglietta, Maria Pia (November 2008). "Reversing opinions on Dollo's Law". Trends in Ecology & Evolution. Cambridge, MA: Cell Press. 23 (11): 602–609. doi:10.1016/j.tree.2008.06.013. ISSN 0169-5347. PMID 18814933.
130. Hoekstra, Hopi E.; Hoekstra, Jonathan M.; Berrigan, David; et al. (July 31, 2001). "Strength and tempo of directional selection in the wild". Proc. Natl. Acad. Sci. U.S.A. Washington, D.C.: National Academy of Sciences. 98 (16): 9157–9160. Bibcode:2001PNAS...98.9157H. doi:10.1073/pnas.161281098. ISSN 0027-8424. PMC 55389 . PMID 11470913.
131. Felsenstein, Joseph (November 1979). "Excursions along the Interface between Disruptive and Stabilizing Selection". Genetics. Bethesda, MD: Genetics Society of America. 93 (3): 773–795. ISSN 0016-6731. PMC 1214112 . PMID 17248980.
132. Andersson, Malte; Simmons, Leigh W. (June 2006). "Sexual selection and mate choice". Trends in Ecology & Evolution. Cambridge, MA: Cell Press. 21 (6): 296–302. doi:10.1016/j.tree.2006.03.015. ISSN 0169-5347. PMID 16769428.
133. Kokko, Hanna; Brooks, Robert; McNamara, John M.; Houston, Alasdair I. (July 7, 2002). "The sexual selection continuum". Proceedings of the Royal Society B. London: Royal Society. 269 (1498): 1331–1340. doi:10.1098/rspb.2002.2020. ISSN 0962-8452. PMC 1691039 . PMID 12079655.
134. Quinn, Thomas P.; Hendry, Andrew P.; Buck, Gregory B. (2001). "Balancing natural and sexual selection in sockeye salmon: interactions between body size, reproductive opportunity and vulnerability to predation by bears" (PDF). Evolutionary Ecology Research. 3: 917–937. ISSN 1522-0613. Retrieved 2014-12-15.
135. Hunt, John; Brooks, Robert; Jennions, Michael D.; et al. (December 23, 2004). "High-quality male field crickets invest heavily in sexual display but die young". Nature. London: Nature Publishing Group. 432 (7020): 1024–1027. Bibcode:2004Natur.432.1024H. doi:10.1038/nature03084. ISSN 0028-0836. PMID 15616562.
136. Odum 1971, p. 8
137. Okasha 2006
138. Gould, Stephen Jay (February 28, 1998). "Gulliver's further travels: the necessity and difficulty of a hierarchical theory of selection". Philosophical Transactions of the Royal Society B: Biological Sciences. London: Royal Society. 353 (1366): 307–314. doi:10.1098/rstb.1998.0211. ISSN 0962-8436. PMC 1692213 . PMID 9533127.
139. Mayr, Ernst (March 18, 1997). "The objects of selection". Proc. Natl. Acad. Sci. U.S.A. Washington, D.C.: National Academy of Sciences. 94 (6): 2091–2094.

Bibcode:1997PNAS...94.2091M. doi:10.1073/pnas.94.6.2091. ISSN 0027-8424. PMC 33654 . PMID 9122151.

140. Maynard Smith 1998, pp. 203–211; discussion 211–217

141. Hickey, Donal A. (1992). "Evolutionary dynamics of transposable elements in prokaryotes and eukaryotes". Genetica. Dordrecht, the Netherlands: Kluwer Academic Publishers. 86 (1–3): 269–274. doi:10.1007/BF00133725. ISSN 0016-6707. PMID 1334911.

142. Gould, Stephen Jay; Lloyd, Elisabeth A. (October 12, 1999). "Individuality and adaptation across levels of selection: how shall we name and generalise the unit of Darwinism?". Proc. Natl. Acad. Sci. U.S.A. Washington, D.C.: National Academy of Sciences. 96 (21): 11904–11909. Bibcode:1999PNAS...9611904G. doi:10.1073/pnas.96.21.11904. ISSN 0027-8424. PMC 18385 . PMID 10518549.

143. Lynch, Michael (May 15, 2007). "The frailty of adaptive hypotheses for the origins of organismal complexity". Proc. Natl. Acad. Sci. U.S.A. Washington, D.C.: National Academy of Sciences. 104 (Suppl. 1): 8597–8604. Bibcode:2007PNAS..104.8597L. doi:10.1073/pnas.0702207104. ISSN 0027-8424. PMC 1876435 . PMID 17494740.

144. Smith, Nick G.C.; Webster, Matthew T.; Ellegren, Hans (September 2002). "Deterministic Mutation Rate Variation in the Human Genome". Genome Research. Cold Spring Harbor, NY: Cold Spring Harbor Laboratory Press. 12 (9): 1350–1356. doi:10.1101/gr.220502. ISSN 1088-9051. PMC 186654 . PMID 12213772.

145. Petrov, Dmitri A.; Sangster, Todd A.; Johnston, J. Spencer; et al. (February 11, 2000). "Evidence for DNA Loss as a Determinant of Genome Size". Science. Washington, D.C.: American Association for the Advancement of Science. 287 (5455): 1060–1062. Bibcode:2000Sci...287.1060P. doi:10.1126/science.287.5455.1060. ISSN 0036-8075. PMID 10669421.

146. Petrov, Dmitri A. (May 2002). "DNA loss and evolution of genome size in Drosophila". Genetica. Dordrecht, the Netherlands: Kluwer Academic Publishers. 115 (1): 81–91. doi:10.1023/A:1016076215168. ISSN 0016-6707. PMID 12188050.

147. Kiontke, Karin; Barriere, Antoine; Kolotuev, Irina; et al. (November 2007). "Trends, Stasis, and Drift in the Evolution of Nematode Vulva Development". Current Biology. Cambridge, MA: Cell Press. 17 (22): 1925–1937. doi:10.1016/j.cub.2007.10.061. ISSN 0960-9822. PMID 18024125.

148. Braendle, Christian; Baer, Charles F.; Félix, Marie-Anne (March 12, 2010). Barsh, Gregory S., ed. "Bias and Evolution of the Mutationally Accessible Phenotypic Space in a Developmental System". PLOS Genetics. San Francisco, CA: Public Library of Science. 6 (3): e1000877. doi:10.1371/journal.pgen.1000877. ISSN 1553-7390. PMC 2837400 . PMID 20300655.

149. Palmer, A. Richard (October 29, 2004). "Symmetry breaking and the evolution of development". Science. Washington, D.C.: American Association for the Advancement of Science. 306 (5697): 828–833. Bibcode:2004Sci...306..828P.

doi:10.1126/science.1103707. ISSN 0036-8075. PMID 15514148.

150. West-Eberhard 2003, pp. 140

151. Pocheville, Arnaud; Danchin, Etienne (January 1, 2017). "Chapter 3: Genetic assimilation and the paradox of blind variation". In Huneman, Philippe; Walsh, Denis. Challenging the Modern Synthesis. Oxford University Press.

152. Stoltzfus, Arlin; Yampolsky, Lev Y. (September–October 2009). "Climbing Mount Probable: Mutation as a Cause of Nonrandomness in Evolution". Journal of Heredity. Oxford: Oxford University Press on behalf of the American Genetic Association. 100 (5): 637–647. doi:10.1093/jhered/esp048. ISSN 0022-1503. PMID 19625453.

153. Yampolsky, Lev Y.; Stoltzfus, Arlin (March 2001). "Bias in the introduction of variation as an orienting factor in evolution". Evolution & Development. Hoboken, NJ: Wiley-Blackwell on behalf of the Society for Integrative and Comparative Biology. 3 (2): 73–83. doi:10.1046/j.1525-142x.2001.003002073.x. ISSN 1520-541X. PMID 11341676.

154. Haldane, J. B. S. (January–February 1933). "The Part Played by Recurrent Mutation in Evolution". The American Naturalist. Chicago, IL: University of Chicago Press on behalf of the American Society of Naturalists. 67 (708): 5–19. doi:10.1086/280465. ISSN 0003-0147. JSTOR 2457127.

155. Protas, Meredith; Conrad, Melissa; Gross, Joshua B.; et al. (March 6, 2007). "Regressive Evolution in the Mexican Cave Tetra, Astyanax mexicanus". Current Biology. Cambridge, MA: Cell Press. 17 (5): 452–454. doi:10.1016/j.cub.2007.01.051. ISSN 0960-9822. PMC 2570642 . PMID 17306543.

156. Maughan, Heather; Masel, Joanna; Birky, C. William, Jr.; Nicholson, Wayne L. (October 2007). "The Roles of Mutation Accumulation and Selection in Loss of Sporulation in Experimental Populations of Bacillus subtilis". Genetics. Bethesda, MD: Genetics Society of America. 177 (2): 937–948. doi:10.1534/genetics.107.075663. ISSN 0016-6731. PMC 2034656 . PMID 17720926.

157. Masel, Joanna; King, Oliver D.; Maughan, Heather (January 2007). "The Loss of Adaptive Plasticity during Long Periods of Environmental Stasis". The American Naturalist. Chicago, IL: University of Chicago Press on behalf of the American Society of Naturalists. 169 (1): 38–46. doi:10.1086/510212. ISSN 0003-0147. PMC 1766558 . PMID 17206583.

158. Masel, Joanna (October 25, 2011). "Genetic drift". Current Biology. Cambridge, MA: Cell Press. 21 (20): R837–R838. doi:10.1016/j.cub.2011.08.007. ISSN 0960-9822. PMID 22032182.

159. Lande, Russell (1989). "Fisherian and Wrightian theories of speciation". Genome. Ottawa: National Research Council of Canada. 31 (1): 221–227. doi:10.1139/g89-037. ISSN 0831-2796. PMID 2687093.

160. Mitchell-Olds, Thomas; Willis, John H.; Goldstein, David B. (November 2007). "Which evolutionary processes influence natural genetic variation for phenotypic traits?". Nature Reviews Genetics. London: Nature Publishing Group. 8 (11): 845–856. doi:10.1038/nrg2207. ISSN 1471-0056. PMID 17943192.

161. Nei, Masatoshi (December 2005). "Selectionism and Neutralism in Molecular Evolution". Molecular Biology and Evolution. Oxford: Oxford University Press on behalf of the Society for Molecular Biology and Evolution. 22 (12): 2318–2342. doi:10.1093/molbev/msi242. ISSN 0737-4038. PMC 1513187 . PMID 16120807.

162. Nei, Masatoshi (May 2006). "Selectionism and Neutralism in Molecular Evolution". Molecular Biology and Evolution (Erratum). Oxford: Oxford University Press on behalf of the Society for Molecular Biology and Evolution. 23 (5): 1095. doi:10.1093/molbev/msk009. ISSN 0737-4038.

163. Kimura, Motoo (1991). "The neutral theory of molecular evolution: a review of recent evidence". The Japanese Journal of Human Genetics. Mishima, Japan: Genetics Society of Japan. 66 (4): 367–386. doi:10.1266/jjg.66.367. ISSN 0021-504X. PMID 1954033.

164. Kimura, Motoo (1989). "The neutral theory of molecular evolution and the world view of the neutralists". Genome. Ottawa: National Research Council of Canada. 31 (1): 24–31. doi:10.1139/g89-009. ISSN 0831-2796. PMID 2687096.

165. Kreitman, Martin (August 1996). "The neutral theory is dead. Long live the neutral theory". BioEssays. Hoboken, NJ: John Wiley & Sons. 18 (8): 678–683; discussion 683. doi:10.1002/bies.950180812. ISSN 0265-9247. PMID 8760341.

166. Leigh, E. G., Jr. (November 2007). "Neutral theory: a historical perspective". Journal of Evolutionary Biology. Hoboken, NJ: Wiley-Blackwell on behalf of the European Society for Evolutionary Biology. 20 (6): 2075–2091. doi:10.1111/j.1420-9101.2007.01410.x. ISSN 1010-061X. PMID 17956380.

167. Gillespie, John H. (November 2001). "Is the population size of a species relevant to its evolution?". Evolution. Hoboken, NJ: John Wiley & Sons on behalf of the Society for the Study of Evolution. 55 (11): 2161–2169. doi:10.1111/j.0014-3820.2001.tb00732.x. ISSN 0014-3820. PMID 11794777.

168. Neher, Richard A.; Shraiman, Boris I. (August 2011). "Genetic Draft and Quasi-Neutrality in Large Facultatively Sexual Populations". Genetics. Bethesda, MD: Genetics Society of America. 188 (4): 975–996. doi:10.1534/genetics.111.128876. ISSN 0016-6731. PMC 3176096 . PMID 21625002.

169. Otto, Sarah P.; Whitlock, Michael C. (June 1997). "The Probability of Fixation in Populations of Changing Size" (PDF). Genetics. Bethesda, MD: Genetics Society of America. 146 (2): 723–733. ISSN 0016-6731. PMC 1208011 . PMID 9178020. Retrieved 2014-12-18.

170. Charlesworth, Brian (March 2009). "Fundamental concepts in genetics: effective population size and patterns of molecular evolution and variation". Nature Reviews Genetics. London: Nature Publishing Group. 10 (3): 195–205. doi:10.1038/nrg2526. ISSN 1471-0056. PMID 19204717.

171. Cutter, Asher D.; Choi, Jae Young (August 2010). "Natural selection shapes nucleotide polymorphism across the genome of the nematodeCaenorhabditis briggsae". Genome Research. Cold Spring Harbor, NY: Cold Spring Harbor Laboratory Press. 20 (8): 1103–1111. doi:10.1101/gr.104331.109. ISSN 1088-9051. PMC 2909573 . PMID 20508143.

172. Lien, Sigbjørn; Szyda, Joanna; Schechinger, Birgit; et al. (February 2000). "Evidence for Heterogeneity in Recombination in the Human Pseudoautosomal Region: High Resolution Analysis by Sperm Typing and Radiation-Hybrid Mapping". American Journal of Human Genetics. Cambridge, MA: Cell Press on behalf of the American Society of Human Genetics. 66 (2): 557–566. doi:10.1086/302754. ISSN 0002-9297. PMC 1288109 . PMID 10677316.

173. Barton, Nicholas H. (November 29, 2000). "Genetic hitchhiking". Philosophical Transactions of the Royal Society B: Biological Sciences. London: Royal Society. 355 (1403): 1553–1562. doi:10.1098/rstb.2000.0716. ISSN 0962-8436. PMC 1692896 . PMID 11127900.

174. Wright, Sewall (1932). "The roles of mutation, inbreeding, crossbreeding and selection in evolution". Proceedings of the VI International Congress of Genetrics. 1: 356–366. Retrieved 2014-12-18.

175. Coyne, Jerry A.; Barton, Nicholas H.; Turelli, Michael (June 1997). "Perspective: A Critique of Sewall Wright's Shifting Balance Theory of Evolution". Evolution. Hoboken, NJ: John Wiley & Sons on behalf of the Society for the Study of Evolution. 51 (3): 643–671. doi:10.2307/2411143. ISSN 0014-3820.

176. Baym, Michael; Lieberman, Tami D.; Kelsic, Eric D.; Chait, Remy; Gross, Rotem; Yelin, Idan; Kishony, Roy (2016-09-09). "Spatiotemporal microbial evolution on antibiotic landscapes". Science. 353 (6304): 1147–1151. doi:10.1126/science.aag0822. ISSN 0036-8075. PMID 27609891.

177. Scott, Eugenie C.; Matzke, Nicholas J. (May 15, 2007). "Biological design in science classrooms". Proc. Natl. Acad. Sci. U.S.A. Washington, D.C.: National Academy of Sciences. 104 (Suppl. 1): 8669–8676. Bibcode:2007PNAS..104.8669S. doi:10.1073/pnas.0701505104. ISSN 0027-8424. PMC 1876445 . PMID 17494747.

178. Hendry, Andrew Paul; Kinnison, Michael T. (November 2001). "An introduction to microevolution: rate, pattern, process". Genetica. Dordrecht, the Netherlands: Kluwer Academic Publishers. 112–113 (1): 1–8. doi:10.1023/A:1013368628607. ISSN 0016-6707. PMID 11838760.

179. Leroi, Armand M. (March–April 2000). "The scale independence of evolution". Evolution & Development. Hoboken, NJ: Wiley-Blackwell on behalf of the Society for Integrative and Comparative Biology. 2 (2): 67–77. doi:10.1046/j.1525-142x.2000.00044.x. ISSN 1520-541X. PMID 11258392.

180. Gould 2002, pp. 657–658.

181. Gould, Stephen Jay (July 19, 1994). "Tempo and mode in the macroevolutionary reconstruction of Darwinism". Proc. Natl. Acad. Sci. U.S.A. Washington, D.C.: National Academy of Sciences. 91 (15): 6764–6771. Bibcode:1994PNAS...91.6764G. doi:10.1073/pnas.91.15.6764. ISSN 0027-8424. PMC 44281 . PMID 8041695.

182. Jablonski, David (2000). "Micro- and macroevolution: scale and hierarchy in evolutionary biology and paleobiology". Paleobiology. Boulder, CO: Paleontological Society. 26 (sp4): 15–52. doi:10.1666/0094-8373(2000)26[15:MAMSAH]2.0.CO;2. ISSN 0094-8373.

183. Dougherty, Michael J. (July 20, 1998). "Is the human race evolving or devolving?". Scientific American. Stuttgart: Georg von Holtzbrinck Publishing Group. ISSN 0036-8733. Retrieved 2015-09-11.

184. Isaak, Mark, ed. (July 22, 2003). "Claim CB932: Evolution of degenerate forms". TalkOrigins Archive. Houston, TX: The TalkOrigins Foundation, Inc. Retrieved 2014-12-19.

185. Lane 1996, p. 61

186. Carroll, Sean B. (February 22, 2001). "Chance and necessity: the evolution of morphological complexity and diversity". Nature. London: Nature Publishing Group. 409 (6823): 1102–1109. Bibcode:2001Natur.409.1102C. doi:10.1038/35059227. ISSN 0028-0836. PMID 11234024.

187. Whitman, William B.; Coleman, David C.; Wiebe, William J. (June 9, 1998). "Prokaryotes: The unseen majority". Proc. Natl. Acad. Sci. U.S.A. Washington, D.C.: National Academy of Sciences. 95 (12): 6578–6583. Bibcode:1998PNAS...95.6578W. doi:10.1073/pnas.95.12.6578. ISSN 0027-8424. PMC 33863 . PMID 9618454.

188. Schloss, Patrick D.; Handelsman, Jo (December 2004). "Status of the Microbial Census". Microbiology and Molecular Biology Reviews. Washington, D.C.: American Society for Microbiology. 68 (4): 686–691. doi:10.1128/MMBR.68.4.686-691.2004. ISSN 1092-2172. PMC 539005 . PMID 15590780.

189. Nealson, Kenneth H. (January 1999). "Post-Viking microbiology: new approaches, new data, new insights". Origins of Life and Evolution of Biospheres. Dordrecht, the Netherlands: Kluwer Academic Publishers. 29 (1): 73–93. doi:10.1023/A:1006515817767. ISSN 0169-6149. PMID 11536899.

190. Buckling, Angus; MacLean, R. Craig; Brockhurst, Michael A.; Colegrave, Nick (February 12, 2009). "The Beagle in a bottle". Nature. London: Nature Publishing Group. 457 (7231): 824–829. Bibcode:2009Natur.457..824B. doi:10.1038/nature07892. ISSN 0028-0836. PMID 19212400.

191. Elena, Santiago F.; Lenski, Richard E. (June 2003). "Evolution experiments with microorganisms: the dynamics and genetic bases of adaptation". Nature Reviews Genetics. London: Nature Publishing Group. 4 (6): 457–469. doi:10.1038/nrg1088. ISSN 1471-0056. PMID 12776215.

192. Mayr 1982, p. 483: "Adaptation... could no longer be considered a static condition, a product of a creative past and became instead a continuing dynamic process."

193. The sixth edition of the Oxford Dictionary of Science (2010) defines adaptation as "Any change in the structure or functioning of successive generations of a population that makes it better suited to its environment."

194. Orr, H. Allen (February 2005). "The genetic theory of adaptation: a brief history". Nature Reviews Genetics. London: Nature Publishing Group. 6 (2): 119–127. doi:10.1038/nrg1523. ISSN 1471-0056. PMID 15716908.

195. Dobzhansky 1968, pp. 1–34

196. Dobzhansky 1970, pp. 4–6, 79–82, 84–87

197. Dobzhansky, Theodosius (March 1956). "Genetics of Natural Populations. XXV. Genetic Changes in Populations of Drosophila pseudoobscura and Drosophila persimilis in Some Localities in California". Evolution. Hoboken, NJ: John Wiley & Sons on behalf of the Society for the Study of Evolution. 10 (1): 82–92. doi:10.2307/2406099. ISSN 0014-3820. JSTOR 2406099.

198. Nakajima, Akira; Sugimoto, Yohko; Yoneyama, Hiroshi; Nakae, Taiji (June 2002). "High-Level Fluoroquinolone Resistance in Pseudomonas aeruginosa Due to Interplay of the MexAB-OprM Efflux Pump and the DNA Gyrase Mutation". Microbiology and Immunology. Tokyo: Center for Academic Publications Japan. 46 (6): 391–395. doi:10.1111/j.1348-0421.2002.tb02711.x. ISSN 1348-0421. PMID 12153116.

199. Blount, Zachary D.; Borland, Christina Z.; Lenski, Richard E. (June 10, 2008). "Inaugural Article: Historical contingency and the evolution of a key innovation in an experimental population of Escherichia coli". Proc. Natl. Acad. Sci. U.S.A. Washington, D.C.: National Academy of Sciences. 105 (23): 7899–7906. Bibcode:2008PNAS..105.7899B. doi:10.1073/pnas.0803151105. ISSN 0027-8424. PMC 2430337 . PMID 18524956.

200. Okada, Hirosuke; Negoro, Seiji; Kimura, Hiroyuki; Nakamura, Shunichi (November 10, 1983). "Evolutionary adaptation of plasmid-encoded enzymes for degrading nylon oligomers". Nature. London: Nature Publishing Group. 306 (5939): 203–206. Bibcode:1983Natur.306..203O. doi:10.1038/306203a0. ISSN 0028-0836. PMID 6646204.

201. Ohno, Susumu (April 1984). "Birth of a unique enzyme from an alternative reading frame of the preexisted, internally repetitious coding sequence". Proc. Natl. Acad. Sci. U.S.A. Washington, D.C.: National Academy of Sciences. 81 (8): 2421–2425. Bibcode:1984PNAS...81.2421O. doi:10.1073/pnas.81.8.2421. ISSN 0027-8424. PMC 345072 . PMID 6585807.

202. Copley, Shelley D. (June 2000). "Evolution of a metabolic pathway for degradation of a toxic xenobiotic: the patchwork approach". Trends in Biochemical Sciences. Cambridge, MA: Cell Press. 25 (6): 261–265. doi:10.1016/S0968-0004(00)01562-0. ISSN 0968-0004. PMID 10838562.

203. Crawford, Ronald L.; Jung, Carina M.; Strap, Janice L. (October 2007). "The recent evolution of pentachlorophenol (PCP)-4-monooxygenase (PcpB) and associated pathways for bacterial degradation of PCP". Biodegradation. Dordrecht, the Netherlands: Springer Netherlands. 18 (5): 525–539. doi:10.1007/s10532-006-9090-6. ISSN 0923-9820. PMID 17123025.

204. Eshel, Ilan (December 1973). "Clone-Selection and Optimal Rates of Mutation". Journal of Applied Probability. Sheffield, England. 10 (4): 728–738. doi:10.2307/3212376. ISSN 1475-6072. JSTOR 3212376.

205. Altenberg 1995, pp. 205–259

206. Masel, Joanna; Bergman, Aviv (July 2003). "The evolution of the evolvability properties of the yeast prion [PSI+]". Evolution. Hoboken, NJ: John Wiley & Sons on behalf of the Society for the Study of Evolution. 57 (7): 1498–1512.

doi:10.1111/j.0014-3820.2003.tb00358.x. ISSN 0014-3820. PMID 12940355.

207. Lancaster, Alex K.; Bardill, J. Patrick; True, Heather L.; Masel, Joanna (February 2010). "The Spontaneous Appearance Rate of the Yeast Prion [PSI+] and Its Implications for the Evolution of the Evolvability Properties of the [PSI+] System". Genetics. Bethesda, MD: Genetics Society of America. 184 (2): 393–400. doi:10.1534/genetics.109.110213. ISSN 0016-6731. PMC 2828720 . PMID 19917766.

208. Draghi, Jeremy; Wagner, Günter P. (February 2008). "Evolution of evolvability in a developmental model". Evolution. Hoboken, NJ: John Wiley & Sons on behalf of the Society for the Study of Evolution. 62 (2): 301–315. doi:10.1111/j.1558-5646.2007.00303.x. ISSN 0014-3820. PMID 18031304.

209. Bejder, Lars; Hall, Brian K. (November 2002). "Limbs in whales and limblessness in other vertebrates: mechanisms of evolutionary and developmental transformation and loss". Evolution & Development. Hoboken, NJ: Wiley-Blackwell on behalf of the Society for Integrative and Comparative Biology. 4 (6): 445–458. doi:10.1046/j.1525-142X.2002.02033.x. ISSN 1520-541X. PMID 12492145.

210. Young, Nathan M.; Hallgrímsson, Benedikt (December 2005). "Serial homology and the evolution of mammalian limb covariation structure". Evolution. Hoboken, NJ: John Wiley & Sons on behalf of the Society for the Study of Evolution. 59 (12): 2691–2704. doi:10.1554/05-233.1. ISSN 0014-3820. PMID 16526515.

211. Penny, David; Poole, Anthony (December 1999). "The nature of the last universal common ancestor". Current Opinion in Genetics & Development. Amsterdam, the Netherlands: Elsevier. 9 (6): 672–677. doi:10.1016/S0959-437X(99)00020-9. ISSN 0959-437X. PMID 10607605.

212. Hall, Brian K. (August 2003). "Descent with modification: the unity underlying homology and homoplasy as seen through an analysis of development and evolution". Biological Reviews. Cambridge: Cambridge University Press on behalf of the Cambridge Philosophical Society. 78 (3): 409–433. doi:10.1017/S1464793102006097. ISSN 1464-7931. PMID 14558591.

213. Shubin, Neil; Tabin, Clifford J.; Carroll, Sean (February 12, 2009). "Deep homology and the origins of evolutionary novelty". Nature. London: Nature Publishing Group. 457 (7231): 818–823. Bibcode:2009Natur.457..818S. doi:10.1038/nature07891. ISSN 0028-0836. PMID 19212399.

214. Fong, Daniel F.; Kane, Thomas C.; Culver, David C. (November 1995). "Vestigialization and Loss of Nonfunctional Characters". Annual Review of Ecology and Systematics. Palo Alto, CA: Annual Reviews. 26: 249–268. doi:10.1146/annurev.es.26.110195.001341. ISSN 1545-2069.

215. ZhaoLei Zhang; Gerstein, Mark (August 2004). "Large-scale analysis of pseudogenes in the human genome". Current Opinion in Genetics & Development. Amsterdam, the Netherlands: Elsevier. 14 (4): 328–335. doi:10.1016/j.gde.2004.06.003. ISSN 0959-437X. PMID 15261647.

216. Jeffery, William R. (May–June 2005). "Adaptive Evolution of Eye Degeneration in the Mexican Blind Cavefish". Journal of Heredity. Oxford: Oxford University

Press on behalf of the American Genetic Association. 96 (3): 185–196. doi:10.1093/jhered/esi028. ISSN 0022-1503. PMID 15653557.

217. Maxwell, Erin E.; Larsson, Hans C.E. (May 2007). "Osteology and myology of the wing of the Emu (Dromaius novaehollandiae) and its bearing on the evolution of vestigial structures". Journal of Morphology. Hoboken, NJ: John Wiley & Sons. 268 (5): 423–441. doi:10.1002/jmor.10527. ISSN 0362-2525. PMID 17390336.
218. van der Kooi, Casper J.; Schwander, Tanja (November 2014). "On the fate of sexual traits under asexuality" (PDF). Biological Reviews. Cambridge: Cambridge University Press on behalf of the Cambridge Philosophical Society. 89 (4): 805–819. doi:10.1111/brv.12078. ISSN 1464-7931. PMID 24443922. Retrieved 2015-08-05.
219. Silvestri, Anthony R., Jr.; Singh, Iqbal (April 2003). "The unresolved problem of the third molar: Would people be better off without it?". Journal of the American Dental Association. Chicago, IL: American Dental Association. 134 (4): 450–455. doi:10.14219/jada.archive.2003.0194. ISSN 0002-8177. PMID 12733778.
220. Coyne 2009, p. 62
221. Darwin 1872, pp. 101, 103
222. Gray 2007, p. 66
223. Coyne 2009, pp. 85–86
224. Stevens 1982, p. 87
225. Gould 2002, pp. 1235–1236.
226. Pallen, Mark J.; Matzke, Nicholas J. (October 2006). "From The Origin of Species to the origin of bacterial flagella" (PDF). Nature Reviews Microbiology (PDF). London: Nature Publishing Group. 4 (10): 784–790. doi:10.1038/nrmicro1493. ISSN 1740-1526. PMID 16953248. Retrieved 2014-12-25.
227. Clements, Abigail; Bursac, Dejan; Gatsos, Xenia; et al. (September 15, 2009). "The reducible complexity of a mitochondrial molecular machine". Proc. Natl. Acad. Sci. U.S.A. Washington, D.C.: National Academy of Sciences. 106 (37): 15791–15795. Bibcode:2009PNAS..10615791C. doi:10.1073/pnas.0908264106. ISSN 0027-8424. PMC 2747197 . PMID 19717453.
228. Piatigorsky et al. 1994, pp. 241–250
229. Wistow, Graeme (August 1993). "Lens crystallins: gene recruitment and evolutionary dynamism". Trends in Biochemical Sciences. Cambridge, MA: Cell Press. 18 (8): 301–306. doi:10.1016/0968-0004(93)90041-K. ISSN 0968-0004. PMID 8236445.
230. Johnson, Norman A.; Porter, Adam H. (November 2001). "Toward a new synthesis: population genetics and evolutionary developmental biology". Genetica. Dordrecht, the Netherlands: Kluwer Academic Publishers. 112–113 (1): 45–58. doi:10.1023/A:1013371201773. ISSN 0016-6707. PMID 11838782.
231. Baguñà, Jaume; Garcia-Fernàndez, Jordi (2003). "Evo-Devo: the long and

winding road". The International Journal of Developmental Biology. Bilbao, Spain: University of the Basque Country Press. 47 (7–8): 705–713. ISSN 0214-6282. PMID 14756346.

232. Love, Alan C. (March 2003). "Evolutionary Morphology, Innovation and the Synthesis of Evolutionary and Developmental Biology". Biology and Philosophy. Dordrecht, the Netherlands: Kluwer Academic Publishers. 18 (2): 309–345. doi:10.1023/A:1023940220348. ISSN 0169-3867.

233. Allin, Edgar F. (December 1975). "Evolution of the mammalian middle ear". Journal of Morphology. Hoboken, NJ: John Wiley & Sons. 147 (4): 403–437. doi:10.1002/jmor.1051470404. ISSN 0362-2525. PMID 1202224.

234. Harris, Matthew P.; Hasso, Sean M.; Ferguson, Mark W. J.; Fallon, John F. (February 21, 2006). "The Development of Archosaurian First-Generation Teeth in a Chicken Mutant". Current Biology. Cambridge, MA: Cell Press. 16 (4): 371–377. doi:10.1016/j.cub.2005.12.047. ISSN 0960-9822. PMID 16488870.

235. Carroll, Sean B. (July 11, 2008). "Evo-Devo and an Expanding Evolutionary Synthesis: A Genetic Theory of Morphological Evolution". Cell. Cambridge, MA: Cell Press. 134 (1): 25–36. doi:10.1016/j.cell.2008.06.030. ISSN 0092-8674. PMID 18614008.

236. Wade, Michael J. (March 2007). "The co-evolutionary genetics of ecological communities". Nature Reviews Genetics. London: Nature Publishing Group. 8 (3): 185–195. doi:10.1038/nrg2031. ISSN 1471-0056. PMID 17279094.

237. Geffeney, Shana; Brodie, Edmund D., Jr.; Ruben, Peter C.; Brodie, Edmund D., III (August 23, 2002). "Mechanisms of Adaptation in a Predator-Prey Arms Race: TTX-Resistant Sodium Channels". Science. Washington, D.C.: American Association for the Advancement of Science. 297 (5585): 1336–1339. Bibcode:2002Sci...297.1336G. doi:10.1126/science.1074310. ISSN 0036-8075. PMID 12193784.

238. Brodie, Edmund D., Jr.; Ridenhour, Benjamin J.; Brodie, Edmund D., III (October 2002). "The evolutionary response of predators to dangerous prey: hotspots and coldspots in the geographic mosaic of coevolution between garter snakes and newts". Evolution. Hoboken, NJ: John Wiley & Sons on behalf of the Society for the Study of Evolution. 56 (10): 2067–2082. doi:10.1554/0014-3820(2002)056[2067:teropt]2.0.co;2. ISSN 0014-3820. PMID 12449493.

239. Carroll, Sean B. (December 21, 2009). "Whatever Doesn't Kill Some Animals Can Make Them Deadly". The New York Times. New York: The New York Times Company. ISSN 0362-4331. Retrieved 2014-12-26.

240. Sachs, Joel L. (September 2006). "Cooperation within and among species". Journal of Evolutionary Biology. Hoboken, NJ: Wiley-Blackwell on behalf of the European Society for Evolutionary Biology. 19 (5): 1415–1418; discussion 1426–1436. doi:10.1111/j.1420-9101.2006.01152.x. ISSN 1010-061X. PMID 16910971.

241. Nowak, Martin A. (December 8, 2006). "Five Rules for the Evolution of Cooperation". Science. Washington, D.C.: American Association for the

Advancement of Science. 314 (5805): 1560–1563. Bibcode:2006Sci...314.1560N. doi:10.1126/science.1133755. ISSN 0036-8075. PMC 3279745 . PMID 17158317.

242. Paszkowski, Uta (August 2006). "Mutualism and parasitism: the yin and yang of plant symbioses". Current Opinion in Plant Biology. Amsterdam, the Netherlands: Elsevier. 9 (4): 364–370. doi:10.1016/j.pbi.2006.05.008. ISSN 1369-5266. PMID 16713732.

243. Hause, Bettina; Fester, Thomas (May 2005). "Molecular and cell biology of arbuscular mycorrhizal symbiosis". Planta. Berlin: Springer-Verlag. 221 (2): 184–196. doi:10.1007/s00425-004-1436-x. ISSN 0032-0935. PMID 15871030.

244. Bertram, John S. (December 2000). "The molecular biology of cancer". Molecular Aspects of Medicine. Amsterdam, the Netherlands: Elsevier on behalf of the International Union of Biochemistry and Molecular Biology. 21 (6): 167–223. doi:10.1016/S0098-2997(00)00007-8. ISSN 0098-2997. PMID 11173079.

245. Reeve, H. Kern; Hölldobler, Bert (June 5, 2007). "The emergence of a superorganism through intergroup competition". Proc. Natl. Acad. Sci. U.S.A. Washington, D.C.: National Academy of Sciences. 104 (23): 9736–9740. Bibcode:2007PNAS..104.9736R. doi:10.1073/pnas.0703466104. ISSN 0027-8424. PMC 1887545 . PMID 17517608.

246. Axelrod, Robert; Hamilton, W. D. (March 27, 1981). "The evolution of cooperation". Science. Washington, D.C.: American Association for the Advancement of Science. 211 (4489): 1390–1396. Bibcode:1981Sci...211.1390A. doi:10.1126/science.7466396. ISSN 0036-8075. PMID 7466396.

247. Wilson, Edward O.; Hölldobler, Bert (September 20, 2005). "Eusociality: Origin and consequences". Proc. Natl. Acad. Sci. U.S.A. Washington, D.C.: National Academy of Sciences. 102 (38): 13367–1371. Bibcode:2005PNAS..10213367W. doi:10.1073/pnas.0505858102. ISSN 0027-8424. PMC 1224642 . PMID 16157878.

248. Gavrilets, Sergey (October 2003). "Perspective: models of speciation: what have we learned in 40 years?". Evolution. Hoboken, NJ: John Wiley & Sons on behalf of the Society for the Study of Evolution. 57 (10): 2197–2215. doi:10.1554/02-727. ISSN 0014-3820. PMID 14628909.

249. de Queiroz, Kevin (May 3, 2005). "Ernst Mayr and the modern concept of species". Proc. Natl. Acad. Sci. U.S.A. Washington, D.C.: National Academy of Sciences. 102 (Suppl. 1): 6600–6607. Bibcode:2005PNAS..102.6600D. doi:10.1073/pnas.0502030102. ISSN 0027-8424. PMC 1131873 . PMID 15851674.

250. Ereshefsky, Marc (December 1992). "Eliminative pluralism". Philosophy of Science. Chicago, IL: University of Chicago Press. 59 (4): 671–690. doi:10.1086/289701. ISSN 0031-8248. JSTOR 188136.

251. Mayr 1942, p. 120

252. Fraser, Christophe; Alm, Eric J.; Polz, Martin F.; et al. (February 6, 2009). "The Bacterial Species Challenge: Making Sense of Genetic and Ecological Diversity". Science. Washington, D.C.: American Association for the Advancement of Science. 323 (5915): 741–746. Bibcode:2009Sci...323..741F. doi:10.1126/science.1159388. ISSN 0036-8075. PMID 19197054.

253. Short, Roger Valentine (October 1975). "The contribution of the mule to scientific thought". Journal of Reproduction and Fertility. Supplement. Society for Reproduction and Fertility (23): 359–364. ISSN 0449-3087. OCLC 1639439. PMID 1107543.

254. Gross, Briana L.; Rieseberg, Loren H. (May–June 2005). "The Ecological Genetics of Homoploid Hybrid Speciation". Journal of Heredity. Oxford: Oxford University Press on behalf of the American Genetic Association. 96 (3): 241–252. doi:10.1093/jhered/esi026. ISSN 0022-1503. PMC 2517139 . PMID 15618301.

255. Burke, John M.; Arnold, Michael L. (December 2001). "Genetics and the fitness of hybrids". Annual Review of Genetics. Palo Alto, CA: Annual Reviews. 35: 31–52. doi:10.1146/annurev.genet.35.102401.085719. ISSN 0066-4197. PMID 11700276.

256. Vrijenhoek, Robert C. (April 4, 2006). "Polyploid Hybrids: Multiple Origins of a Treefrog Species". Current Biology. Cambridge, MA: Cell Press. 16 (7): R245–R247. doi:10.1016/j.cub.2006.03.005. ISSN 0960-9822. PMID 16581499.

257. Rice, William R.; Hostert, Ellen E. (December 1993). "Laboratory Experiments on Speciation: What Have We Learned in 40 Years?". Evolution. Hoboken, NJ: John Wiley & Sons on behalf of the Society for the Study of Evolution. 47 (6): 1637–1653. doi:10.2307/2410209. ISSN 0014-3820.

258. Jiggins, Chris D.; Bridle, Jon R. (March 2004). "Speciation in the apple maggot fly: a blend of vintages?". Trends in Ecology & Evolution. Cambridge, MA: Cell Press. 19 (3): 111–114. doi:10.1016/j.tree.2003.12.008. ISSN 0169-5347. PMID 16701238.

259. Boxhorn, Joseph (September 1, 1995). "Observed Instances of Speciation". TalkOrigins Archive. Houston, TX: The TalkOrigins Foundation, Inc. Retrieved 2008-12-26.

260. Weinberg, James R.; Starczak, Victoria R.; Jörg, Daniele (August 1992). "Evidence for Rapid Speciation Following a Founder Event in the Laboratory". Evolution. Hoboken, NJ: John Wiley & Sons on behalf of the Society for the Study of Evolution. 46 (4): 1214–1220. doi:10.2307/2409766. ISSN 0014-3820. JSTOR 2409766.

261. Herrel, Anthony; Huyghe, Katleen; Vanhooydonck, Bieke; et al. (March 25, 2008). "Rapid large-scale evolutionary divergence in morphology and performance associated with exploitation of a different dietary resource". Proc. Natl. Acad. Sci. U.S.A. Washington, D.C.: National Academy of Sciences. 105 (12): 4792–4795. Bibcode:2008PNAS..105.4792H. doi:10.1073/pnas.0711998105. ISSN 0027-8424. PMC 2290806 . PMID 18344323.

262. Losos, Jonathan B.; Warhelt, Kenneth I.; Schoener, Thomas W. (May 1, 1997). "Adaptive differentiation following experimental island colonization in Anolis lizards". Nature. London: Nature Publishing Group. 387 (6628): 70–73. Bibcode:1997Natur.387...70L. doi:10.1038/387070a0. ISSN 0028-0836.

263. Hoskin, Conrad J.; Higgle, Megan; McDonald, Keith R.; Moritz, Craig (October 27, 2005). "Reinforcement drives rapid allopatric speciation". Nature. London:

Nature Publishing Group. 437 (7063): 1353–1356. Bibcode:2005Natur.437.1353H. doi:10.1038/nature04004. ISSN 0028-0836. PMID 16251964.

264. Templeton, Alan R. (April 1980). "The Theory of Speciation VIA the Founder Principle" (PDF). Genetics. Bethesda, MD: Genetics Society of America. 94 (4): 1011–1038. ISSN 0016-6731. PMC 1214177 . PMID 6777243. Retrieved 2014-12-29.

265. Antonovics, Janis (July 2006). "Evolution in closely adjacent plant populations X: long-term persistence of prereproductive isolation at a mine boundary". Heredity. London: Nature Publishing Group for The Genetics Society. 97 (1): 33–37. doi:10.1038/sj.hdy.6800835. ISSN 0018-067X. PMID 16639420. Retrieved 2014-12-29.

266. Nosil, Patrik; Crespi, Bernard J.; Gries, Regine; Gries, Gerhard (March 2007). "Natural selection and divergence in mate preference during speciation". Genetica. Dordrecht, the Netherlands: Kluwer Academic Publishers. 129 (3): 309–327. doi:10.1007/s10709-006-0013-6. ISSN 0016-6707. PMID 16900317.

267. Savolainen, Vincent; Anstett, Marie-Charlotte; Lexer, Christian; et al. (May 11, 2006). "Sympatric speciation in palms on an oceanic island". Nature. London: Nature Publishing Group. 441 (7090): 210–213. Bibcode:2006Natur.441..210S. doi:10.1038/nature04566. ISSN 0028-0836. PMID 16467788.

268. Barluenga, Marta; Stölting, Kai N.; Salzburger, Walter; et al. (February 9, 2006). "Sympatric speciation in Nicaraguan crater lake cichlid fish". Nature. London: Nature Publishing Group. 439 (7077): 719–23. Bibcode:2006Natur.439..719B. doi:10.1038/nature04325. ISSN 0028-0836. PMID 16467837.

269. Gavrilets, Sergey (March 21, 2006). "The Maynard Smith model of sympatric speciation". Journal of Theoretical Biology. Amsterdam, the Netherlands: Elsevier. 239 (2): 172–182. doi:10.1016/j.jtbi.2005.08.041. ISSN 0022-5193. PMID 16242727.

270. Wood, Troy E.; Takebayashi, Naoki; Barker, Michael S.; et al. (August 18, 2009). "The frequency of polyploid speciation in vascular plants". Proc. Natl. Acad. Sci. U.S.A. Washington, D.C.: National Academy of Sciences. 106 (33): 13875–13879. Bibcode:2009PNAS..10613875W. doi:10.1073/pnas.0811575106. ISSN 0027-8424. PMC 2728988 . PMID 19667210.

271. Hegarty, Matthew J.; Hiscock, Simon J. (May 20, 2008). "Genomic Clues to the Evolutionary Success of Polyploid Plants". Current Biology. Cambridge, MA: Cell Press. 18 (10): R435–R444. doi:10.1016/j.cub.2008.03.043. ISSN 0960-9822. PMID 18492478.

272. Jakobsson, Mattias; Hagenblad, Jenny; Tavaré, Simon; et al. (June 2006). "A Unique Recent Origin of the Allotetraploid Species Arabidopsis suecica: Evidence from Nuclear DNA Markers". Molecular Biology and Evolution. Oxford: Oxford University Press on behalf of the Society for Molecular Biology and Evolution. 23 (6): 1217–1231. doi:10.1093/molbev/msk006. ISSN 0737-4038. PMID 16549398.

273. Säll, Torbjörn; Jakobsson, Mattias; Lind-Halldén, Christina; Halldén, Christer (September 2003). "Chloroplast DNA indicates a single origin of the allotetraploid

Arabidopsis suecica". Journal of Evolutionary Biology. Hoboken, NJ: Wiley-Blackwell on behalf of the European Society for Evolutionary Biology. 16 (5): 1019–1029. doi:10.1046/j.1420-9101.2003.00554.x. ISSN 1010-061X. PMID 14635917.

274. Bomblies, Kirsten; Weigel, Detlef (December 2007). "Arabidopsis—a model genus for speciation". Current Opinion in Genetics & Development. Amsterdam, the Netherlands: Elsevier. 17 (6): 500–504. doi:10.1016/j.gde.2007.09.006. ISSN 0959-437X. PMID 18006296.

275. Sémon, Marie; Wolfe, Kenneth H. (December 2007). "Consequences of genome duplication". Current Opinion in Genetics & Development. Amsterdam, the Netherlands: Elsevier. 17 (6): 505–512. doi:10.1016/j.gde.2007.09.007. ISSN 0959-437X. PMID 18006297.

276. Eldredge & Gould 1972, pp. 82–115

277. Benton, Michael J. (April 7, 1995). "Diversification and extinction in the history of life". Science. Washington, D.C.: American Association for the Advancement of Science. 268 (5207): 52–58. Bibcode:1995Sci...268...52B. doi:10.1126/science.7701342. ISSN 0036-8075. PMID 7701342.

278. Raup, David M. (March 28, 1986). "Biological extinction in Earth history". Science. Washington, D.C.: American Association for the Advancement of Science. 231 (4745): 1528–1533. Bibcode:1986Sci...231.1528R. doi:10.1126/science.11542058. ISSN 0036-8075. PMID 11542058.

279. Avise, John C.; Hubbell, Stephen P.; Ayala, Francisco J. (August 12, 2008). "In the light of evolution II: Biodiversity and extinction". Proc. Natl. Acad. Sci. U.S.A. Washington, D.C.: National Academy of Sciences. 105 (Suppl. 1): 11453–11457. Bibcode:2008PNAS..10511453A. doi:10.1073/pnas.0802504105. ISSN 0027-8424. PMC 2556414 . PMID 18695213.

280. Raup, David M. (July 19, 1994). "The role of extinction in evolution". Proc. Natl. Acad. Sci. U.S.A. Washington, D.C.: National Academy of Sciences. 91 (15): 6758–6763. Bibcode:1994PNAS...91.6758R. doi:10.1073/pnas.91.15.6758. ISSN 0027-8424. PMC 44280 . PMID 8041694.

281. Novacek, Michael J.; Cleland, Elsa E. (May 8, 2001). "The current biodiversity extinction event: scenarios for mitigation and recovery". Proc. Natl. Acad. Sci. U.S.A. Washington, D.C.: National Academy of Sciences. 98 (10): 5466–5470. Bibcode:2001PNAS...98.5466N. doi:10.1073/pnas.091093698. ISSN 0027-8424. PMC 33235 . PMID 11344295.

282. Pimm, Stuart; Raven, Peter; Peterson, Alan; et al. (July 18, 2006). "Human impacts on the rates of recent, present and future bird extinctions". Proc. Natl. Acad. Sci. U.S.A. Washington, D.C.: National Academy of Sciences. 103 (29): 10941–10946. Bibcode:2006PNAS..10310941P. doi:10.1073/pnas.0604181103. ISSN 0027-8424. PMC 1544153 . PMID 16829570.

283. Barnosky, Anthony D.; Koch, Paul L.; Feranec, Robert S.; et al. (October 1, 2004). "Assessing the Causes of Late Pleistocene Extinctions on the Continents". Science. Washington, D.C.: American Association for the Advancement of Science.

306 (5693): 70–75. Bibcode:2004Sci...306...70B. doi:10.1126/science.1101476. ISSN 0036-8075. PMID 15459379.

284. Lewis, Owen T. (January 29, 2006). "Climate change, species–area curves and the extinction crisis". Philosophical Transactions of the Royal Society B: Biological Sciences. London: Royal Society. 361 (1465): 163–171. doi:10.1098/rstb.2005.1712. ISSN 0962-8436. PMC 1831839 . PMID 16553315.
285. Jablonski, David (May 8, 2001). "Lessons from the past: Evolutionary impacts of mass extinctions". Proc. Natl. Acad. Sci. U.S.A. Washington, D.C.: National Academy of Sciences. 98 (10): 5393–5398. Bibcode:2001PNAS...98.5393J. doi:10.1073/pnas.101092598. ISSN 0027-8424. PMC 33224 . PMID 11344284.
286. "Age of the Earth". United States Geological Survey. July 9, 2007. Retrieved 2015-05-31.
287. Dalrymple 2001, pp. 205–221
288. Manhesa, Gérard; Allègre, Claude J.; Dupréa, Bernard; Hamelin, Bruno (May 1980). "Lead isotope study of basic-ultrabasic layered complexes: Speculations about the age of the earth and primitive mantle characteristics". Earth and Planetary Science Letters. Amsterdam, the Netherlands: Elsevier. 47 (3): 370–382. Bibcode:1980E&PSL..47..370M. doi:10.1016/0012-821X(80)90024-2. ISSN 0012-821X.
289. Raven & Johnson 2002, p. 68
290. McKinney 1997, p. 110
291. Peretó, Juli (March 2005). "Controversies on the origin of life"(PDF). International Microbiology. Barcelona: Spanish Society for Microbiology. 8 (1): 23–31. ISSN 1139-6709. PMID 15906258.
292. Joyce, Gerald F. (July 11, 2002). "The antiquity of RNA-based evolution". Nature. London: Nature Publishing Group. 418 (6894): 214–221. Bibcode:2002Natur.418..214J. doi:10.1038/418214a. ISSN 0028-0836. PMID 12110897.
293. Trevors, Jack T.; Psenner, Roland (December 2001). "From self-assembly of life to present-day bacteria: a possible role for nanocells". FEMS Microbiology Reviews. Amsterdam, the Netherlands: Elsevier on behalf of the Federation of European Microbiological Societies. 25 (5): 573–582. doi:10.1111/j.1574-6976.2001.tb00592.x. ISSN 1574-6976. PMID 11742692.
294. Theobald, Douglas L. (May 13, 2010). "A formal test of the theory of universal common ancestry". Nature. London: Nature Publishing Group. 465 (7295): 219–222. Bibcode:2010Natur.465..219T. doi:10.1038/nature09014. ISSN 0028-0836. PMID 20463738.
295. Bapteste, Eric; Walsh, David A. (June 2005). "Does the 'Ring of Life' ring true?". Trends in Microbiology. Cambridge, MA: Cell Press. 13 (6): 256–261. doi:10.1016/j.tim.2005.03.012. ISSN 0966-842X. PMID 15936656.
296. Darwin 1859, p. 1
297. Doolittle, W. Ford; Bapteste, Eric (February 13, 2007). "Pattern pluralism and the

Tree of Life hypothesis". Proc. Natl. Acad. Sci. U.S.A. Washington, D.C.: National Academy of Sciences. 104 (7): 2043–2049. Bibcode:2007PNAS..104.2043D. doi:10.1073/pnas.0610699104. ISSN 0027-8424. PMC 1892968 . PMID 17261804.

298. Kunin, Victor; Goldovsky, Leon; Darzentas, Nikos; Ouzounis, Christos A. (July 2005). "The net of life: Reconstructing the microbial phylogenetic network". Genome Research. Cold Spring Harbor, NY: Cold Spring Harbor Laboratory Press. 15 (7): 954–959. doi:10.1101/gr.3666505. ISSN 1088-9051. PMC 1172039 . PMID 15965028.

299. Jablonski, David (June 25, 1999). "The Future of the Fossil Record". Science. Washington, D.C.: American Association for the Advancement of Science. 284 (5423): 2114–2116. doi:10.1126/science.284.5423.2114. ISSN 0036-8075. PMID 10381868.

300. Mason, Stephen F. (September 6, 1984). "Origins of biomolecular handedness". Nature. London: Nature Publishing Group. 311 (5981): 19–23. Bibcode:1984Natur.311...19M. doi:10.1038/311019a0. ISSN 0028-0836. PMID 6472461.

301. Wolf, Yuri I.; Rogozin, Igor B.; Grishin, Nick V.; Koonin, Eugene V. (September 1, 2002). "Genome trees and the tree of life". Trends in Genetics. Cambridge, MA: Cell Press. 18 (9): 472–479. doi:10.1016/S0168-9525(02)02744-0. ISSN 0168-9525. PMID 12175808.

302. Varki, Ajit; Altheide, Tasha K. (December 2005). "Comparing the human and chimpanzee genomes: searching for needles in a haystack". Genome Research. Cold Spring Harbor, NY: Cold Spring Harbor Laboratory Press. 15 (12): 1746–1758. doi:10.1101/gr.3737405. ISSN 1088-9051. PMID 16339373.

303. Ciccarelli, Francesca D.; Doerks, Tobias; von Mering, Christian; et al. (March 3, 2006). "Toward Automatic Reconstruction of a Highly Resolved Tree of Life". Science. Washington, D.C.: American Association for the Advancement of Science. 311 (5765): 1283–1287. Bibcode:2006Sci...311.1283C. doi:10.1126/science.1123061. ISSN 0036-8075. PMID 16513982.

304. Cavalier-Smith, Thomas (June 29, 2006). "Cell evolution and Earth history: stasis and revolution". Philosophical Transactions of the Royal Society B: Biological Sciences. London: Royal Society. 361 (1470): 969–1006. doi:10.1098/rstb.2006.1842. ISSN 0962-8436. PMC 1578732 . PMID 16754610.

305. Schopf, J. William (June 29, 2006). "Fossil evidence of Archaean life". Philosophical Transactions of the Royal Society B: Biological Sciences. London: Royal Society. 361 (1470): 869–885. doi:10.1098/rstb.2006.1834. ISSN 0962-8436. PMC 1578735 . PMID 16754604.

306. Altermann, Wladyslaw; Kazmierczak, Józef (November 2003). "Archean microfossils: a reappraisal of early life on Earth". Research in Microbiology. Amsterdam, the Netherlands: Elsevier for the Pasteur Institute. 154 (9): 611–617. doi:10.1016/j.resmic.2003.08.006. ISSN 0923-2508. PMID 14596897.

307. Schopf, J. William (July 19, 1994). "Disparate rates, differing fates: tempo and mode of evolution changed from the Precambrian to the Phanerozoic". Proc.

Natl. Acad. Sci. U.S.A. Washington, D.C.: National Academy of Sciences. 91 (15): 6735–6742. Bibcode:1994PNAS...91.6735S. doi:10.1073/pnas.91.15.6735. ISSN 0027-8424. PMC 44277 . PMID 8041691.

308. Poole, Anthony M.; Penny, David (January 2007). "Evaluating hypotheses for the origin of eukaryotes". BioEssays. Hoboken, NJ: John Wiley & Sons. 29 (1): 74–84. doi:10.1002/bies.20516. ISSN 0265-9247. PMID 17187354.
309. Dyall, Sabrina D.; Brown, Mark T.; Johnson, Patricia J. (April 9, 2004). "Ancient Invasions: From Endosymbionts to Organelles". Science. Washington, D.C.: American Association for the Advancement of Science. 304 (5668): 253–257. Bibcode:2004Sci...304..253D. doi:10.1126/science.1094884. ISSN 0036-8075. PMID 15073369.
310. Martin, William (October 2005). "The missing link between hydrogenosomes and mitochondria". Trends in Microbiology. Cambridge, MA: Cell Press. 13 (10): 457–459. doi:10.1016/j.tim.2005.08.005. ISSN 0966-842X. PMID 16109488.
311. Lang, B. Franz; Gray, Michael W.; Burger, Gertraud (December 1999). "Mitochondrial genome evolution and the origin of eukaryotes". Annual Review of Genetics. Palo Alto, CA: Annual Reviews. 33: 351–397. doi:10.1146/annurev.genet.33.1.351. ISSN 0066-4197. PMID 10690412.
312. McFadden, Geoffrey Ian (December 1, 1999). "Endosymbiosis and evolution of the plant cell". Current Opinion in Plant Biology. Amsterdam, the Netherlands: Elsevier. 2 (6): 513–519. doi:10.1016/S1369-5266(99)00025-4. ISSN 1369-5266. PMID 10607659.
313. DeLong, Edward F.; Pace, Norman R. (August 1, 2001). "Environmental Diversity of Bacteria and Archaea". Systematic Biology. Oxford: Oxford University Press on behalf of the Society of Systematic Biologists. 50 (4): 470–478. doi:10.1080/106351501750435040. ISSN 1063-5157. PMID 12116647.
314. Kaiser, Dale (December 2001). "Building a multicellular organism". Annual Review of Genetics. Palo Alto, CA: Annual Reviews. 35: 103–123. doi:10.1146/annurev.genet.35.102401.090145. ISSN 0066-4197. PMID 11700279.
315. Zimmer, Carl (January 7, 2016). "Genetic Flip Helped Organisms Go From One Cell to Many". The New York Times. Retrieved January 7, 2016.
316. Valentine, James W.; Jablonski, David; Erwin, Douglas H. (March 1, 1999). "Fossils, molecules and embryos: new perspectives on the Cambrian explosion" (PDF). Development. Cambridge: The Company of Biologists. 126 (5): 851–859. ISSN 0950-1991. PMID 9927587. Retrieved 2014-12-30.
317. Ohno, Susumu (January 1997). "The reason for as well as the consequence of the Cambrian explosion in animal evolution". Journal of Molecular Evolution. New York: Springer-Verlag New York. 44 (Suppl. 1): S23–S27. doi:10.1007/PL00000055. ISSN 0022-2844. PMID 9071008.
318. Valentine, James W.; Jablonski, David (2003). "Morphological and developmental macroevolution: a paleontological perspective". The International Journal of Developmental Biology. Bilbao, Spain: University of the Basque Country Press. 47 (7–8): 517–522. ISSN 0214-6282. PMID 14756327. Retrieved 2014-12-30.

319. Waters, Elizabeth R. (December 2003). "Molecular adaptation and the origin of land plants". Molecular Phylogenetics and Evolution. Academic Press. 29 (3): 456–463. doi:10.1016/j.ympev.2003.07.018. ISSN 1055-7903. PMID 14615186.

320. Mayhew, Peter J. (August 2007). "Why are there so many insect species? Perspectives from fossils and phylogenies". Biological Reviews. Cambridge: Cambridge University Press on behalf of the Cambridge Philosophical Society. 82 (3): 425–454. doi:10.1111/j.1469-185X.2007.00018.x. ISSN 1464-7931. PMID 17624962.

321. Carroll, Robert L. (May 2007). "The Palaeozoic Ancestry of Salamanders, Frogs and Caecilians". Zoological Journal of the Linnean Society. Hoboken, NJ: Wiley-Blackwell. 150 (Supplement s1): 1–140. doi:10.1111/j.1096-3642.2007.00246.x. ISSN 1096-3642.

322. Wible, John R.; Rougier, Guillermo W.; Novacek, Michael J.; Asher, Robert J. (June 21, 2007). "Cretaceous eutherians and Laurasian origin for placental mammals near the K/T boundary". Nature. London: Nature Publishing Group. 447 (7147): 1003–1006. Bibcode:2007Natur.447.1003W. doi:10.1038/nature05854. ISSN 0028-0836. PMID 17581585.

323. Witmer, Lawrence M. (July 28, 2011). "Palaeontology: An icon knocked from its perch". Nature. London: Nature Publishing Group. 475 (7357): 458–459. doi:10.1038/475458a. ISSN 0028-0836. PMID 21796198.

324. Bull, James J.; Wichman, Holly A. (November 2001). "Applied evolution". Annual Review of Ecology and Systematics. Palo Alto, CA: Annual Reviews. 32: 183–217. doi:10.1146/annurev.ecolsys.32.081501.114020. ISSN 1545-2069.

325. Doebley, John F.; Gaut, Brandon S.; Smith, Bruce D. (December 29, 2006). "The Molecular Genetics of Crop Domestication". Cell. Cambridge, MA: Cell Press. 127 (7): 1309–1321. doi:10.1016/j.cell.2006.12.006. ISSN 0092-8674. PMID 17190597.

326. Jäckel, Christian; Kast, Peter; Hilvert, Donald (June 2008). "Protein Design by Directed Evolution". Annual Review of Biophysics. Palo Alto, CA: Annual Reviews. 37: 153–173. doi:10.1146/annurev.biophys.37.032807.125832. ISSN 1936-122X. PMID 18573077.

327. Maher, Brendan (April 8, 2009). "Evolution: Biology's next top model?". Nature. London: Nature Publishing Group. 458 (7239): 695–698. doi:10.1038/458695a. ISSN 0028-0836. PMID 19360058.

328. Borowsky, Richard (January 8, 2008). "Restoring sight in blind cavefish". Current Biology. Cambridge, MA: Cell Press. 18 (1): R23–R24. doi:10.1016/j.cub.2007.11.023. ISSN 0960-9822. PMID 18177707.

329. Gross, Joshua B.; Borowsky, Richard; Tabin, Clifford J. (January 2, 2009). Barsh, Gregory S., ed. "A novel role for Mc1r in the parallel evolution of depigmentation in independent populations of the cavefishAstyanax mexicanus". PLOS Genetics. San Francisco, CA: Public Library of Science. 5 (1): e1000326. doi:10.1371/journal.pgen.1000326. ISSN 1553-7390. PMC 2603666 . PMID 19119422.

330. Merlo, Lauren M.F.; Pepper, John W.; Reid, Brian J.; Maley, Carlo C. (December

2006). "Cancer as an evolutionary and ecological process.". Nature Reviews Cancer. London: Nature Publishing Group. 6 (12): 924–935. doi:10.1038/nrc2013. ISSN 1474-175X. PMID 17109012.

331. Pan, Dabo; Weiwei Xue; Wenqi Zhang; et al. (October 2012). "Understanding the drug resistance mechanism of hepatitis C virus NS3/4A to ITMN-191 due to R155K, A156V, D168A/E mutations: a computational study.". Biochimica et Biophysica Acta (BBA) - General Subjects. Amsterdam, the Netherlands: Elsevier. 1820 (10): 1526–1534. doi:10.1016/j.bbagen.2012.06.001. ISSN 0304-4165. PMID 22698669.
332. Woodford, Neil; Ellington, Matthew J. (January 2007). "The emergence of antibiotic resistance by mutation.". Clinical Microbiology and Infection. Amsterdam, the Netherlands: Elsevier for the European Society of Clinical Microbiology and Infectious Diseases. 13 (1): 5–18. doi:10.1111/j.1469-0691.2006.01492.x. ISSN 1198-743X. PMID 17184282.
333. Labbé, Pierrick; Berticat, Claire; Berthomieu, Arnaud; et al. (November 16, 2007). "Forty Years of Erratic Insecticide Resistance Evolution in the Mosquito Culex pipiens". PLOS Genetics. San Francisco, CA: Public Library of Science. 3 (11): e205. doi:10.1371/journal.pgen.0030205. ISSN 1553-7390. PMC 2077897 . PMID 18020711.
334. Neve, Paul (October 2007). "Challenges for herbicide resistance evolution and management: 50 years after Harper". Weed Research. Hoboken, NJ: Wiley-Blackwell on behalf of the European Weed Research Society. 47 (5): 365–369. doi:10.1111/j.1365-3180.2007.00581.x. ISSN 0043-1737.
335. Rodríguez-Rojas, Alexandro; Rodríguez-Beltrán, Jerónimo; Couce, Alejandro; Blázquez, Jesús (August 2013). "Antibiotics and antibiotic resistance: A bitter fight against evolution". International Journal of Medical Microbiology. Amsterdam, the Netherlands: Elsevier. 303 (6–7): 293–297. doi:10.1016/j.ijmm.2013.02.004. ISSN 1438-4221. PMID 23517688.
336. Schenk, Martijn F.; Szendro, Ivan G.; Krug, Joachim; de Visser, J. Arjan G. M. (June 28, 2012). "Quantifying the Adaptive Potential of an Antibiotic Resistance Enzyme". PLOS Genetics. San Francisco, CA: Public Library of Science. 8 (6): e1002783. doi:10.1371/journal.pgen.1002783. ISSN 1553-7390. PMC 3386231 . PMID 22761587.
337. Read, Andrew F.; Lynch, Penelope A.; Thomas, Matthew B. (April 7, 2009). "How to Make Evolution-Proof Insecticides for Malaria Control". PLOS Biology. San Francisco, CA: Public Library of Science. 7 (4): e1000058. doi:10.1371/journal.pbio.1000058. ISSN 1545-7885. PMC 3279047 . PMID 19355786.
338. Fraser, Alex S. (January 18, 1958). "Monte Carlo Analyses of Genetic Models". Nature. London: Nature Publishing Group. 181 (4603): 208–209. Bibcode:1958Natur.181..208F. doi:10.1038/181208a0. ISSN 0028-0836. PMID 13504138.
339. Rechenberg 1973
340. Holland 1975

341. Koza 1992

342. Jamshidi, Mo (August 15, 2003). "Tools for intelligent control: fuzzy controllers, neural networks and genetic algorithms". Philosophical Transactions of the Royal Society A. London: Royal Society. 361 (1809): 1781–1808. Bibcode:2003RSPTA.361.1781J. doi:10.1098/rsta.2003.1225. ISSN 1364-503X. PMID 12952685.

343. For an overview of the philosophical, religious and cosmological controversies, see: Dennett 1995

344. For the scientific and social reception of evolution in the 19th and early 20th centuries, see:

345. Johnston, Ian C. (1999). "Section Three: The Origins of Evolutionary Theory". . . . And Still We Evolve: A Handbook for the Early History of Modern Science (3rd revised ed.). Nanaimo, BC: Liberal Studies Department, Malaspina University-College. Retrieved 2015-01-01.

346. Bowler 2003

347. Zuckerkandl, Emile (December 30, 2006). "Intelligent design and biological complexity". Gene. Amsterdam, the Netherlands: Elsevier. 385: 2–18. doi:10.1016/j.gene.2006.03.025. ISSN 0378-1119. PMID 17011142.

348. Ross, Marcus R. (May 2005). "Who Believes What? Clearing up Confusion over Intelligent Design and Young-Earth Creationism"(PDF). Journal of Geoscience Education. National Association of Geoscience Teachers. 53 (3): 319. ISSN 1089-9995. Retrieved 2008-04-28.

349. Hameed, Salman (December 12, 2008). "Bracing for Islamic Creationism" (PDF). Science. Washington, D.C.: American Association for the Advancement of Science. 322 (5908): 1637–1638. doi:10.1126/science.1163672. ISSN 0036-8075. PMID 19074331.

350. Bowler 2003

351. Miller, Jon D.; Scott, Eugenie C.; Okamoto, Shinji (August 11, 2006). "Public Acceptance of Evolution". Science. Washington, D.C.: American Association for the Advancement of Science. 313 (5788): 765–766. doi:10.1126/science.1126746. ISSN 0036-8075. PMID 16902112.

352. Spergel, David Nathaniel; Verde, Licia; Peiris, Hiranya V.; et al. (2003). "First-Year Wilkinson Microwave Anisotropy Probe (WMAP) Observations: Determination of Cosmological Parameters". The Astrophysical Journal Supplement Series. Chicago, IL: University of Chicago Press on behalf of the American Astronomical Society. 148 (1): 175–194. arXiv:astro-ph/0302209 . Bibcode:2003ApJS..148..175S. doi:10.1086/377226.

353. Wilde, Simon A.; Valley, John W.; Peck, William H.; Graham, Colin M. (January 11, 2001). "Evidence from detrital zircons for the existence of continental crust and oceans on the Earth 4.4 Gyr ago". Nature. London: Nature Publishing Group. 409 (6817): 175–178. doi:10.1038/35051550. ISSN 0028-0836. PMID 11196637.

354. Branch, Glenn (March 2007). "Understanding Creationism after Kitzmiller". BioScience. Oxford: Oxford University Press on behalf of the American Institute of Biological Sciences. 57 (3): 278–284. doi:10.1641/B570313. ISSN 0006-3568.

CHAPTER 3

Ecology

Chapter Contents

- **Introduction**

Ecology is a branch of biology which is focused on the examination of living organisms in the natural environment. Ecologists look at how organisms interact with the environment and each other, and they study the complex and interconnected systems which influence life on Earth. Ecology is also sometimes known as environmental biology, and there are a number of subdisciplines within this branch of the sciences which deal with specific topics of interest, such as the relationship between humans and the natural environment.

Ecology is sometimes known as environmental biology.

INTRODUCTION

Ecology is the scientific analysis and study of interactions among organisms and their environment. It is an interdisciplinary field that includes biology, geography, and Earth science. Ecology includes the study of interactions organisms have with each other, other organisms, and with abiotic components of their environment. Topics of interest to ecologists include the diversity, distribution, amount (biomass), and number (population) of particular organisms, as well as cooperation and competition between organisms, both within and among ecosystems. Ecosystems are composed of dynamically interacting parts including organisms, the communities they make up, and the non-living components of their environment. Ecosystem processes, such as primary production, pedogenesis, nutrient cycling, and various niche construction activities, regulate the flux of energy and matter through an environment. These processes are sustained by organisms with specific life history traits, and the variety of organisms is called biodiversity. Biodiversity, which refers to the varieties of species, genes, and ecosystems, enhances certain ecosystem services.

Ecology is not synonymous with environment, environmentalism, natural history, or environmental science. It is closely related to evolutionary biology, genetics, and ethology. An important focus for ecologists is to improve the understanding of how biodiversity affects ecological function. Ecologists seek to explain:

- Life processes, interactions, and adaptations
- The movement of materials and energy through living communities
- The successional development of ecosystems
- The abundance and distribution of organisms and biodiversity in the context of the environment.

Ecology is a human science as well. There are many practical applications of ecology in conservation biology, wetland management, natural resource management (agroecology, agriculture, forestry, agroforestry, fisheries), city planning (urban ecology), community health, economics, basic and applied science, and human social interaction (human ecology). For example, the *Circles of Sustainability* approach treats ecology as more than the environment 'out there'. It is not treated as separate from humans. Organisms (including humans) and resources compose ecosystems which, in turn, maintain biophysical feedback mechanisms that moderate processes acting on living (biotic) and non-living (abiotic) components of the planet. Ecosystems sustain life-supporting functions and produce natural capital like biomass production (food, fuel, fiber, and medicine), the regulation of climate, global biogeochemical cycles, water filtration, soil formation, erosion control, flood protection, and many other natural features of scientific, historical, economic, or intrinsic value.

Ecology addresses the full scale of life, from tiny bacteria to processes that span the entire planet. Ecologists study many diverse and complex relations among species, such as predation and pollination. The diversity of life is organized into different habitats, from terrestrial (middle) to aquatic ecosystems.

The word "ecology" ("Ökologie») was coined in 1866 by the German scientist Ernst Haeckel (1834–1919). Ecological thought is derivative of established currents in philosophy, particularly from ethics and politics. Ancient Greek philosophers such as Hippocrates and Aristotle laid the foundations of ecology in their studies on natural history. Modern ecology became a much more rigorous science in the late 19th century. Evolutionary concepts relating to adaptation and natural selection became the cornerstones of modern ecological theory.

Levels, Scope, and Scale of Organization

The scope of ecology contains a wide array of interacting levels of organization spanning micro-level (e.g., cells) to a planetary scale (e.g., biosphere) phenomena. Ecosystems, for example, contain abiotic resources and interacting life forms (i.e., individual organisms that aggregate into populations which aggregate into distinct ecological communities). Ecosystems are dynamic, they do not always follow a linear successional path, but they are always changing, sometimes rapidly and sometimes so slowly that it can take thousands of years for ecological processes to bring about certain successional stages of a forest. An ecosystem's area can vary greatly, from tiny to vast. A single tree is of little consequence to the classification of a forest ecosystem, but critically relevant to organisms living in and on it. Several generations of an aphid population can exist over the lifespan of a single leaf. Each of those aphids, in turn, support diverse bacterial communities. The nature of connections in ecological communities cannot be explained by knowing the details of each species in isolation, because the emergent pattern is neither revealed nor predicted until the ecosystem is studied as an integrated whole. Some ecological principles, however, do exhibit collective properties where the sum of the components explain the properties of the whole, such as birth rates of a population being equal to the sum of individual births over a designated time frame.

Hierarchical Ecology

The scale of ecological dynamics can operate like a closed system, such as aphids migrating on a single tree, while at the same time remain open with regard to broader scale influences, such as atmosphere or climate. Hence, ecologists classify ecosystems hierarchically by analyzing data collected from finer scale units, such as vegetation associations, climate, and soil types, and integrate this information to identify emergent patterns of uniform organization and processes that operate on local to regional, landscape, and chronological scales. To structure the study of ecology into a conceptually manageable framework, the biological world is organized into a nested hierarchy, ranging in scale from genes, to cells, to tissues, to organs, to organisms, to species, to populations, to communities, to ecosystems, to biomes, and up to the level of the biosphere. This framework forms a panarchy and exhibits non-linear behaviors; this means that "effect and cause are disproportionate, so that small changes to critical variables, such as the number of nitrogen fixers, can lead to disproportionate, perhaps irreversible, changes in the system properties."

Biodiversity

Biodiversity (an abbreviation of "biological diversity") describes the diversity of life from genes to ecosystems and spans every level of biological organization. The term has several interpretations, and there are many ways to index, measure, characterize, and represent its complex organization.Biodiversity includes species diversity, ecosystem diversity, and genetic diversity and scientists are interested in the way that this diversity affects the complex ecological processes operating at and among these respective levels. Biodiversity plays an important role in ecosystem services which by definition maintain and improve human quality of life. Preventing species extinctions is one way to preserve biodiversity and that goal rests on techniques that preserve genetic diversity, habitat and the ability for species to migrate. Conservation priorities and management techniques require different approaches and considerations to address the full ecological scope of biodiversity. Natural capital that supports populations is critical for maintaining ecosystem services and species migration (e.g., riverine fish runs and avian insect control) has been implicated as one mechanism by which those service losses are experienced. An understanding of biodiversity has practical applications for species and ecosystem-level conservation planners as they make management recommendations to consulting firms, governments, and industry.

Habitat

The habitat of a species describes the environment over which a species is known to occur and the type of community that is formed as a result. More specifically, "habitats can be defined as regions in environmental space that are composed of multiple dimensions, each representing a biotic or abiotic environmental variable; that is, any component or characteristic of the environment related directly (e.g. forage biomass and quality) or indirectly (e.g. elevation) to the use of a location by the animal."[:745] For example, a habitat might be an aquatic or terrestrial environment that can be further categorized as a montane or alpine ecosystem. Habitat shifts provide important evidence of competition in nature where one population changes relative to the habitats that most other individuals of the species occupy. For example, one population of a species of tropical lizards (*Tropidurus hispidus*) has a flattened body relative to the main populations that live in open savanna. The population that lives in an isolated rock outcrop hides in crevasses where its flattened body offers a selective advantage. Habitat shifts also occur in the developmental life history of amphibians, and in insects that transition from aquatic to terrestrial habitats. Biotope and habitat are sometimes used interchangeably, but the former applies to a community's environment, whereas the latter applies to a species' environment. Additionally, some species are ecosystem engineers, altering the environment within a localized region. For instance, beavers manage water levels by building dams which improves their habitat in a landscape.

Niche

Definitions of the niche date back to 1917, but G. Evelyn Hutchinson made conceptual advances in 1957 by introducing a widely adopted definition: "the set of biotic and abiotic conditions in which a species is able to persist and maintain stable population sizes."[:] The ecological niche is a central concept in the ecology of organisms and is sub-divided into the *fundamental* and the *realized* niche. The fundamental niche is the set of environmental conditions under which a species is able to persist. The realized niche is the set of environmental plus ecological conditions under which a species persists. The Hutchinsonian niche is defined more technically as a "Euclidean hyperspace whose *dimensions* are defined as environmental variables and whose *size* is a function of the number of values that the environmental values may assume for which an organism has *positive fitness*."[:]

Termite mounds with varied heights of chimneys regulate gas exchange, temperature and other environmental parameters that are needed to sustain the internal physiology of the entire colony.

Biogeographical patterns and range distributions are explained or predicted through knowledge of a species' traits and niche requirements. Species have functional traits that are uniquely adapted to the ecological niche. A trait is a measurable property, phenotype, or characteristic of an organism that may influence its survival. Genes play an important role in the interplay of development and environmental expression of traits. Resident species evolve traits that are fitted to the selection pressures of their local environment. This tends to afford them a competitive advantage and discourages similarly adapted species from having an overlapping geographic range. The competitive exclusion principle states that two species cannot coexist indefinitely by living off the same limiting resource; one will always out-compete the other. When similarly adapted species overlap geographically, closer inspection reveals subtle ecological differences in their habitat or dietary requirements. Some models and empirical studies, however, suggest that disturbances can stabilize the co-evolution and shared niche occupancy of similar species inhabiting species-rich communities. The habitat plus the niche is called the ecotope, which is defined as the full range of environmental and biological variables affecting an entire species.

Niche Construction

Organisms are subject to environmental pressures, but they also modify their habitats. The regulatory feedback between organisms and their environment can affect conditions from local (e.g., a beaver pond) to global scales, over time and even after death, such as decaying logs or silica skeleton deposits from marine organisms. The process and concept of ecosystem engineering is related to niche construction, but the former relates only to the physical modifications of the habitat whereas the latter also considers the evolutionary implications of physical changes to the environment and the feedback this causes on the process of natural selection. Ecosystem engineers are defined as: "organisms that directly or indirectly modulate the availability of resources to other species, by causing physical state changes in biotic or abiotic materials. In so doing they modify, maintain and create habitats."[:373]

The ecosystem engineering concept has stimulated a new appreciation for the influence that organisms have on the ecosystem and evolutionary process. The term "niche construction" is more often used in reference to the under-appreciated feedback mechanisms of natural selection imparting forces on the abiotic niche. An example of natural selection through ecosystem engineering occurs in the nests of social insects, including ants, bees, wasps, and termites. There is an emergent homeostasis or homeorhesis in the structure of the nest that regulates, maintains and defends the physiology of the entire colony. Termite mounds, for example, maintain a constant internal temperature through the design of air-conditioning chimneys. The structure of the nests themselves are subject to the forces of natural selection. Moreover, a nest can survive over successive generations, so that progeny inherit both genetic material and a legacy niche that was constructed before their time.

Biome

Biomes are larger units of organization that categorize regions of the Earth's ecosystems, mainly according to the structure and composition of vegetation. There are different methods to define the continental boundaries of biomes dominated by different functional types of vegetative communities that are limited in distribution by climate, precipitation, weather and other environmental variables. Biomes include tropical rainforest, temperate broadleaf and mixed forest, temperate deciduous forest, taiga, tundra, hot desert, and polar desert. Other researchers have recently categorized other biomes, such as the human and oceanic microbiomes. To a microbe, the human body is a habitat and a landscape.Microbiomes were discovered largely through advances in molecular genetics, which have revealed a hidden richness of microbial diversity on the planet. The oceanic microbiome plays a significant role in the ecological biogeochemistry of the planet's oceans.

Biosphere

The largest scale of ecological organization is the biosphere: the total sum of ecosystems on the planet. Ecological relationships regulate the flux of energy, nutrients, and climate all the way up to the planetary scale. For example, the dynamic history of the planetary atmosphere's CO_2 and O_2composition has been affected by the biogenic flux of gases coming from respiration and photosynthesis, with levels fluctuating over time in relation to the ecology and evolution of plants and animals. Ecological theory has also been used to explain self-emergent regulatory phenomena at the planetary scale: for example, the Gaia hypothesis is an example of holism applied in ecological theory. The Gaia hypothesis states that there is an emergent feedback loop generated by the metabolism of living organisms that maintains the core temperature of the Earth and atmospheric conditions within a narrow self-regulating range of tolerance.

Individual Ecology

Understanding traits of individual organisms helps explain patterns and processes at other levels of organization including populations, communities, and ecosystems. Several areas of ecology of evolution that focus on such traits are life history theory, ecophysiology, metabolic theory of ecology, and Ethology. Examples of such traits include features of an organisms life cycle such as age to maturity, life span, or metabolic costs of reproduction. Other traits may be related to structure, such as the spines of a cactus or dorsal spines of a bluegill sunfish, or behaviors such as courtship displays or pair bonding. Other traits include emergent properties that are the result at least in part of interactions with the surrounding environment such as growth rate, resource uptake rate, winter, and deciduous vs. drought deciduous trees and shrubs.

One set of characteristics relate to body size and temperature. The metabolic theory of ecology provides a predictive qualitative set of relationships between an organism's body size and temperature and metabolic processes. In general, smaller, warmer organisms have higher metabolic rates and this results in a variety of predictions regarding individual somatic growth rates, reproduction and population growth rates, population size, and resource uptake rates.

The traits of organisms are subject to change through acclimation, development, and evolution. For this reason, individuals form a shared focus for ecology and for evolutionary ecology.

Population Ecology

Population ecology studies the dynamics of specie populations and how these populations interact with the wider environment. A population consists of individuals of the same species that live, interact, and migrate through the same niche and habitat.

A primary law of population ecology is the Malthusian growth model which states, "a population will grow (or decline) exponentially as long as the environment experienced by all individuals in the population remains constant."[:18] Simplified population models usually start with four variables: death, birth, immigration, and emigration.

An example of an introductory population model describes a closed population, such as on an island, where immigration and emigration does not take place. Hypotheses are evaluated with reference to a null hypothesis which states that random processes create the observed data. In these island models, the rate of population change is described by:

$$\frac{\mathrm{d}N}{\mathrm{d}T} = bN - dN = (b - d)N = rN,$$

where N is the total number of individuals in the population, b and d are the per capita rates of birth and death respectively, and r is the per capita rate of population change.

Using these modelling techniques, Malthus' population principle of growth was later transformed into a model known as the logistic equation:

$$\frac{dN}{dT} = aN\left(1 - \frac{N}{K}\right),$$

where N is the number of individuals measured as biomass density, a is the maximum per-capita rate of change, and K is the carrying capacity of the population. The formula states that the rate of change in population size (dN/dT) is equal to growth (aN) that is limited by carrying capacity ($1 - N/K$).

Population ecology builds upon these introductory models to further understand demographic processes in real study populations. Commonly used types of data include life history, fecundity, and survivorship, and these are analysed using mathematical techniques such as matrix algebra. The information is used for managing wildlife stocks and setting harvest quotas. In cases where basic models are insufficient, ecologists may adopt different kinds of statistical methods, such as the Akaike information criterion, or use models that can become mathematically complex as "several competing hypotheses are simultaneously confronted with the data."

Metapopulations and Migration

The concept of metapopulations was defined in 1969 as "a population of populations which go extinct locally and recolonize".Metapopulation ecology is another statistical approach that is often used in conservation research. Metapopulation models simplify the landscape into patches of varying levels of quality, and metapopulations are linked by the migratory behaviours of organisms. Animal migration is set apart from other kinds of movement; because, it involves the seasonal departure and return of individuals from a habitat. Migration is also a population-level phenomenon, as with the migration routes followed by plants as they occupied northern post-glacial environments. Plant ecologists use pollen records that accumulate and stratify in wetlands to reconstruct the timing of plant migration and dispersal relative to historic and contemporary climates. These migration routes involved an expansion of the range as plant populations expanded from one area to another. There is a larger taxonomy of movement, such as commuting, foraging, territorial behaviour, stasis, and ranging. Dispersal is usually distinguished from migration; because, it involves the one way permanent movement of individuals from their birth population into another population.

In metapopulation terminology, migrating individuals are classed as emigrants (when they leave a region) or immigrants (when they enter a region), and sites are classed either as sources or sinks. A site is a generic term that refers to places where ecologists sample populations, such as ponds or defined sampling areas in a forest. Source patches are productive sites that generate a seasonal supply of juveniles that migrate to other patch locations. Sink patches are unproductive sites that only receive migrants; the population at the site will disappear unless rescued by an adjacent source patch or environmental conditions become more favourable. Metapopulation models examine patch dynamics over time to answer potential questions about spatial and demographic ecology. The ecology of metapopulations is a dynamic process of extinction and colonization. Small patches of lower quality (i.e., sinks) are maintained or rescued by a seasonal influx of new immigrants. A dynamic metapopulation structure evolves from year to year, where some patches are sinks in dry years and are sources when conditions are more favourable. Ecologists use a mixture of computer models and field studies to explain metapopulation structure.

Community Ecology

Community ecology is the study of the interactions among a collections of species that inhabit the same geographic area. Community ecologists study the determinants of patterns and processes for two or more interacting species. Research in community ecology might measure species diversity in grasslands in relation to soil fertility. It might also include the analysis of predator-prey dynamics, competition among similar plant species, or mutualistic interactions between crabs and corals.

Interspecific interactions such as predation are a key aspect of community ecology.

Ecosystem Ecology

Ecosystems may be habitats within biomes that form an integrated whole and a dynamically responsive system having both physical and biological complexes. Ecosystem ecology is the science of determining the fluxes of materials (e.g. carbon, phosphorus) between different pools (e.g., tree biomass, soil organic material). Ecosystem ecologist attempt to determine the underlying causes of these fluxes. Research in ecosystem ecology might measure primary production (g C/m^2) in a wetland in relation to decomposition and consumption rates (g C/m^2/y). This requires an understanding of the community connections between plants (i.e., primary producers) and the decomposers (e.g., fungi and bacteria),

The underlying concept of ecosystem can be traced back to 1864 in the published work of George Perkins Marsh ("Man and Nature"). Within an ecosystem, organisms are linked to the physical and biological components of their environment to which they are adapted. Ecosystems are complex adaptive systems where the interaction of life processes form self-organizing patterns across different scales of time and space. Ecosystems are broadly categorized as terrestrial, freshwater, atmospheric, or marine. Differences stem from the nature of the unique physical environments that shapes the biodiversity within each. A more recent addition to ecosystem ecology are technoecosystems, which are affected by or primarily the result of human activity.

A riparian forest in the White Mountains, New Hampshire (USA) is an example of ecosystem ecology

Food Webs

A food web is the archetypal ecological network. Plants capture solar energy and use it to synthesize simple sugars during photosynthesis. As plants grow, they accumulate nutrients and are eaten by grazing herbivores, and the energy is transferred through a chain of organisms by consumption. The simplified linear feeding pathways that move from a basal trophic species to a top consumer is called the food chain. The larger interlocking pattern of food chains in an ecological community creates a complex food web. Food webs are a type of concept map or a heuristic device that is used to illustrate and study pathways of energy and material flows.

Food webs are often limited relative to the real world. Complete empirical measurements are generally restricted to a specific habitat, such as a cave or a pond, and principles gleaned from food web microcosm studies are extrapolated to larger systems. Feeding relations require extensive investigations into the gut contents of organisms, which can be difficult to decipher, or stable isotopes can be used to trace the flow of nutrient diets and energy through a food web. Despite these limitations, food webs remain a valuable tool in understanding community ecosystems.

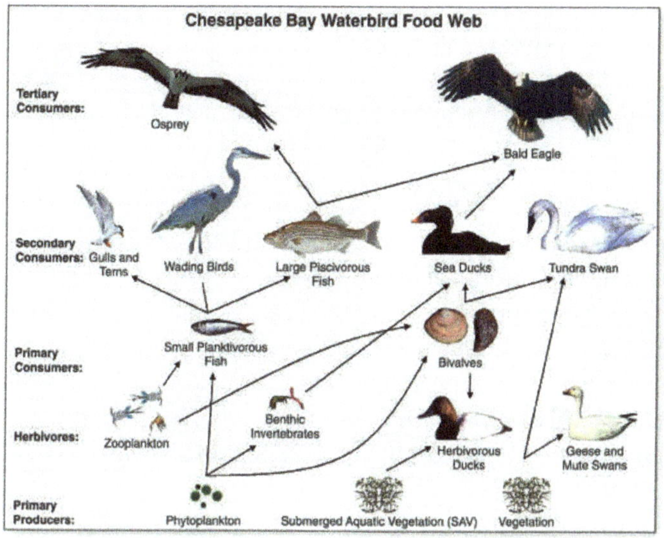

Generalized food web of waterbirds from Chesapeake Bay

Food webs exhibit principles of ecological emergence through the nature of trophic relationships: some species have many weak feeding links (e.g., omnivores) while some are more specialized with fewer stronger feeding links (e.g., primary predators). Theoretical and empirical studies identify non-random emergent patterns of few strong and many weak linkages that explain how ecological communities remain stable over time. Food webs are composed of subgroups where members in a community are linked by strong interactions, and the weak interactions occur between these subgroups. This increases food web stability. Step by step lines or relations are drawn until a web of life is illustrated.

Trophic Levels

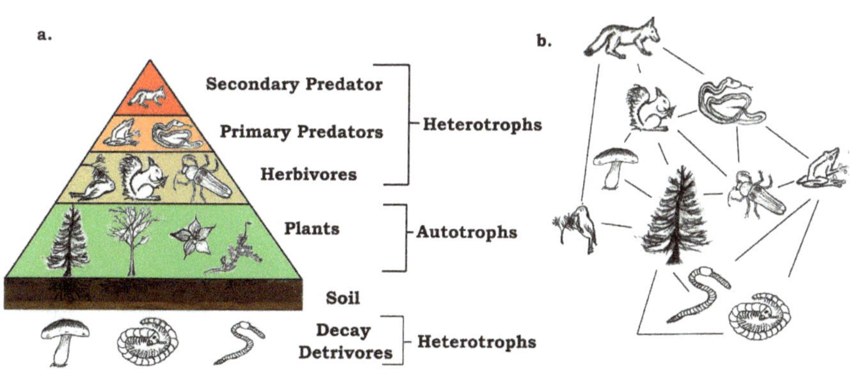

A trophic pyramid (a) and a food-web (b) illustrating ecological relationships among creatures that are typical of a northern boreal terrestrial ecosystem. The trophic pyramid roughly represents the biomass (usually measured as total dry-weight) at each level. Plants generally have the greatest biomass. Names of trophic categories are shown to the right of the pyramid. Some ecosystems, such as many wetlands, do not organize as a strict pyramid, because aquatic plants are not as productive as long-lived terrestrial plants such as trees. Ecological trophic pyramids are typically one of three kinds: 1) pyramid of numbers, 2) pyramid of biomass, or 3) pyramid of energy.

A trophic level (from Greek *troph*, τροφή, trophē, meaning "food" or "feeding") is "a group of organisms acquiring a considerable majority of its energy from the adjacent level nearer the abiotic source." Links in food webs primarily connect feeding relations or trophism among species. Biodiversity within ecosystems can be organized into trophic pyramids, in which the vertical dimension represents feeding relations that become further removed from the base of the food chain up toward top predators, and the horizontal dimension represents the abundance or biomass at each level. When the relative abundance or biomass of each species is sorted into its respective trophic level, they naturally sort into a 'pyramid of numbers'.

Species are broadly categorized as autotrophs (or primary producers), heterotrophs (or consumers), and Detritivores (or decomposers). Autotrophs are organisms that produce their own food (production is greater than respiration) by photosynthesis or chemosynthesis. Heterotrophs are organisms that must feed on others for nourishment and energy (respiration exceeds production). Heterotrophs can be further sub-divided into different functional groups, including primary consumers (strict herbivores), secondary consumers (carnivorous predators that feed exclusively on herbivores), and tertiary consumers (predators that feed on a mix of herbivores and predators). Omnivores do not fit neatly into a functional category because they eat both plant and animal tissues. It has been suggested that omnivores have a greater functional influence as predators, because compared to herbivores, they are relatively inefficient at grazing.

Trophic levels are part of the holistic or complex systems view of ecosystems. Each trophic level contains unrelated species that are grouped together because they share common ecological functions, giving a macroscopic view of the system. While the notion of trophic levels provides insight into energy flow and top-down control within food webs, it is troubled by the prevalence of omnivory in real ecosystems. This has led some ecologists to "reiterate that the notion that species clearly aggregate into discrete, homogeneous trophic levels is fiction." Nonetheless, recent studies have shown that real trophic levels do exist, but "above the herbivore trophic level, food webs are better characterized as a tangled web of omnivores."

Keystone Species

A keystone species is a species that is connected to a disproportionately large number of other species in the food-web. Keystone species have lower levels of biomass in the trophic pyramid relative to the importance of their role. The many connections that a keystone species holds means that it maintains the organization and structure of entire communities. The loss of a keystone species results in a range of dramatic cascading effects that alters trophic dynamics, other food web connections, and can cause the extinction of other species.

Sea otters (*Enhydra lutris*) are commonly cited as an example of a keystone species; because, they limit the density of sea urchins that feed on kelp. If sea otters are removed from the system, the urchins graze until the kelp beds disappear, and this has a dramatic effect on community structure. Hunting of sea otters, for example, is thought to have led indirectly to the extinction of the Steller's sea cow (*Hydrodamalis gigas*). While the keystone species concept has been used extensively as a conservation tool, it has been criticized for being poorly defined from an operational stance. It is difficult to experimentally determine what species may hold a keystone role in each ecosystem. Furthermore, food web theory suggests that keystone species may not be common, so it is unclear how generally the keystone species model can be applied.

Sea otters, an example of a keystone species

Ecological Complexity

Complexity is understood as a large computational effort needed to piece together numerous interacting parts exceeding the iterative memory capacity of the human mind. Global patterns of biological diversity are complex. This biocomplexity stems from the interplay among ecological processes that operate and influence patterns at different scales that grade into each other, such as transitional areas or ecotones spanning landscapes. Complexity stems from the interplay among levels of biological organization as energy, and matter is integrated into larger units that superimpose onto the smaller parts. "What were wholes on one level become parts on a higher one." Small scale patterns do not necessarily explain large scale phenomena, otherwise captured in the expression (coined by Aristotle) 'the sum is greater than the parts'.

"Complexity in ecology is of at least six distinct types: spatial, temporal, structural, process, behavioral, and geometric." From these principles, ecologists have identified emergent and self-organizing phenomena that operate at different environmental scales of influence, ranging from molecular to planetary, and these require different explanations at each integrative level. Ecological complexity relates to the dynamic resilience of ecosystems that transition to multiple shifting steady-states directed by random fluctuations of history. Long-term ecological studies provide important track records to better understand the complexity and resilience of ecosystems over longer temporal and broader spatial scales. These studies are managed by the International Long Term Ecological Network (LTER).

The longest experiment in existence is the Park Grass Experiment, which was initiated in 1856. Another example is the Hubbard Brook study, which has been in operation since 1960.

Holism

Holism remains a critical part of the theoretical foundation in contemporary ecological studies. Holism addresses the biological organization of life that self-organizes into layers of emergent whole systems that function according to non-reducible properties. This means that higher order patterns of a whole functional system, such as an ecosystem, cannot be predicted or understood by a simple summation of the parts. "New properties emerge because the components interact, not because the basic nature of the components is changed."

Ecological studies are necessarily holistic as opposed to reductionistic. Holism has three scientific meanings or uses that identify with ecology: 1) the mechanistic complexity of ecosystems, 2) the practical description of patterns in quantitative reductionist terms where correlations may be identified but nothing is understood about the causal relations without reference to the whole system, which leads to 3) a metaphysical hierarchy whereby the causal relations of larger systems are understood without reference to the smaller parts. Scientific holism differs from mysticism that has appropriated the same term. An example of metaphysical holism is identified in the trend of increased exterior thickness in shells of different species. The reason for a thickness increase can be understood through reference to principles of natural selection via predation without need to reference or understand the biomolecular properties of the exterior shells.

Relation to Evolution

Ecology and evolution are considered sister disciplines of the life sciences. Natural selection, life history, development, adaptation, populations, and inheritance are examples of concepts that thread equally into ecological and evolutionary theory. Morphological, behavioural, and genetic traits, for example, can be mapped onto evolutionary trees to study the historical development of a species in relation to their functions and roles in different ecological circumstances. In this framework, the analytical tools of ecologists and evolutionists overlap as they organize, classify, and investigate life through common systematic principals, such as phylogenetics or the Linnaean system of taxonomy. The two disciplines often appear together, such as in the title of the journal *Trends in Ecology and Evolution*. There is no sharp boundary separating ecology from evolution, and they differ more in their areas of applied focus. Both disciplines discover and explain emergent and unique properties and processes operating across different spatial or temporal scales of organization. While the boundary between ecology and evolution is not always clear, ecologists study the abiotic and biotic factors that influence evolutionary processes, and evolution can be rapid, occurring on ecological timescales as short as one generation.

Behavioural Ecology

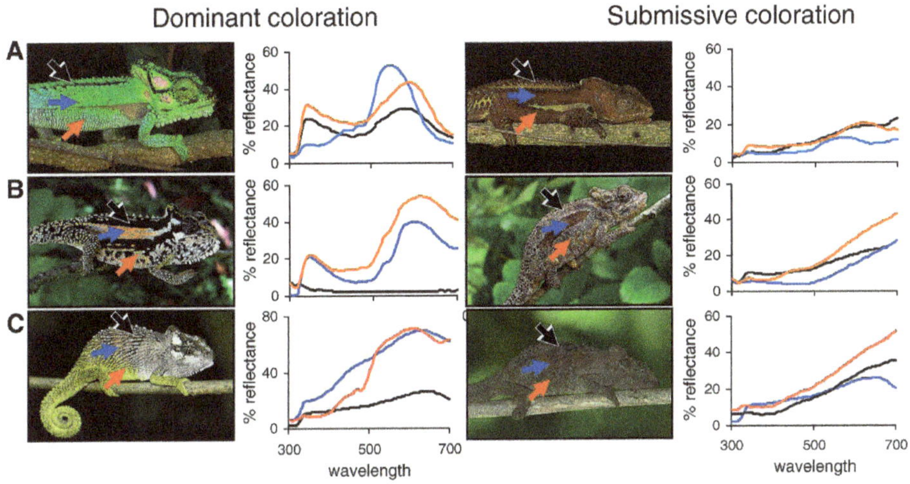

Social display and colour variation in differently adapted species of chameleons (*Bradypodion* spp.). Chameleons change their skin colour to match their background as a behavioural defence mechanism and also use colour to communicate with other members of their species, such as dominant (left) versus submissive (right) patterns shown in the three species (A-C) above.

All organisms can exhibit behaviours. Even plants express complex behaviour, including memory and communication. Behavioural ecology is the study of an organism's behaviour in its environment and its ecological and evolutionary implications. Ethology is the study of observable movement or behaviour in animals. This could include investigations of motile sperm of plants, mobile phytoplankton, zooplankton swimming toward the female egg, the cultivation of fungi by weevils, the mating dance of a salamander, or social gatherings of amoeba.

Adaptation is the central unifying concept in behavioural ecology. Behaviours can be recorded as traits and inherited in much the same way that eye and hair colour can. Behaviours can evolve by means of natural selection as adaptive traits conferring functional utilities that increases reproductive fitness.

Predator-prey interactions are an introductory concept into food-web studies as well as behavioural ecology. Prey species can exhibit different kinds of behavioural adaptations to predators, such as avoid, flee, or defend. Many prey species are faced with multiple predators that differ in the degree of danger posed. To be adapted to their environment and face predatory threats, organisms must balance their energy budgets as they invest in different aspects of their life history, such as growth, feeding, mating, socializing, or modifying their habitat. Hypotheses posited in behavioural ecology are generally based on adaptive principles of conservation, optimization, or efficiency. For example, "[t]he threat-sensitive predator avoidance hypothesis predicts that prey should assess the degree of threat posed by different predators and match their behaviour according to current levels of risk" or "[t]he optimal flight initiation distance occurs where expected postencounter fitness is maximized, which depends on the prey's initial fitness, benefits obtainable by not fleeing, energetic escape costs, and expected fitness loss due to predation risk."

Symbiosis: Leafhoppers (*Eurymela fenestrata*) are protected by ants (*Iridomyrmex purpureus*) in a symbiotic relationship. The ants protect the leafhoppers from predators and in return the leafhoppers feeding on plants exude honeydew from their anus that provides energy and nutrients to tending ants.

Elaborate sexual displays and posturing are encountered in the behavioural ecology of animals. The birds-of-paradise, for example, sing and display elaborate ornaments during courtship. These displays serve a dual purpose of signalling healthy or well-adapted individuals and desirable genes. The displays are driven by sexual selection as an advertisement of quality of traits among suitors.

Cognitive Ecology

Cognitive ecology integrates theory and observations from evolutionary ecology and neurobiology, primarily cognitive science, in order to understand the effect that animal interaction with their habitat has on their cognitive systems and how those systems restrict behavior within an ecological and evolutionary framework. "Until recently, however, cognitive scientists have not paid sufficient attention to the fundamental fact that cognitive traits evolved under particular natural settings. With consideration of the selection pressure on cognition, cognitive ecology can contribute intellectual coherence to the multidisciplinary study of cognition." As a study involving the 'coupling' or interactions between organism and environment, cognitive ecology is closely related to enactivism, a field based upon the view that "...we must see the organism and environment as bound together in reciprocal specification and selection...".

Social Ecology

Social ecological behaviours are notable in the social insects, slime moulds, social spiders, human society, and naked mole-rats where eusocialism has evolved. Social behaviours include reciprocally beneficial behaviours among kin and nest mates and evolve from kin and group selection. Kin selection explains altruism through genetic relationships, whereby an altruistic behaviour leading to death is rewarded by the survival of genetic copies distributed among surviving relatives. The social insects, including ants, bees, and wasps are most famously studied for this type of relationship because the male drones are clones that share the same genetic make-up as every other male in the colony. In contrast, group selectionists find examples of altruism among non-genetic relatives and explain this through selection acting on the group; whereby, it becomes selectively advantageous for groups if their members express altruistic behaviours to one another. Groups with predominantly altruistic members beat groups with predominantly selfish members.

Coevolution

Bumblebees and the flowers they pollinate have coevolved so that both have become dependent on each other for survival.

Ecological interactions can be classified broadly into a host and an associate relationship. A host is any entity that harbours another that is called the associate. Relationships within a species that are mutually or reciprocally beneficial are called mutualisms. Examples of mutualism include fungus-growing ants employing agricultural symbiosis, bacteria living in the guts of insects and other organisms, the fig wasp and yucca moth pollination complex, lichens with fungi and photosynthetic algae, and corals with photosynthetic algae. If there is a physical connection between host and associate, the relationship is called symbiosis. Approximately 60% of all plants, for example, have a symbiotic relationship with arbuscular mycorrhizal fungi living in their roots forming an exchange network of carbohydrates for mineral nutrients.

Indirect mutualisms occur where the organisms live apart. For example, trees living in the equatorial regions of the planet supply oxygen into the atmosphere that sustains species living in distant polar regions of the planet. This relationship is called commensalism; because, many others receive the benefits of clean air at no cost or harm to trees supplying the oxygen. If the associate benefits while the host suffers, the relationship is called parasitism. Although parasites impose a cost to their host (e.g., via damage to their reproductive organs or propagules, denying the services of a beneficial partner), their net effect on host fitness is not necessarily negative and, thus, becomes difficult to forecast. Co-evolution is also driven by competition among species or among members of the same species under the banner of reciprocal antagonism, such as grasses competing for growth space. The Red Queen Hypothesis, for example, posits that parasites track down and specialize on the locally common genetic defense systems of its host that drives the evolution of sexual reproduction to diversify the genetic constituency of populations responding to the antagonistic pressure.

Parasitism: A harvestman arachnid being parasitized by mites. The harvestman is being consumed, while the mites benefit from traveling on and feeding off of their host.

Biogeography

Biogeography (an amalgamation of *biology* and *geography*) is the comparative study of the geographic distribution of organisms and the corresponding evolution of their traits in space and time. The *Journal of Biogeography* was established in 1974. Biogeography and ecology share many of their disciplinary roots. For example, the theory of island biogeography, published by the mathematician Robert MacArthur and ecologist Edward O. Wilson in 1967 is considered one of the fundamentals of ecological theory.

Biogeography has a long history in the natural sciences concerning the spatial distribution of plants and animals. Ecology and evolution provide the explanatory context for biogeographical studies. Biogeographical patterns result from ecological processes that influence range distributions, such as migration and dispersal. and from historical processes that split populations or species into different areas. The biogeographic processes that result in the natural splitting of species explains much of the modern distribution of the Earth's biota. The splitting of lineages in a species is called vicariance biogeography and it is a sub-discipline of biogeography. There are also practical applications in the field of biogeography concerning ecological systems and processes. For example, the range and distribution of biodiversity and invasive species responding to climate change is a serious concern and active area of research in the context of global warming.

r/K-Selection theory

A population ecology concept is r/K selection theory, one of the first predictive models in ecology used to explain life-history evolution. The premise behind the r/K selection model is that natural selection pressures change according to population density. For example, when an island is first colonized, density of individuals is low. The initial increase in population size is not limited by competition, leaving an abundance of available resources for rapid population growth. These early phases of population growth experience *density-independent* forces of natural selection, which is called *r*-selection. As the population becomes more crowded, it approaches the island's carrying capacity, thus forcing individuals to compete more heavily for fewer available resources. Under crowded conditions, the population experiences density-dependent forces of natural selection, called *K*-selection.

In the *r/K*-selection model, the first variable *r* is the intrinsic rate of natural increase in population size and the second variable *K* is the carrying capacity of a population. Different species evolve different life-history strategies spanning a continuum between these two selective forces. An *r*-selected species is one that has high birth rates, low levels of parental investment, and high rates of mortality before individuals reach maturity. Evolution favours high rates of fecundity in *r*-selected species. Many kinds of insects and invasive species exhibit *r*-selected characteristics. In contrast, a *K*-selected species has low rates of fecundity, high levels of parental investment in the young, and low rates of mortality as individuals mature. Humans and elephants are examples of species exhibiting *K*-selected characteristics, including longevity and efficiency in the conversion of more resources into fewer offspring.

Molecular Ecology

The important relationship between ecology and genetic inheritance predates modern techniques for molecular analysis. Molecular ecological research became more feasible with the development of rapid and accessible genetic technologies, such as the polymerase chain reaction (PCR). The rise of molecular technologies and influx of research questions into this new ecological field resulted in the publication *Molecular Ecology* in 1992. Molecular ecology uses various analytical techniques to study genes in an evolutionary and ecological context. In 1994, John Avise also played a leading role in this area of science with the publication of his book, *Molecular Markers, Natural History and Evolution*. Newer technologies opened a wave of genetic analysis into organisms once difficult to study from an ecological or evolutionary standpoint, such as bacteria, fungi, and nematodes. Molecular ecology engendered a new research paradigm for investigating ecological questions considered otherwise intractable. Molecular investigations revealed previously obscured details in the tiny intricacies of nature and improved resolution into probing questions about behavioural and biogeographical ecology. For example, molecular ecology revealed promiscuous sexual behaviour and multiple male partners in tree swallows previously thought to be socially monogamous. In a biogeographical context, the marriage between genetics, ecology, and evolution resulted in a new sub-discipline called phylogeography.

Human Ecology

Ecology is as much a biological science as it is a human science. Human ecology is an interdisciplinary investigation into the ecology of our species. "Human ecology may be defined: (1) from a bio-ecological standpoint as the study of man as the ecological dominant in plant and animal communities and systems; (2) from a bio-ecological standpoint as simply another animal affecting and being affected by his physical environment; and (3) as a human being, somehow different from animal life in general, interacting with physical and modified environments in a distinctive and creative way. A truly interdisciplinary human ecology will most likely address itself to all three.":3 The term was formally introduced in 1921, but many sociologists, geographers, psychologists, and other disciplines were interested in human relations to natural systems centuries prior, especially in the late 19th century.

The ecological complexities human beings are facing through the technological transformation of the planetary biome has brought on the Anthropocene. The unique set of circumstances has generated the need for a new unifying science called coupled human and natural systems that builds upon, but moves beyond the field of human ecology. Ecosystems tie into human societies through the critical and all encompassing life-supporting functions they sustain. In recognition of these functions and the incapability of traditional economic valuation methods to see the value in ecosystems, there has been a surge of interest in social-natural capital, which provides the means to put a value on the stock and use of information and materials stemming from ecosystem goods and services. Ecosystems produce, regulate, maintain, and supply services of critical necessity and beneficial to human health (cognitive and physiological), economies, and they even provide an information or reference function as a living library giving opportunities for science and cognitive development in children engaged in the complexity of the natural world. Ecosystems relate importantly to human ecology as they are the ultimate base foundation of global economics as every commodity, and the capacity for exchange ultimately stems from the ecosystems on Earth.

Restoration and Management

Ecology is an employed science of restoration, repairing disturbed sites through human intervention, in natural resource management, and in environmental impact assessments. Edward O. Wilson predicted in 1992 that the 21st century "will be the era of restoration in ecology". Ecological science has boomed in the industrial investment of restoring ecosystems and their processes in abandoned sites after disturbance. Natural resource managers, in forestry, for example, employ ecologists to develop, adapt, and implement ecosystem based methods into the planning, operation, and restoration phases of land-use. Ecological science is used in the methods of sustainable harvesting, disease, and fire outbreak management, in fisheries stock management, for integrating land-use with protected areas and communities, and conservation in complex geo-political landscapes.

Relation to the Environment

The environment of ecosystems includes both physical parameters and biotic attributes. It is dynamically interlinked, and contains resources for organisms at any time throughout their life cycle. Like "ecology", the term "environment" has different conceptual meanings and overlaps with the concept of "nature". Environment "includes the physical world, the social world of human relations and the built world of human creation."[:62]The physical environment is external to the level of biological organization under investigation, including abiotic factors such as temperature, radiation, light, chemistry, climate and geology. The biotic environment includes genes, cells, organisms, members of the same species (conspecifics) and other species that share a habitat.

The distinction between external and internal environments, however, is an abstraction parsing life and environment into units or facts that are inseparable in reality. There is an interpenetration of cause and effect between the environment and life. The laws of thermodynamics, for example, apply to ecology by means of its physical state. With an understanding of metabolic and thermodynamic principles, a complete accounting of energy and material flow can be traced through an ecosystem. In this way, the environmental and ecological relations are studied through reference to conceptually manageable and isolated material parts. After the effective environmental components are understood through reference to their causes; however, they conceptually link back together as an integrated whole, or *holocoenotic* system as it was once called. This is known as the dialectical approach to ecology. The dialectical approach examines the parts, but integrates the organism and the environment into a dynamic whole (or umwelt). Change in one ecological or environmental factor can concurrently affect the dynamic state of an entire ecosystem.

Disturbance and Resilience

Ecosystems are regularly confronted with natural environmental variations and disturbances over time and geographic space. A disturbance is any process that removes biomass from a community, such as a fire, flood, drought, or predation. Disturbances occur over vastly different ranges in terms of magnitudes as well as distances and time periods, and are both the cause and product of natural fluctuations in death rates, species assemblages, and biomass densities within an ecological community. These disturbances create places of renewal where new directions emerge from the patchwork of natural experimentation and opportunity. Ecological resilience is a cornerstone theory in ecosystem management.

Biodiversity fuels the resilience of ecosystems acting as a kind of regenerative insurance.

Metabolism and the early Atmosphere

The Earth was formed approximately 4.5 billion years ago. As it cooled and a crust and oceans formed, its atmosphere transformed from being dominated by hydrogen to one composed mostly of methane and ammonia. Over the next billion years, the metabolic activity of life transformed the atmosphere into a mixture of carbon dioxide, nitrogen, and water vapor. These gases changed the way that light from the sun hit the Earth's surface and greenhouse effects trapped heat. There were untapped sources of free energy within the mixture of reducing and oxidizing gasses that set the stage for primitive ecosystems to evolve and, in turn, the atmosphere also evolved.

The leaf is the primary site of photosynthesis in most plants.

Throughout history, the Earth's atmosphere and biogeochemical cycles have been in a dynamic equilibrium with planetary ecosystems. The history is characterized by periods of significant transformation followed by millions of years of stability. The evolution of the earliest organisms, likely anaerobic methanogen microbes, started the process by converting atmospheric hydrogen into methane ($4H_2 + CO_2 \rightarrow CH_4 + 2H_2O$). Anoxygenic photosynthesis reduced hydrogen concentrations and increased atmospheric methane, by converting hydrogen sulfide into water or other sulfur compounds (for example, $2H_2S + CO_2 + h\nu \rightarrow CH_2O + H_2O + 2S$). Early forms of fermentation also increased levels of atmospheric methane. The transition to an oxygen-dominant atmosphere (the *Great Oxidation*) did not begin until approximately 2.4–2.3 billion years ago, but photosynthetic processes started 0.3 to 1 billion years prior.

Radiation: Heat, Temperature and Light

The biology of life operates within a certain range of temperatures. Heat is a form of energy that regulates temperature. Heat affects growth rates, activity, behaviour, and primary production. Temperature is largely dependent on the incidence of solar radiation. The latitudinal and longitudinal spatial variation of temperature greatly affects climates and consequently the distribution of biodiversity and levels of primary production in different ecosystems or biomes across the planet. Heat and temperature relate importantly to metabolic activity. Poikilotherms, for example, have a body temperature that is largely regulated and dependent on the temperature of the external environment. In contrast, homeotherms regulate their internal body temperature by expending metabolic energy.

There is a relationship between light, primary production, and ecological energy budgets. Sunlight is the primary input of energy into the planet's ecosystems. Light is composed of electromagnetic energy of different wavelengths. Radiant energy from the sun generates heat, provides photons of light measured as active energy in the chemical reactions of life, and also acts as a catalyst for genetic mutation. Plants, algae, and some bacteria absorb light and assimilate the energy through photosynthesis. Organisms capable of assimilating energy by photosynthesis or through inorganic fixation of H_2S are autotrophs. Autotrophs — responsible for primary production — assimilate light energy which becomes metabolically stored as potential energy in the form of biochemical enthalpic bonds.

Physical Environments

Water

Diffusion of carbon dioxide and oxygen is approximately 10,000 times slower in water than in air. When soils are flooded, they quickly lose oxygen, becoming hypoxic (an environment with O_2 concentration below 2 mg/liter) and eventually completely anoxic where anaerobic bacteria thrive among the roots. Water also influences the intensity and spectral composition of light as it reflects off the water surface and submerged particles. Aquatic plants exhibit a wide variety of morphological and physiological adaptations that allow them to survive, compete, and diversify in these environments. For example, their roots and stems contain large air spaces (aerenchyma) that regulate the efficient transportation of gases (for example, CO_2 and O_2) used in respiration and photosynthesis. Salt water plants (halophytes) have additional specialized adaptations, such as the development of special organs for shedding salt and osmoregulating their internal salt (NaCl) concentrations, to live in estuarine, brackish, or oceanic environments. Anaerobic soil microorganisms in aquatic environments use nitrate, manganese ions, ferric ions, sulfate, carbon dioxide, and some organic compounds; other microorganisms are facultative anaerobes and use oxygen during respiration when the soil becomes drier. The activity of soil microorganisms and the chemistry of the water reduces the oxidation-reduction potentials of the water. Carbon dioxide, for example, is reduced to methane (CH_4) by methanogenic bacteria. The physiology of fish is also specially adapted to compensate for environmental salt levels through osmoregulation. Their gills form electrochemical gradients that mediate salt excretion in salt water and uptake in fresh water.

Gravity

The shape and energy of the land is significantly affected by gravitational forces. On a large scale, the distribution of gravitational forces on the earth is uneven and influences the shape and movement of tectonic plates as well as influencing geomorphic processes such as orogeny and erosion. These forces govern many of the geophysical properties and distributions of ecological biomes across the Earth. On the organismal scale, gravitational forces provide directional cues for plant and fungal growth (gravitropism), orientation cues for animal migrations, and influence the biomechanics and size of animals. Ecological traits, such as allocation of biomass in trees during growth are subject to mechanical failure as gravitational forces influence the position and structure of branches and leaves. The cardiovascular systems of animals are functionally adapted to overcome pressure and gravitational forces that change according to the features of organisms (e.g., height, size, shape), their behaviour (e.g., diving, running, flying), and the habitat occupied (e.g., water, hot deserts, cold tundra).

Pressure

Climatic and osmotic pressure places physiological constraints on organisms, especially those that fly and respire at high altitudes, or dive to deep ocean depths. These constraints influence vertical limits of ecosystems in the biosphere, as organisms are physiologically sensitive and adapted to atmospheric and osmotic water pressure differences. For example, oxygen levels decrease with decreasing pressure and are a limiting factor for life at higher altitudes. Water transportation by plants is another important ecophysiological process affected by osmotic pressure gradients. Water pressure in the depths of oceans requires that organisms adapt to these conditions. For example, diving animals such as whales, dolphins, and seals are specially adapted to deal with changes in sound due to water pressure differences. Differences between hagfish species provide another example of adaptation to deep-sea pressure through specialized protein adaptations.

Wind and Turbulence

The architecture of the inflorescence in grasses is subject to the physical pressures of wind and shaped by the forces of natural selection facilitating wind-pollination (anemophily).

Turbulent forces in air and water affect the environment and ecosystem distribution, form and dynamics. On a planetary scale, ecosystems are affected by circulation patterns in the global trade winds. Wind power and the turbulent forces it creates can influence heat, nutrient, and biochemical profiles of ecosystems.For example, wind running over the surface of a lake creates turbulence, mixing the water column and influencing the environmental profile to create thermally layered zones, affecting how fish, algae, and other parts of the aquatic ecosystem are structured. Wind speed and turbulence also influence evapotranspiration rates and energy budgets in plants and animals. Wind speed, temperature and moisture content can vary as winds travel across different land features and elevations. For example, the westerlies come into contact with the coastal and interior mountains of western North America to produce a rain shadow on the leeward side of the mountain. The air expands and moisture condenses as the winds increase in elevation; this is called orographic lift and can cause precipitation. This environmental process produces spatial divisions in biodiversity, as species adapted to wetter conditions are range-restricted to the coastal mountain valleys and unable to migrate across the xeric ecosystems (e.g., of the Columbia Basin in western North America) to intermix with sister lineages that are segregated to the interior mountain systems.

Fire

Forest fires modify the land by leaving behind an environmental mosaic that diversifies the landscape into different seral stages and habitats of varied quality (left). Some species are adapted to forest fires, such as pine trees that open their cones only after fire exposure (right).

Plants convert carbon dioxide into biomass and emit oxygen into the atmosphere. By approximately 350 million years ago (the end of the Devonian period), photosynthesis had brought the concentration of atmospheric oxygen above 17%, which allowed combustion to occur. Fire releases CO_2 and converts fuel into ash and tar. Fire is a significant ecological parameter that raises many issues pertaining to its control and suppression.While the issue of fire in relation to ecology and plants has been recognized for a long time, Charles Cooper brought attention to the issue of forest fires in relation to the ecology of forest fire suppression and management in the 1960s.

Native North Americans were among the first to influence fire regimes by controlling their spread near their homes or by lighting fires to stimulate the production of herbaceous foods and basketry materials. Fire creates a heterogeneous ecosystem age and canopy structure, and the altered soil nutrient supply and cleared canopy structure opens new ecological niches for seedling establishment. Most ecosystems are adapted to natural fire cycles. Plants, for example, are equipped with a variety of adaptations to deal with forest fires. Some species (e.g., *Pinus halepensis*) cannot germinate until after their seeds have lived through a fire or been exposed to certain compounds from smoke. Environmentally triggered germination of seeds is called serotiny. Fire plays a major role in the persistence and resilience of ecosystems.

Soils

Soil is the living top layer of mineral and organic dirt that covers the surface of the planet. It is the chief organizing centre of most ecosystem functions, and it is of critical importance in agricultural science and ecology. The decomposition of dead organic matter (for example, leaves on the forest floor), results in soils containing minerals and nutrients that feed into plant production. The whole of the planet's soil ecosystems is called the pedosphere where a large biomass of the Earth's biodiversity organizes into trophic levels. Invertebrates that feed and shred larger leaves, for example, create smaller bits for smaller organisms in the feeding chain. Collectively, these organisms are the detritivores that regulate soil formation. Tree roots, fungi, bacteria, worms, ants, beetles, centipedes, spiders, mammals, birds, reptiles, amphibians, and other less familiar creatures all work to create the trophic web of life in soil ecosystems. Soils form composite phenotypes where inorganic matter is enveloped into the physiology of a whole community. As organisms feed and migrate through soils they physically displace materials, an ecological process called bioturbation. This aerates soils and stimulates heterotrophic growth and production. Soil microorganisms are influenced by and feed back into the trophic dynamics of the ecosystem. No single axis of causality can be discerned to segregate the biological from geomorphological systems in soils. Paleoecological studies of soils places the origin for bioturbation to a time before the Cambrian period. Other events, such as the evolution of trees and the colonization of land in the Devonian period played a significant role in the early development of ecological trophism in soils.

Biogeochemistry and Climate

Ecologists study and measure nutrient budgets to understand how these materials are regulated, flow, and recycled through the environment. This research has led to an understanding that there is global feedback between ecosystems and the physical parameters of this planet, including minerals, soil, pH, ions, water, and atmospheric gases. Six major elements (hydrogen, carbon, nitrogen, oxygen, sulfur, and phosphorus; H, C, N, O, S, and P) form the constitution of all biological macromolecules and feed into the Earth's geochemical processes. From the smallest scale of biology, the combined effect of billions upon billions of ecological processes amplify and ultimately regulate the biogeochemical cycles of the Earth. Understanding the relations and cycles mediated between these elements and their ecological pathways has significant bearing toward understanding global biogeochemistry.

The ecology of global carbon budgets gives one example of the linkage between biodiversity and biogeochemistry. It is estimated that the Earth's oceans hold 40,000 gigatonnes (Gt) of carbon, that vegetation and soil hold 2070 Gt, and that fossil fuel emissions are 6.3 Gt carbon per year.There have been major restructurings in these global carbon budgets during the Earth's history, regulated to a large extent by the ecology of the land. For example, through the early-mid Eocene volcanic outgassing, the oxidation of methane stored in wetlands, and seafloor gases increased atmospheric CO_2 (carbon dioxide) concentrations to levels as high as 3500 ppm.

In the Oligocene, from twenty-five to thirty-two million years ago, there was another significant restructuring of the global carbon cycle as grasses evolved a new mechanism of photosynthesis, C_4 photosynthesis, and expanded their ranges. This new pathway evolved in response to the drop in atmospheric CO_2 concentrations below 550 ppm. The relative abundance and distribution of biodiversity alters the dynamics between organisms and their environment such that ecosystems can be both cause and effect in relation to climate change. Human-driven modifications to the planet's ecosystems (e.g., disturbance, biodiversity loss, agriculture) contributes to rising atmospheric greenhouse gas levels. Transformation of the global carbon cycle in the next century is projected to raise planetary temperatures, lead to more extreme fluctuations in weather, alter species distributions, and increase extinction rates. The effect of global warming is already being registered in melting glaciers, melting mountain ice caps, and rising sea levels. Consequently, species distributions are changing along waterfronts and in continental areas where migration patterns and breeding grounds are tracking the prevailing shifts in climate. Large sections of permafrost are also melting to create a new mosaic of flooded areas having increased rates of soil decomposition activity that raises methane (CH_4) emissions. There is concern over increases in atmospheric methane in the context of the global carbon cycle, because methane is a greenhouse gas that is 23 times more effective at absorbing long-wave radiation than CO_2 on a 100-year time scale. Hence, there is a relationship between global warming, decomposition and respiration in soils and wetlands producing significant climate feedbacks and globally altered biogeochemical cycles.

Early Beginnings

Ecology has a complex origin, due in large part to its interdisciplinary nature. Ancient Greek philosophers such as Hippocrates and Aristotle were among the first to record observations on natural history. However, they viewed life in terms of essentialism, where species were conceptualized as static unchanging things while varieties were seen as aberrations of an idealized type. This contrasts against the modern understanding of ecological theory where varieties are viewed as the real phenomena of interest and having a role in the origins of adaptations by means of natural selection. Early conceptions of ecology, such as a balance and regulation in nature can be traced to Herodotus (died *c.* 425 BC), who described one of the earliest accounts of mutualism in his observation of "natural dentistry". Basking Nile crocodiles, he noted, would open their mouths to give sandpipers safe access to pluck leeches out, giving nutrition to the sandpiper and oral hygiene for the crocodile.Aristotle was an early influence on the philosophical development of ecology. He and his student Theophrastus made extensive observations on plant and animal migrations, biogeography, physiology, and on their behaviour, giving an early analogue to the modern concept of an ecological niche.

Ecological concepts such as food chains, population regulation, and productivity were first developed in the 1700s, through the published works of microscopist Antoni van Leeuwenhoek (1632–1723) and botanist Richard Bradley (1688?–1732). Biogeographer Alexander von Humboldt (1769–1859) was an early pioneer in ecological thinking and was among the first to recognize ecological gradients, where species are replaced or altered in form along environmental gradients, such as a cline forming along a rise in elevation. Humboldt drew inspiration from Isaac Newton as he developed a form of "terrestrial physics". In Newtonian fashion, he brought a scientific exactitude for measurement into natural history and even alluded to concepts that are the foundation of a modern ecological law on species-to-area relationships. Natural historians, such as Humboldt, James Hutton, and Jean-Baptiste Lamarck (among others) laid the foundations of the modern ecological sciences. The term "ecology" (German: *Oekologie, Ökologie*) is of a more recent origin and was first coined by the German biologist Ernst Haeckel in his book *Generelle Morphologie der Organismen* (1866). Haeckel was a zoologist, artist, writer, and later in life a professor of comparative anatomy.

Ernst Haeckel (left) and Eugenius Warming (right), two founders of ecology

Opinions differ on who was the founder of modern ecological theory. Some mark Haeckel's definition as the beginning;others say it was Eugenius Warming with the writing of Oecology of Plants: An Introduction to the Study of Plant Communities (1895), or Carl Linnaeus' principles on the economy of nature that matured in the early 18th century. Linnaeus founded an early branch of ecology that he called the economy of nature. His works influenced Charles Darwin, who adopted Linnaeus' phrase on the *economy or polity of nature* in *The Origin of Species*. Linnaeus was the first to frame the balance of nature as a testable hypothesis. Haeckel, who admired Darwin's work, defined ecology in reference to the economy of nature, which has led some to question whether ecology and the economy of nature are synonymous.

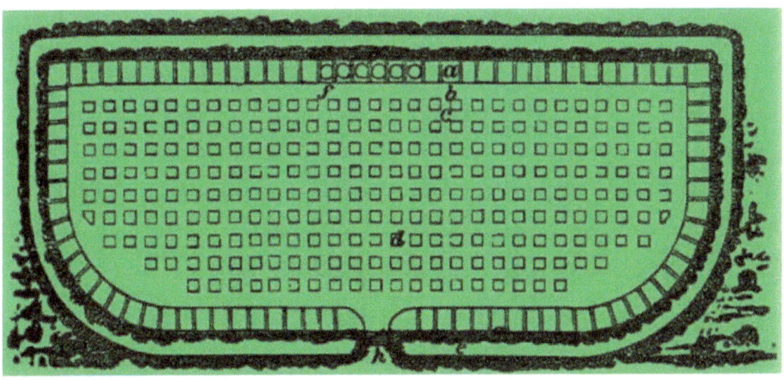

The layout of the first ecological experiment, carried out in a grass garden at Woburn Abbey in 1816, was noted by Charles Darwin in *The Origin of Species*. The experiment studied the performance of different mixtures of species planted in different kinds of soils.

From Aristotle until Darwin, the natural world was predominantly considered static and unchanging. Prior to *The Origin of Species*, there was little appreciation or understanding of the dynamic and reciprocal relations between organisms, their adaptations, and the environment. An exception is the 1789 publication *Natural History of Selborne* by Gilbert White (1720–1793), considered by some to be one of the earliest texts on ecology. While Charles Darwin is mainly noted for his treatise on evolution, he was one of the founders of soil ecology, and he made note of the first ecological experiment in *The Origin of Species*. Evolutionary theory changed the way that researchers approached the ecological sciences.

Since 1900

Modern ecology is a young science that first attracted substantial scientific attention toward the end of the 19th century (around the same time that evolutionary studies were gaining scientific interest). Notable scientist Ellen Swallow Richards may have first introduced the term "oekology" (which eventually morphed into home economics) in the U.S. as early 1892.

In the early 20th century, ecology transitioned from a more descriptive form of natural history to a more analytical form of *scientific natural history*. Frederic Clements published the first American ecology book in 1905, presenting the idea of plant communities as a superorganism. This publication launched a debate between ecological holism and individualism that lasted until the 1970s. Clements' superorganism concept proposed that ecosystems progress through regular and determined stages of seral development that are analogous to the developmental stages of an organism. The Clementsian paradigm was challenged by Henry Gleason, who stated that ecological communities develop from the unique and coincidental association of individual organisms. This perceptual shift placed the focus back onto the life histories of individual organisms and how this relates to the development of community associations.

The Clementsian superorganism theory was an overextended application of an idealistic form of holism. The term "holism" was coined in 1926 by Jan Christiaan Smuts, a South African general and polarizing historical figure who was inspired by Clements' superorganism concept.[C] Around the same time, Charles Elton pioneered the concept of food chains in his classical book *Animal Ecology*. Elton defined ecological relations using concepts of food chains, food cycles, and food size, and described numerical relations among different functional groups and their relative abundance. Elton's 'food cycle' was replaced by 'food web' in a subsequent ecological text. Alfred J. Lotka brought in many theoretical concepts applying thermodynamic principles to ecology.

In 1942, Raymond Lindeman wrote a landmark paper on the trophic dynamics of ecology, which was published posthumously after initially being rejected for its theoretical emphasis. Trophic dynamics became the foundation for much of the work to follow on energy and material flow through ecosystems. Robert MacArthur advanced mathematical theory, predictions, and tests in ecology in the 1950s, which inspired a resurgent school of theoretical mathematical ecologists. Ecology also has developed through contributions from other nations, including Russia's Vladimir Vernadsky and his founding of the biosphere concept in the 1920s and Japan's Kinji Imanishi and his concepts of harmony in nature and habitat segregation in the 1950s. Scientific recognition of contributions to ecology from non-English-speaking cultures is hampered by language and translation barriers.

Ecology surged in popular and scientific interest during the 1960–1970s environmental movement. There are strong historical and scientific ties between ecology, environmental management, and protection. The historical emphasis and poetic naturalistic writings advocating the protection of wild places by notable ecologists in the history of conservation biology, such as Aldo Leopold and Arthur Tansley, have been seen as far removed from urban centres where, it is claimed, the concentration of pollution and environmental degradation is located. Palamar (2008) notes an overshadowing by mainstream environmentalism of pioneering women in the early 1900s who fought for urban health ecology (then called euthenics) and brought about changes in environmental legislation. Women such as Ellen Swallow Richards and Julia Lathrop, among others, were precursors to the more popularized environmental movements after the 1950s.

In 1962, marine biologist and ecologist Rachel Carson's book *Silent Spring* helped to mobilize the environmental movement by alerting the public to toxic pesticides, such as DDT, bioaccumulating in the environment. Carson used ecological science to link the release of environmental toxins to human and ecosystem health. Since then, ecologists have worked to bridge their understanding of the degradation of the planet's ecosystems with environmental politics, law, restoration, and natural resources management.

REFERENCES

1. Eric Laferrière; Peter J. Stoett (2 September 2003). International Relations Theory and Ecological Thought: Towards a Synthesis. Routledge. pp. 25–. ISBN 978-1-134-71068-3.
2. Stadler, B.; Michalzik, B.; Müller, T. (1998). "Linking aphid ecology with nutrient fluxes in a coniferous forest". Ecology. 79 (5): 1514–1525. doi:10.1890/0012-9658(1998)079[1514:LAEWNF]2.0.CO;2. ISSN 0012-9658.
3. Humphreys, N. J.; Douglas, A. E. (1997). "Partitioning of symbiotic bacteria between generations of an insect: a quantitative study of aBuchnera sp. in the pea aphid (Acyrthosiphon pisum) reared at different temperatures". Applied and Environmental Microbiology. 63 (8): 3294–3296. PMC 1389233 . PMID 16535678.
4. Liere, Heidi; Jackson, Doug; Vandermeer, John; Wilby, Andrew (20 September 2012). "Ecological Complexity in a Coffee Agroecosystem: Spatial Heterogeneity, Population Persistence and Biological Control". PLoS ONE. 7 (9): e45508. Bibcode:2012PLoSO...745508L. doi:10.1371/journal.pone.0045508. PMC 3447771 . PMID 23029061.
5. Odum, E. P.; Barrett, G. W. (2005). Fundamentals of Ecology. Brooks Cole. p. 598. ISBN 978-0-534-42066-6.
6. O'Neill, D. L.; Deangelis, D. L.; Waide, J. B.; Allen, T. F. H. (1986). A Hierarchical Concept of Ecosystems. Princeton University Press. p. 253. ISBN 0-691-08436-X.

7. Nachtomy, Ohad; Shavit, Ayelet; Smith, Justin (2002). "Leibnizian organisms, nested individuals, and units of selection". Theory in Biosciences. 121 (2): 205–230. doi:10.1007/s12064-002-0020-9.
8. Holling, C. S. (2004). "Understanding the complexity of economic, ecological, and social systems". Ecosystems. 4 (5): 390–405. doi:10.1007/s10021-001-0101-5.
9. Levin, S. A. (1999). Fragile Dominion: Complexity and the Commons. Reading, MA: Perseus Books. ISBN 978-0-7382-0319-5.
10. Noss, R. F.; Carpenter, A. Y. (1994). Saving Nature's Legacy: Protecting and Restoring Biodiversity. Island Press. p. 443. ISBN 978-1-55963-248-5.
11. Noss, R. F. (1990). "Indicators for monitoring biodiversity: A hierarchical approach". Conservation Biology. 4 (4): 355–364. doi:10.1111/j.1523-1739.1990.tb00309.x. JSTOR 2385928.
12. Scholes, R. J.; Mace, G. M.; Turner, W.; Geller, G. N.; Jürgens, N.; Larigauderie, A.; Muchoney, D.; Walther, B. A.; Mooney, H. A. (2008). "Toward a global biodiversity observing system" (PDF). Science. 321 (5892): 1044–1045. doi:10.1126/science.1162055. PMID 18719268.
13. Cardinale, Bradley J.; Duffy, J. Emmett; Gonzalez, Andrew; Hooper, David U.; Perrings, Charles; Venail, Patrick; Narwani, Anita; Mace, Georgina M.; Tilman, David; Wardle, David A.; Kinzig, Ann P.; Daily, Gretchen C.; Loreau, Michel; Grace, James B.; Larigauderie, Anne; Srivastava, Diane S.; Naeem, Shahid; Gonzalez, Andrew; Hooper, David U.; Perrings, Charles; Venail, Patrick; Narwani, Anita; Mace, Georgina M.; Tilman, David; Wardle, David A.; Kinzig, Ann P.; Daily, Gretchen C.; Loreau, Michel; Grace, James B.; Larigauderie, Anne; Srivastava, Diane S.; Naeem, Shahid (6 June 2012). "Biodiversity loss and its impact on humanity". Nature. 486 (7401): 59–67. Bibcode:2012Natur.486...59C. doi:10.1038/nature11148. PMID 22678280.
14. Wilson, E. O. (2000). "A global biodiversity map". Science. 289 (5488): 2279. PMID 11041790.
15. Purvis, A.; Hector, A. (2000). "Getting the measure of biodiversity"(PDF). Nature. 405 (6783): 212–218. doi:10.1038/35012221. PMID 10821281.
16. Ostfeld, R. S. (2009). "Biodiversity loss and the rise of zoonotic pathogens" (PDF). Clinical Microbiology and Infection. 15 (s1): 40–43. doi:10.1111/j.1469-0691.2008.02691.x. PMID 19220353.
17. Tierney, Geraldine L.; Faber-Langendoen, Don; Mitchell, Brian R.; Shriver, W. Gregory; Gibbs, James P. (2009). "Monitoring and evaluating the ecological integrity of forest ecosystems" (PDF). Frontiers in Ecology and the Environment. 7 (6): 308–316. doi:10.1890/070176.
18. Ceballos, G.; Ehrlich, P. R. (2002). "Mammal population losses and the extinction crisis" (PDF). Science. 296 (5569): 904–907. Bibcode:2002Sci...296..904C. doi:10.1126/science.1069349. PMID 11988573. Retrieved 2010-03-16.
19. Palumbi, Stephen R; S. R., Paul A; Allan, J. David; Beck, Michael W.; Fautin, Daphne G.; Fogarty, Michael J.; Halpern, Benjamin S.; Incze, Lewis S.; Leong, Jo-Ann; et al. (2009). "Managing for ocean biodiversity to sustain marine

ecosystem services" (PDF). Frontiers in Ecology and the Environment. 7 (4): 204–211. doi:10.1890/070135.

20. Wilcove, D. S.; Wikelski, M. (2008). "Going, going, gone: Is animal migration disappearing". PLoS Biology. 6 (7): e188. doi:10.1371/journal.pbio.0060188. PMC 2486312 . PMID 18666834.
21. Hammond, H. (2009). Maintaining Whole Systems on the Earth's Crown: Ecosystem-based Conservation Planning for the Boreal Forest. Slocan Park, BC: Silva Forest Foundation. p. 380. ISBN 978-0-9734779-0-0.
22. Whittaker, R. H.; Levin, S. A.; Root, R. B. (1973). "Niche, habitat, and ecotope" (PDF). The American Naturalist. 107 (955): 321–338. doi:10.1086/282837.
23. Beyer, Hawthorne, L.; Haydon, Daniel, T.; Morales, Juan M.; Frair, Jacqueline L.; Hebblewhite, Mark; Mitchell, Michael; Matthiopoulos, Jason (2010). "The interpretation of habitat preference metrics under use–availability designs". Philosophical Transactions of the Royal Society B. 365 (1550): 2245–2254. doi:10.1098/rstb.2010.0083. PMC 2894962 . PMID 20566501.
24. Schoener, T. W. (1975). "Presence and absence of habitat shift in some widespread lizard species". Ecological Monographs. 45 (3): 233–258. doi:10.2307/1942423. JSTOR 1942423.
25. Vitt, L. J.; Caldwell, J. P.; Zani, P. A.; Titus, T. A. (1997). "The role of habitat shift in the evolution of lizard morphology: Evidence from tropicalTropidurus". Proceedings of the National Academy of Sciences. 94 (8): 3828–3832. Bibcode:1997PNAS...94.3828V. doi:10.1073/pnas.94.8.3828. PMC 20526 . PMID 9108063.
26. Kiessling, W.; Simpson, C.; Foote, M. (2009). "Reefs as cradles of evolution and sources of biodiversity in the Phanerozoic". Science. 327 (5962): 196–198. Bibcode:2010Sci...327..196K. doi:10.1126/science.1182241. PMID 20056888.
27. Laland, K. N.; Odling-Smee, F. J.; Feldman, M. W. (1999). "Evolutionary consequences of niche construction and their implications for ecology". Proceedings of the National Academy of Sciences. 96 (18): 10242–10247. Bibcode:1999PNAS...9610242L. doi:10.1073/pnas.96.18.10242. PMC 17873 . PMID 10468593.
28. Hughes, D. P.; Pierce, N. E.; Boomsma, J. J. (2008). "Social insect symbionts: evolution in homeostatic fortresses" (PDF). Trends in Ecology & Evolution. 23 (12): 672–677. doi:10.1016/j.tree.2008.07.011. PMID 18951653.
29. Wiens, J. J.; Graham, C. H. (2005). "Niche conservatism: Integrating evolution, ecology, and conservation biology" (PDF). Annual Review of Ecology, Evolution, and Systematics. 36: 519–539. doi:10.1146/annurev.ecolsys.36.102803.095431.
30. Hutchinson, G. E. (1957). A Treatise on Limnology. New York, NY: Wiley. p. 1015. ISBN 0-471-42572-9.
31. Hutchinson, G. E. (1957). "Concluding remarks" (PDF). Cold Spring Harbor Symposia on Quantitative Biology. 22: 415–427. doi:10.1101/SQB.1957.022.01.039.
32. Begon, M.; Townsend, C. R.; Harper, J. L. (2005). Ecology: From Individuals

to Ecosystems (4th ed.). Wiley-Blackwell. p. 752. ISBN 1-4051-1117-8.

33. D. L., Hardesty (1975). "The niche concept: suggestions for its use in human ecology". Human Ecology. 3 (2): 71–85. doi:10.1007/BF01552263. JSTOR 4602315.
34. Pearman, P. B.; Guisan, A.; Broennimann, O.; Randin, C. F. (2008). "Niche dynamics in space and time" (PDF). Trends in Ecology & Evolution. 23 (3): 149–158. doi:10.1016/j.tree.2007.11.005. PMID 18289716.
35. Levins, R.; Lewontin, R. (1980). "Dialectics and reductionism in ecology" (PDF). Synthese. 43: 47–78. doi:10.1007/bf00413856.
36. Hardin, G. (1960). "The competitive exclusion principal". Science. 131 (3409): 1292–1297. Bibcode:1960Sci...131.1292H. doi:10.1126/science.131.3409.1292. PMID 14399717.
37. Scheffer, M.; van Nes, E. H. (2006). "Self-organized similarity, the evolutionary emergence of groups of similar species" (PDF). Proceedings of the National Academy of Sciences. 103 (16): 6230–6235. Bibcode:2006PNAS..103.6230S. doi:10.1073/pnas.0508024103.
38. Hastings, Alan; Byers, James E.; Crooks, Jeffrey A.; Cuddington, Kim; Jones, Clive G.; Lambrinos, John G.; Talley, Theresa S.; Wilson, William G. (2007). "Ecosystem engineering in space and time". Ecology Letters. 10 (2): 153–164. doi:10.1111/j.1461-0248.2006.00997.x. PMID 17257103.
39. Jones, Clive G.; Lawton, John H.; Shachak, Moshe (1994). "Organisms as ecosystem engineers". Oikos. 69 (3): 373–386. doi:10.2307/3545850. JSTOR 3545850.
40. Wright, J. P.; Jones, C.G. (2006). "The concept of organisms as ecosystem engineers ten years on: Progress, limitations, and challenges". BioScience. 56 (3): 203–209. doi:10.1641/0006-3568(2006)056[0203:TCOOAE]2.0.CO;2. ISSN 0006-3568.
41. Palmer, M.; White, P. S. (1994). "On the existence of ecological communities" (PDF). Journal of Vegetation Sciences. 5 (2): 279–282. doi:10.2307/3236162. JSTOR 3236162.
42. Prentice; I. C.; Harrison, S. P.; Leemans, R.; Monserud, R. A.; Solomon, A. M. (1992). "Special paper: A global biome model based on plant physiology and dominance, soil properties and climate". Journal of Biogeography. 19 (2): 117–134. doi:10.2307/2845499. JSTOR 2845499.
43. Turnbaugh, Peter J.; Ley, Ruth E.; Hamady, Micah; Fraser-Liggett, Claire M.; Knight, Rob; Gordon, Jeffrey I. (2007). "The human microbiome project". Nature. 449 (7164): 804–810. Bibcode:2007Natur.449..804T. doi:10.1038/nature06244. PMC 3709439 . PMID 17943116.
44. DeLong, E. F. (2009). "The microbial ocean from genomes to biomes" (PDF). Nature. 459 (7244): 200–206. Bibcode:2009Natur.459..200D. doi:10.1038/nature08059. PMID 19444206.
45. Igamberdiev, Abir U.; Lea, P. J. (2006). "Land plants equilibrate O2and CO2

concentrations in the atmosphere" (PDF). Photosynthesis Research. 87 (2): 177–194. doi:10.1007/s11120-005-8388-2. PMID 16432665.

46. Lovelock, J.; Margulis, Lynn (1973). "Atmospheric homeostasis by and for the biosphere: The Gaia hypothesis". Tellus. 26: 2–10. Bibcode:1974Tell...26....2L. doi:10.1111/j.2153-3490.1974.tb01946.x.
47. Lovelock, J. (2003). "The living Earth". Nature. 426 (6968): 769–770. Bibcode:2003Natur.426..769L. doi:10.1038/426769a. PMID 14685210.
48. Waples, R. S.; Gaggiotti, O. (2006). "What is a population? An empirical evaluation of some genetic methods for identifying the number of gene pools and their degree of connectivity". Molecular Ecology. 15 (6): 1419–1439. doi:10.1111/j.1365-294X.2006.02890.x. PMID 16629801.
49. Turchin, P. (2001). "Does population ecology have general laws?". Oikos. 94 (1): 17–26. doi:10.1034/j.1600-0706.2001.11310.x.
50. Vandermeer, J. H.; Goldberg, D. E. (2003). Population Ecology: First Principles. Woodstock, Oxfordshire: Princeton University Press. ISBN 0-691-11440-4.
51. Berryman, A. A. (1992). "The origins and evolution of predator-prey theory". Ecology. 73 (5): 1530–1535. doi:10.2307/1940005. JSTOR 1940005.
52. Anderson, D. R.; Burnham, K. P.; Thompson, W. L. (2000). "Null hypotheses testing: Problems, prevalence, and an alternative" (PDF). J. Wildl. Mngt. 64 (4): 912–923.
53. Johnson, J. B.; Omland, K. S. (2004). "Model selection in ecology and evolution" (PDF). Trends in Ecology and Evolution. 19 (2): 101–108. doi:10.1016/j.tree.2003.10.013. PMID 16701236.
54. Levins, R. (1969). "Some demographic and genetic consequences of environmental heterogeneity for biological control". Bulletin of the Entomological Society of America. 15: 237–240. ISBN 978-0-231-12680-9.
55. Levins, R. (1970). "Extinction". In Gerstenhaber, M. Some Mathematical Questions in Biology. pp. 77–107. ISBN 978-0-8218-1152-8.
56. Smith, M. A.; Green, D. M. (2005). "Dispersal and the metapopulation paradigm in amphibian ecology and conservation: Are all amphibian populations metapopulations?". Ecography. 28 (1): 110–128. doi:10.1111/j.0906-7590.2005.04042.x.
57. Hanski, I. (1998). "Metapopulation dynamics" (PDF). Nature. 396 (6706): 41–49. Bibcode:1998Natur.396...41H. doi:10.1038/23876.
58. Nebel, S. (2010). "Animal migration". Nature Education Knowledge. 10 (1): 29.
59. Clark, J. S.; Fastie, C.; Hurtt, G.; Jackson, S. T.; Johnson, C.; King, G. A.; Lewis, M.; Lynch, J.; Pacala, S.; et al. (1998). "Reid's paradox of rapid plant migration" (PDF). BioScience. 48 (1): 13–24. doi:10.2307/1313224. JSTOR 1313224.
60. Dingle, H. (1996-01-18). Migration: The Biology of Life on the Move. Oxford University Press. p. 480. ISBN 0-19-509723-8.
61. Hanski, I.; Gaggiotti, O. E., eds. (2004). Ecology, Genetics and Evolution of

Metapopulations. Burlington, MA: Elsevier Academic Press. ISBN 0-12-323448-4.

62. MacKenzie; D.I. (2006). Occupancy Estimation and Modeling: Inferring Patterns and Dynamics of Species Occurrence. London, UK: Elsevier Academic Press. p. 324. ISBN 978-0-12-088766-8.
63. Johnson, M. T.; Strinchcombe, J. R. (2007). "An emerging synthesis between community ecology and evolutionary biology". Trends in Ecology and Evolution. 22 (5): 250–257. doi:10.1016/j.tree.2007.01.014. PMID 17296244.
64. Tansley, A. G. (1935). "The use and abuse of vegetational concepts and terms" (PDF). Ecology. 16 (3): 284–307. doi:10.2307/1930070. JSTOR 1930070.
65. Brinson, M. M.; Lugo, A. E.; Brown, S (1981). "Primary Productivity, Decomposition and Consumer Activity in Freshwater Wetlands". Annual Review of Ecology and Systematics. 12: 123–161. doi:10.1146/annurev.es.12.110181.001011.
66. Marsh, G. P. (1864). Man and Nature: Physical Geography as Modified by Human Action. Cambridge, MA: Belknap Press. p. 560.
67. O'Neil, R. V. (2001). "Is it time to bury the ecosystem concept? (With full military honors, of course!)" (PDF). Ecology. 82 (12): 3275–3284. doi:10.1890/0012-9658(2001)082[3275:IITTBT]2.0.CO;2. ISSN 0012-9658.
68. Levin, S. A. (1998). "Ecosystems and the biosphere as complex adaptive systems". Ecosystems. 1 (5): 431–436. CiteSeerX 10.1.1.83.6318 . doi:10.1007/s100219900037.
69. Pimm, S. (2002). Food Webs. University of Chicago Press. p. 258. ISBN 978-0-226-66832-1.
70. Pimm, S. L.; Lawton, J. H.; Cohen, J. E. (1991). "Food web patterns and their consequences" (PDF). Nature. 350 (6320): 669–674. Bibcode:1991Natur.350..669P. doi:10.1038/350669a0.
71. Worm, B.; Duffy, J. E. (2003). "Biodiversity, productivity and stability in real food webs". Trends in Ecology and Evolution. 18 (12): 628–632. doi:10.1016/j.tree.2003.09.003.
72. McCann, K. (2007). "Protecting biostructure" (PDF). Nature. 446 (7131): 29. Bibcode:2007Natur.446...29M. doi:10.1038/446029a. PMID 17330028.
73. Wilbur, H. W. (1997). "Experimental ecology of food webs: Complex systems in temporary ponds" (PDF). Ecology. 78 (8): 2279–2302. doi:10.1890/0012-9658(1997)078[2279:EEOFWC]2.0.CO;2. ISSN 0012-9658.
74. Emmerson, M.; Yearsley, J. M. (2004). "Weak interactions, omnivory and emergent food-web properties" (PDF). Philosophical Transactions of the Royal Society B. 271 (1537): 397–405. doi:10.1098/rspb.2003.2592.
75. Krause, A. E.; Frank, K. A.; Mason, D. M.; Ulanowicz, R. E.; Taylor, W. W. (2003). "Compartments revealed in food-web structure" (PDF). Nature. 426 (6964): 282–285. Bibcode:2003Natur.426..282K. doi:10.1038/nature02115. PMID 14628050.
76. Egerton, Frank N. (2007). "Understanding food chains and food webs, 1700–1970". Bulletin of the Ecological Society of America. 88: 50–69. doi:10.1890/0012-

9623(2007)88[50:UFCAFW]2.0.CO;2. ISSN 0012-9623.

77. Shurin, J. B.; Gruner, D. S.; Hillebrand, H. (2006). "All wet or dried up? Real differences between aquatic and terrestrial food webs". Proceedings of the Royal Society B. 273 (1582): 1–9. doi:10.1098/rspb.2005.3377. PMC 1560001 . PMID 16519227.
78. Edwards, J.; Fraser, K. (1983). "Concept maps as reflectors of conceptual understanding". Research in Science Education. 13: 19–26. Bibcode:1983RScEd..13...19E. doi:10.1007/BF02356689.
79. Hairston Jr., N. G.; Hairston Sr., N. G. (1993). "Cause-effect relationships in energy flow, trophic structure, and interspecific interactions" (PDF). The American Naturalist. 142 (3): 379–411. doi:10.1086/285546.
80. Duffy, J. Emmett; Cardinale, Bradley J.; France, Kristin E.; McIntyre, Peter B.; Thébault, Elisa; Loreau, Michel (2007). "The functional role of biodiversity in ecosystems: incorporating trophic complexity". Ecology Letters. 10 (6): 522–538. doi:10.1111/j.1461-0248.2007.01037.x. PMID 17498151.
81. Elton, C. S. (1927). Animal Ecology. London, UK.: Sidgwick and Jackson. ISBN 0-226-20639-4.
82. Davic, R. D. (2003). "Linking keystone species and functional groups: a new operational definition of the keystone species concept". Conservation Ecology. 7 (1): r11.
83. Oksanen, L. (1991). "Trophic levels and trophic dynamics: A consensus emerging?". Trends in Ecology and Evolution. 6 (2): 58–60. doi:10.1016/0169-5347(91)90124-G. PMID 21232425.
84. Loehle, C.; Pechmann, Joseph H. K. (1988). "Evolution: The missing ingredient in systems ecology". The American Naturalist. 132 (9): 884–899. doi:10.1086/284895. JSTOR 2462267.
85. Ulanowicz, R. E.; Kemp, W. Michael (1979). "Toward canonical trophic aggregations". The American Naturalist. 114 (6): 871–883. doi:10.1086/283534. JSTOR 2460557.
86. Li, B. (2000). "Why is the holistic approach becoming so important in landscape ecology?". Landscape and Urban Planning. 50 (1–3): 27–41. doi:10.1016/S0169-2046(00)00078-5.
87. Polis, G. A.; Strong, D. R. (1996). "Food web complexity and community dynamics" (PDF). The American Naturalist. 147 (5): 813–846. doi:10.1086/285880.
88. Thompson, R. M.; Hemberg, M.; Starzomski, B. M.; Shurin, J. B. (2007). "Trophic levels and trophic tangles: The prevalence of omnivory in real food webs" (PDF). Ecology. 88 (3): 612–617. doi:10.1890/05-1454. PMID 17503589.
89. Fischer, J.; Lindenmayer, D. B.; Manning, A. D. (2006). "Biodiversity, ecosystem function, and resilience: ten guiding principles for commodity production landscapes" (PDF). Frontiers in Ecology and the Environment. 4 (2): 80–86. doi:10.1890/1540-9295(2006)004[0080:BEFART]2.0.CO;2. ISSN 1540-9295.
90. Libralato, S.; Christensen, V.; Pauly, D. (2006). "A method for identifying

keystone species in food web models" (PDF). Ecological Modelling. 195 (3–4): 153–171. doi:10.1016/j.ecolmodel.2005.11.029.

91. Mills, L. S.; Soule, M. E.; Doak, D. F. (1993). "The keystone-species concept in ecology and conservation". BioScience. 43 (4): 219–224. doi:10.2307/1312122. JSTOR 1312122.
92. Anderson, P. K. (1995). "Competition, predation, and the evolution and extinction of Steller's sea cow, Hydrodamalis gigas". Marine Mammal Science. 11 (3): 391–394. doi:10.1111/j.1748-7692.1995.tb00294.x.
93. Polis, G. A.; Sears, Anna L. W.; Huxel, Gary R.; Strong, Donald R.; Maron, John (2000). "When is a trophic cascade a trophic cascade?"(PDF). Trends in Ecology and Evolution. 15 (11): 473–475. doi:10.1016/S0169-5347(00)01971-6. PMID 11050351.
94. Novikoff, A. B. (1945). "The concept of integrative levels and biology" (PDF). Science. 101 (2618): 209–215. Bibcode:1945Sci...101..209N. doi:10.1126/science.101.2618.209. PMID 17814095.
95. Schneider, D. D. (2001). "The rise of the concept of scale in ecology" (PDF). BioScience. 51 (7): 545–553. doi:10.1641/0006-3568(2001)051[0545:TROTCO]2.0.CO;2. ISSN 0006-3568.
96. Molnar, J.; Marvier, M.; Kareiva, P. (2004). "The sum is greater than the parts" (PDF). Conservation Biology. 18 (6): 1670–1671. doi:10.1111/j.1523-1739.2004.00l07.x.
97. Loehle, C. (2004). "Challenges of ecological complexity". Ecological Complexity. 1 (1): 3–6. doi:10.1016/j.ecocom.2003.09.001.
98. Odum, E. P. (1977). "The emergence of ecology as a new integrative discipline". Science. 195 (4284): 1289–1293. Bibcode:1977Sci...195.1289O. doi:10.1126/science.195.4284.1289. PMID 17738398.
99. Scheffer, M.; Carpenter, S.; Foley, J. A.; Walker, B.; Walker, B. (2001). "Catastrophic shifts in ecosystems" (PDF). Nature. 413 (6856): 591–596. Bibcode:2001Natur.413..591S. doi:10.1038/35098000. PMID 11595939.
100. "Welcome to ILTER — ILTER". International Long Term Ecological Research. Retrieved 2010-03-16.
101. Silverton, Jonathan; Poulton, Paul; Johnston, Edward; Edwards, Grant; Heard, Matthew; Biss, Pamela M. (2006). "The Park Grass Experiment 1856–2006: Its contribution to ecology" (PDF). Journal of Ecology. 94 (4): 801–814. doi:10.1111/j.1365-2745.2006.01145.x.
102. "Hubbard Brook Ecosystem Study Front Page". Retrieved 2010-03-16.
103. Liu, J.; Dietz, Thomas; Carpenter, Stephen R.; Folke, Carl; Alberti, Marina; Redman, Charles L.; Schneider, Stephen H.; Ostrom, Elinor; Pell, Alice N.; et al. (2009). "Coupled human and natural systems" (PDF). AMBIO: A Journal of the Human Environment. 36 (8): 639–649. doi:10.1579/0044-7447(2007)36[639:CHANS]2.0.CO;2. ISSN 0044-7447.
104. Mikkelson, G. M. (2010). "Part-whole relationships and the unity of ecology".

In Skipper, R. A.; Allen, C.; Ankeny, R.; Craver, C. F.; Darden, L.; Richardson, R.C. Philosophy Across the Life Sciences (PDF). Cambridge, MA: MIT Press.

105. Wilson, D. S. (1988). "Holism and reductionism in evolutionary ecology". Oikos. 53 (2): 269–273. doi:10.2307/3566073. JSTOR 3566073.

106. Miles, D. B.; Dunham, A. E. (1993). "Historical perspectives in ecology and evolutionary biology: The use of phylogenetic comparative analyses". Annual Review of Ecology and Systematics. 24: 587–619. doi:10.1146/annurev.es.24.110193.003103.

107. Craze, P., ed. (2 August 2012). "Trends in Ecology and Evolution". Cell Press, Elsevier, Inc.

108. Allee, W. C.; Park, O.; Emerson, A. E.; Park, T.; Schmidt, K. P. (1949). Principles of Animal Ecology. W. B. Sunders, Co. p. 837. ISBN 0-7216-1120-6.

109. Rickleffs, Robert, E. (1996). The Economy of Nature. University of Chicago Press. p. 678. ISBN 0-7167-3847-3.

110. Yoshida, T. "Rapid evolution drives ecological dynamics in a predator–prey system". Nature Publishing Group.

111. Stuart-Fox, D.; Moussalli, A. (2008). "Selection for social signalling drives the evolution of chameleon colour change". PLoS Biology. 6 (1): e25. doi:10.1371/journal.pbio.0060025. PMC 2214820 . PMID 18232740.

112. Karban, R. (2008). "Plant behaviour and communication". Ecology Letters. 11 (7): 727–739. doi:10.1111/j.1461-0248.2008.01183.x. PMID 18400016.

113. Tinbergen, N. (1963). "On aims and methods of ethology" (PDF). Zeitschrift für Tierpsychologie. 20 (4): 410–433. doi:10.1111/j.1439-0310.1963.tb01161.x.

114. Hamner, W. M. (1985). "The importance of ethology for investigations of marine zooplankton". Bulletin of Marine Science. 37 (2): 414–424.

115. Strassmann, J. E.; Zhu, Y.; Queller, D. C. (2000). "Altruism and social cheating in the social amoeba Dictyostelium discoideum". Nature. 408 (6815): 965–967. doi:10.1038/35050087. PMID 11140681.

116. Sakurai, K. (1985). "An attelabid weevil (Euops splendida) cultivates fungi". Journal of Ethology. 3 (2): 151–156. doi:10.1007/BF02350306.

117. Anderson, J. D. (1961). "The courtship behaviour of Ambystoma macrodactylum croceum". Copeia. 2 (2): 132–139. doi:10.2307/1439987. JSTOR 1439987.

118. "Behavioral Ecology". International Society for Behavioral Ecology. Retrieved 15 April 2011.

119. Gould, Stephen J.; Vrba, Elizabeth S. (1982). "Exaptation–a missing term in the science of form". Paleobiology. 8 (1): 4–15.

120. Wilson, E. O. (2000). Sociobiology: The New Synthesis (25th anniversary ed.). President and Fellows of Harvard College. ISBN 978-0-674-00089-6. Ives, A. R.; Cardinale, B. J.; Snyder, W. E. (2004). "A synthesis of subdisciplines: Predator–prey interactions, and biodiversity and ecosystem functioning" (PDF). Ecology Letters. 8 (1): 102–116. doi:10.1111/j.1461-0248.2004.00698.x.

121. Krebs, J. R.; Davies, N. B. (1993). An Introduction to Behavioural Ecology. Wiley-Blackwell. p. 432. ISBN 978-0-632-03546-5.

122. Webb, J. K.; Pike, D. A.; Shine, R. (2010). "Olfactory recognition of predators by nocturnal lizards: safety outweighs thermal benefits". Behavioural Ecology. 21 (1): 72–77. doi:10.1093/beheco/arp152.

123. Cooper, W. E.; Frederick, W. G. (2010). "Predator lethality, optimal escape behavior, and autotomy". Behavioral Ecology. 21 (1): 91–96. doi:10.1093/beheco/arp151.

124. Eastwood, R. (2004). "Successive replacement of tending ant species at aggregations of scale insects (Hemiptera: Margarodidae and Eriococcidae) on Eucalyptus in south-east Queensland" (PDF). Australian Journal of Entomology. 43: 1–4. doi:10.1111/j.1440-6055.2003.00371.x.

125. Kodric-Brown, A.; Brown, J. H. (1984). "Truth in advertising: The kinds of traits favored by sexual selection" (PDF). The American Naturalist. 124 (3): 309–323. doi:10.1086/284275.[dead link]

126. Adrian G Palacios, Francisco Bozinovic; Bozinovic (2003). "An "enactive" approach to integrative and comparative biology: Thoughts on the table" (PDF). Biology Research. 36 (1): 95–99. doi:10.4067/S0716-97602003000100008. PMID 12795209.

127. Reuven Dukas (1998). "§1.3 Why study cognitive ecology?". In Reuven Dukas. Cognitive Ecology: The Evolutionary Ecology of Information Processing and Decision Making. University of Chicago Press. p. 4. ISBN 9780226169323.

128. Reuven Dukas; John M. Ratcliffe (2009). "Introduction". In Reuven Dukas; John M. Ratcliffe. Cognitive Ecology II. University of Chicago Press. pp. 1 ff. ISBN 9780226169378. Cognitive ecology focuses on he ecology and evolution of "cognition" defined as the neuronal processes concerned with the acquisition, retention, and use of information....we ought to rely on ecological and evolutionary knowledge for studying cognition.

129. Francisco J Varela; Evan Thompson; Eleanor Rosch (1993). The Embodied Mind: Cognitive Science and Human Experience(Paperback ed.). MIT Press. p. 174. ISBN 9780262261234.

130. Sherman, P. W.; Lacey, E. A.; Reeve, H. K.; Keller, L. (1995). "The eusociality continuum" (PDF). Behavioural Ecology. 6 (1): 102–108. doi:10.1093/beheco/6.1.102.

131. Wilson, D. S.; Wilson, E. O. (2007). "Rethinking the theoretical foundation of sociobiology". The Quarterly Review of Biology. 82 (4): 327–348. doi:10.1086/522809. PMID 18217526.

132. Page, R. D. M. (1991). "Clocks, clades, and cospeciation: Comparing rates of evolution and timing of cospeciation events in host-parasite assemblages". Systematic Zoology. 40 (2): 188–198. doi:10.2307/2992256. JSTOR 2992256.

133. Herre, E. A.; Knowlton, N.; Mueller, U. G.; Rehner, S. A. (1999). "The evolution of mutualisms: exploring the paths between conflict and cooperation" (PDF). Trends

in Ecology and Evolution. 14 (2): 49–53. doi:10.1016/S0169-5347(98)01529-8. PMID 10234251.

134. Gilbert, F. S. (1990). Insect life cycles: Genetics, evolution, and co-ordination. New York, NY: Springer-Verlag. p. 258. ISBN 0-387-19550-5.
135. Kiers, E. T.; van der Heijden, M. G. A. (2006). "Mutualistic stability in the arbuscular mycorrhizal symbiosis: Exploring hypotheses of evolutionary cooperation" (PDF). Ecology. 87 (7): 1627–1636. doi:10.1890/0012-9658(2006)87[1627:MSITAM]2.0.CO;2. ISSN 0012-9658. PMID 16922314.
136. Strain, B. R. (1985). "Physiological and ecological controls on carbon sequestering in terrestrial ecosystems". Biogeochemistry. 1 (3): 219–232. doi:10.1007/BF02187200.
137. Bronstein, J. L. (2001). "The exploitation of mutualisms". Ecology Letters. 4 (3): 277–287. doi:10.1046/j.1461-0248.2001.00218.x.
138. Irwin, Rebecca E.; Bronstein, Judith L.; Manson, Jessamyn S.; Richardson, Leif (2010). "Nectar robbing: Ecological and evolutionary perspectives". Annual Review of Ecology, Evolution, and Systematics. 41 (2): 271–292. doi:10.1146/annurev.ecolsys.110308.120330.
139. Boucher, D. H.; James, S.; Keeler, K. H. (1982). "The ecology of mutualism". Annual Review of Ecology and Systematics. 13: 315–347. doi:10.1146/annurev.es.13.110182.001531.
140. King, K. C.; Delph, L. F.; Jokela, J.; Lively, C. M. (2009). "The geographic mosaic of sex and the Red Queen". Current Biology. 19 (17): 1438–1441. doi:10.1016/j.cub.2009.06.062. PMID 19631541.
141. Parenti, L. R.; Ebach, M. C. (2009). Comparative Biogeography: Discovering and Classifying Biogeographical Patterns of a Dynamic Earth. London, England: University of California Press. ISBN 978-0-520-25945-4.
142. "Journal of Biogeography". Wiley. Retrieved 3 August 2012.
143. MacArthur, R.; Wilson, E. O. (1967). "The Theory of Island Biogeography". Princeton, NJ: Princeton University Press.
144. Wiens, J. J.; Donoghue, M. J. (2004). "Historical biogeography, ecology and species richness" (PDF). Trends in Ecology and Evolution. 19 (12): 639–644. doi:10.1016/j.tree.2004.09.011. PMID 16701326.
145. Morrone, J. J.; Crisci, J. V. (1995). "Historical biogeography: Introduction to methods". Annual Review of Ecology and Systematics. 26: 373–401. doi:10.1146/annurev.es.26.110195.002105.
146. Svenning, Jens-Christian; Condi, R. (2008). "Biodiversity in a warmer world". Science. 322 (5899): 206–207. doi:10.1126/science.1164542. PMID 18845738.
147. Landhäusser, Simon M.; Deshaies, D.; Lieffers, V. J. (2009). "Disturbance facilitates rapid range expansion of aspen into higher elevations of the Rocky Mountains under a warming climate". Journal of Biogeography. 37 (1): 68–76. doi:10.1111/j.1365-2699.2009.02182.x.

148. Reznick, D.; Bryant, M. J.; Bashey, F. (2002). "r- and K-selection revisited: The role of population regulation in life-history evolution"(PDF). Ecology. 83 (6): 1509–1520. doi:10.1890/0012-9658(2002)083[1509:RAKSRT]2.0.CO;2. ISSN 0012-9658.

149. Pianka, E. R. (1972). "r and K Selection or b and d Selection?". The American Naturalist. 106 (951): 581–588. doi:10.1086/282798.

150. Rieseberg, L. (ed.). "Molecular Ecology". Molecular Ecology. Wiley. doi:10.1111/(ISSN)1365-294X.

151. Avise, J. (1994). Molecular Markers, Natural History and Evolution. Kluwer Academic Publishers. ISBN 0-412-03771-8.

152. O'Brian, E.; Dawson, R. (2007). "Context-dependent genetic benefits of extra-pair mate choice in a socially monogamous passerine"(PDF). Behavioral Ecology and Sociobiology. 61 (5): 775–782. doi:10.1007/s00265-006-0308-8.

153. Avise, J. (2000). Phylogeography: The History and Formation of Species. President and Fellows of Harvard College. ISBN 0-674-66638-0.

154. Rachel Carson (1962). ""Silent Spring" (excerpt)". Houghton Miffin. Retrieved 4 October 2012.

155. Young, G. L. (1974). "Human ecology as an interdisciplinary concept: A critical inquiry". Advances in Ecological Research. Advances in Ecological Research. 8: 1–105. doi:10.1016/S0065-2504(08)60277-9. ISBN 978-0-12-013908-8.

156. Gross, M. (2004). "Human geography and ecological sociology: the unfolding of human ecology, 1890 to 1930 – and beyond". Social Science History. 28 (4): 575–605. doi:10.1215/01455532-28-4-575.

157. "Millennium Ecosystem Assessment – Synthesis Report". United Nations. 2005. Retrieved 4 February 2010.

158. de Groot, R. S.; Wilson, M. A.; Boumans, R. M. J. (2002). "A typology for the classification, description and valuation of ecosystem functions, goods and services" (PDF). Ecological Economics. 41 (3): 393–408. doi:10.1016/S0921-8009(02)00089-7.

159. Aguirre, A. A. (2009). "Biodiversity and human health". EcoHealth. 6: 153–156. doi:10.1007/s10393-009-0242-0.

160. Grumbine, R. E. (1994). "What is ecosystem management?"(PDF). Conservation Biology. 8 (1): 27–38. doi:10.1046/j.1523-1739.1994.08010027.x.

161. Wilson, E. O. (1992). The Diversity of Life. Harvard University Press. p. 440. ISBN 978-0-674-05817-0.

162. Slocombe, D. S. (1993). "Implementing ecosystem-based management". BioScience. 43 (9): 612–622. doi:10.2307/1312148. JSTOR 1312148.

163. Hobss, R. J.; Harris, J. A. (2001). "Restoration ecology: Repairing the Earth's ecosystems in the new millennium" (PDF). Restoration Ecology. 9 (2): 239–246. doi:10.1046/j.1526-100x.2001.009002239.x.

164. Mason, H. L.; Langenheim, J. H. (1957). "Language analysis and the concept "environment"". Ecology. 38 (2): 325–340. doi:10.2307/1931693. JSTOR 1931693.

165. Kleese, D. A. (2001). "Nature and nature in Psychology". Journal of Theoretical and Philosophical Psychology. 21: 61–79. doi:10.1037/h0091199.

166. Campbell, Neil A.; Williamson, Brad; Heyden, Robin J. (2006). Biology: Exploring Life. Boston, Massachusetts: Pearson Prentice Hall. ISBN 0-13-250882-6.

167. Kormondy, E. E. (1995). Concepts of Ecology (4th ed.). Benjamin Cummings. ISBN 0-13-478116-3.

168. Hughes, A. R. "Disturbance and diversity: an ecological chicken and egg problem". Nature Education Knowledge. 1 (8): 26.

169. Levin, S. A. (1992). "The problem of pattern and scale in ecology: The Robert H. MacArthur Award" (PDF). Ecology. 73 (6): 1943–1967. doi:10.2307/1941447. JSTOR 1941447. Retrieved 2010-03-16.

170. Holling, C. S. (1973). "Resilience and stability of ecological systems". Annual Review of Ecology and Systematics. 4 (1): 1–23. doi:10.1146/annurev.es.04.110173.000245. JSTOR 2096802.

171. Folke, C.; Carpenter, S.; Walker, B.; Scheffer, M.; Elmqvist, T.; Gunderson, L.; Holling, C.S. (2004). "Regime shifts, resilience, and biodiversity in ecosystem management" (PDF). Annual Review of Ecology and Systematics. 35: 557–581. doi:10.1146/annurev.ecolsys.35.021103.105711. JSTOR 2096802.

172. Morgan Ernest, S. K.; Enquist, Brian J.; Brown, James H.; Charnov, Eric L.; Gillooly, James F.; Savage, Van M.; White, Ethan P.; Smith, Felisa A.; Hadly, Elizabeth A.; Haskell, John P.; Lyons, S. Kathleen; Maurer, Brian A.; Niklas, Karl J.; Tiffney, Bruce (2003). "Thermodynamic and metabolic effects on the scaling of production and population energy use" (PDF). Ecology Letters. 6 (11): 990–995. doi:10.1046/j.1461-0248.2003.00526.x.

173. Allègre, Claude J.; Manhès, Gérard; Göpel, Christa (1995). "The age of the Earth". Geochimica et Cosmochimica Acta. 59 (8): 1455–1456. Bibcode:1995GeCoA..59.1445A. doi:10.1016/0016-7037(95)00054-4.

174. Wills, C.; Bada, J. (2001). The Spark of Life: Darwin and the Primeval Soup. Cambridge, MA: Perseus Publishing. ISBN 978-0-7382-0493-2.

175. Goldblatt, Colin; Lenton, Timothy M.; Watson, Andrew J. (2006). "Bistability of atmospheric oxygen and the Great Oxidation" (PDF). Nature. 443 (7112): 683–686. Bibcode:2006Natur.443..683G. doi:10.1038/nature05169. PMID 17036001.

176. Catling, D. C.; Claire, M. W. (2005). "How Earth's atmosphere evolved to an oxic state: A status report" (PDF). Earth and Planetary Science Letters. 237: 1–20. Bibcode:2005E&PSL.237....1C. doi:10.1016/j.epsl.2005.06.013.

177. Cronk, J. K.; Fennessy, M. S. (2001). Wetland Plants: Biology and Ecology. Washington, D.C.: Lewis Publishers. ISBN 1-56670-372-7.

178. Evans, D. H.; Piermarini, P. M.; Potts, W. T. W. (1999). "Ionic transport in the fish gill epithelium" (PDF). Journal of Experimental Zoology. 283 (7): 641–652. doi:10.1002/(SICI)1097-010X(19990601)283:7<641::AID-JEZ3>3.0.CO;2-W.

179. Swenson, N. G.; Enquist, B. J. (2008). "The relationship between stem and branch wood specific gravity and the ability of each measure to predict leaf

area". American Journal of Botany. 95 (4): 516–519. doi:10.3732/ajb.95.4.516. PMID 21632377.

180. Gartner, Gabriel E.A.; Hicks, James W.; Manzani, Paulo R.; et al. (2010). "Phylogeny, ecology, and heart position in snakes" (PDF). Physiological and Biochemical Zoology. 83 (1): 43–54. doi:10.1086/648509. PMID 19968564.
181. Jacobsen, D. (2008). "Low oxygen pressure as a driving factor for the altitudinal decline in taxon richness of stream macroinvertebrates". Oecologia. 154 (4): 795–807. doi:10.1007/s00442-007-0877-x. PMID 17960424.
182. Wheeler, T. D.; Stroock, A. D. (2008). "The transpiration of water at negative pressures in a synthetic tree". Nature. 455 (7210): 208–212. Bibcode:2008Natur.455..208W. doi:10.1038/nature07226. PMID 18784721.
183. Pockman, W. T.; Sperry, J. S.; O'Leary, J. W. (1995). "Sustained and significant negative water pressure in xylem". Nature. 378 (6558): 715–716. Bibcode:1995Natur.378..715P. doi:10.1038/378715a0.
184. Zimmermann, U.; Schneider, H.; Wegner, L. H.; Wagner, M.; Szimtenings, A.; Haase, F.; Bentrup, F. W. (2002). "What are the driving forces for water lifting in the xylem conduit?". Physiologia Plantarum. 114 (3): 327–335. doi:10.1034/j.1399-3054.2002.1140301.x. PMID 12060254.
185. Kastak, D.; Schusterman, R. J. (1998). "Low-frequency amphibious hearing in pinnipeds: Methods, measurements, noise, and ecology". Journal of the Acoustical Society of America. 103 (4): 2216–2228. Bibcode:1998ASAJ..103.2216K. doi:10.1121/1.421367. PMID 9566340.
186. Nishiguchi, Y.; Ito, I.; Okada, M. (2010). "Structure and function of lactate dehydrogenase from hagfish". Marine Drugs. 8 (3): 594–607. doi:10.3390/md8030594. PMC 2857353 . PMID 20411117.
187. Friedman, J.; Harder, L. D. (2004). "Inflorescence architecture and wind pollination in six grass species" (PDF). Functional Ecology. 18 (6): 851–860. doi:10.1111/j.0269-8463.2004.00921.x.
188. Harder, L. D.; Johnson, S. D. (2009). "Darwin's beautiful contrivances: evolutionary and functional evidence for floral adaptation" (PDF). New Phytologist. 183 (3): 530–545. doi:10.1111/j.1469-8137.2009.02914.x. PMID 19552694.
189. Shimeta, J.; Jumars, P. A.; Lessard, E. J. (1995). "Influences of turbulence on suspension feeding by planktonic protozoa; experiments in laminar shear fields" (PDF). Limnolology and Oceanography. 40 (5): 845–859. doi:10.4319/lo.1995.40.5.0845.
190. Etemad-Shahidi, A.; Imberger, J. (2001). "Anatomy of turbulence in thermally stratified lakes" (PDF). Limnolology and Oceanography. 46 (5): 1158–1170. doi:10.4319/lo.2001.46.5.1158.
191. Wolf, B. O.; Walsberg, G. E. (2006). "Thermal effects of radiation and wind on a small bird and implications for microsite selection". Ecology. 77 (7): 2228–2236. doi:10.2307/2265716. JSTOR 2265716.

192. Daubenmire, R. (1975). "Floristic plant geography of eastern Washington and northern Idaho". Journal of Biogeography. 2 (1): 1–18. doi:10.2307/3038197. JSTOR 3038197.

193. Steele, C. A.; Carstens, B. C.; Storfer, A.; Sullivan, J. (2005). "Testing hypotheses of speciation timing in Dicamptodon copei and Dicamptodon aterrimus (Caudata: Dicamptodontidae)" (PDF). Molecular Phylogenetics and Evolution. 36 (1): 90–100. doi:10.1016/j.ympev.2004.12.001. PMID 15904859.

194. Lenton, T. M.; Watson, A. (2000). "Redfield revisited. 2. What regulates the oxygen content of the atmosphere" (PDF). Global Biogeochemical Cycles. 14 (1): 249–268. Bibcode:2000GBioC..14..249L. doi:10.1029/1999GB900076.

195. Lobert, J. M.; Warnatz, J. (1993). "Emissions from the combustion process in vegetation". In Crutzen, P. J.; Goldammer, J. G. Fire in the Environment: The Ecological, Atmospheric and Climatic Importance of Vegetation Fires (PDF). Wiley. ISBN 978-0-471-93604-6.

196. Garren, K. H. (1943). "Effects of fire on vegetation of the southeastern United States". Botanical Review. 9 (9): 617–654. doi:10.1007/BF02872506.

197. Cooper, C. F. (1960). "Changes in vegetation, structure, and growth of southwestern pine forests since white settlement". Ecological Monographs. 30 (2): 130–164. doi:10.2307/1948549.

198. Cooper, C. F. (1961). "The ecology of fire". Scientific American. 204 (4): 150–160. doi:10.1038/scientificamerican0461-150.

199. van Wagtendonk, Jan W. (2007). "History and evolution of wildland fire use" (PDF). Fire Ecology Special Issue. 3 (2): 3–17. doi:10.4996/fireecology.0302003.

200. Boerner, R. E. J. (1982). "Fire and nutrient cycling in temperate ecosystems". BioScience. 32 (3): 187–192. doi:10.2307/1308941. JSTOR 1308941.

201. Goubitz, S.; Werger, M. J. A.; Ne'eman, G. (2009). "Germination response to fire-related factors of seeds from non-serotinous and serotinous cones". Plant Ecology. 169 (2): 195–204. doi:10.1023/A:1026036332277.

202. Ne'eman, G.; Goubitz, S.; Nathan, R. (2004). "Reproductive traits of Pinus halepensis in the light of fire: a critical review". Plant Ecology. 171 (1/2): 69–79. doi:10.1023/B:VEGE.0000029380.04821.99.

203. Flematti, Gavin R.; Ghisalberti, Emilio L.; Dixon, Kingsley W.; Trengove, R. D. (2004). "A compound from smoke that promotes seed germination" (PDF). Science. 305 (5686): 977. doi:10.1126/science.1099944. PMID 15247439.

204. Coleman, D. C.; Corssley, D. A.; Hendrix, P. F. (2004). Fundamentals of Soil Ecology (2nd ed.). Academic Press. ISBN 0-12-179726-0.

205. Wilkinson, M. T.; Richards, P. J.; Humphreys, G. S. (2009). "Breaking ground: Pedological, geological, and ecological implications of soil bioturbation" (PDF). Earth-Science Reviews. 97 (1–4): 257–272. Bibcode:2009ESRv...97..257W. doi:10.1016/j.earscirev.2009.09.005.

206. Phillips, J. D. (2009). "Soils as extended composite phenotypes". Geoderma. 149 (1–2): 143–151. doi:10.1016/j.geoderma.2008.11.028.

207. Reinhardt, L.; Jerolmack, D.; Cardinale, B. J.; Vanacker, V.; Wright, J. "Dynamic interactions of life and its landscape: Feedbacks at the interface of geomorphology and ecology" (PDF). Earth Surf. Process. Landforms. 35: 78–101. Bibcode:2010ESPL...35...78R. doi:10.1002/esp.1912.

208. Davic, R. D.; Welsh, H. H. (2004). "On the ecological role of salamanders" (PDF). Annual Review of Ecology and Systematics. 35: 405–434. doi:10.1146/annurev.ecolsys.35.112202.130116.

209. Hasiotis, S. T. (2003). "Complex ichnofossils of solitary and social soil organisms: Understanding their evolution and roles in terrestrial paleoecosystems". Palaeogeography, Palaeoclimatology, Palaeoecology. 192 (2): 259–320. doi:10.1016/S0031-0182(02)00689-2.

210. Falkowski, P. G.; Fenchel, T.; Delong, E. F. (2008). "The microbial engines that drive Earth's biogeochemical cycles". Science. 320 (5879): 1034–1039. Bibcode:2008Sci...320.1034F. doi:10.1126/science.1153213. PMID 18497287.

211. Grace, J. (2004). "Understanding and managing the global carbon cycle". Journal of Ecology. 92 (2): 189–202. doi:10.1111/j.0022-0477.2004.00874.x.

212. Pearson, P. N.; Palmer, M. R. (2000). "Atmospheric carbon dioxide concentrations over the past 60 million years" (PDF). Nature. 406 (6797): 695–699. doi:10.1038/35021000. PMID 10963587.

213. Pagani, M.; Zachos, J. C.; Freeman, K. H.; Tipple, B.; Bohaty, S. (2005). "Marked decline in atmospheric carbon dioxide concentrations during the Paleogene". Science. 309 (5734): 600–603. Bibcode:2005Sci...309..600P. doi:10.1126/science.1110063. PMID 15961630.

214. Zhuan, Q.; Melillo, J. M.; McGuire, A. D.; Kicklighter, D. W.; Prinn, R. G.; Steudler, P. A.; Felzer, B. S.; Hu, S. (2007). "Net emission of CH4and CO2 in Alaska: Implications for the region's greenhouse gas budget" (PDF). Ecological Applications. 17 (1): 203–212. doi:10.1890/1051-0761(2007)017[0203:NEOCAC]2.0.CO;2. ISSN 1051-0761. PMID 17479846.

215. Cox, Peter M.; Betts, Richard A.; Jones, Chris D.; Spall, Steven A.; Totterdell, Ian J. (2000). "Acceleration of global warming due to carbon-cycle feedbacks in a coupled climate model" (PDF). Nature. 408 (6809): 184–187. doi:10.1038/35041539. PMID 11089968.

216. Erwin, D. H. (2009). "Climate as a driver of evolutionary change". Current Biology. 19 (14): R575–R583. doi:10.1016/j.cub.2009.05.047. PMID 19640496.

217. Bamber, J. (2012). "Shrinking glaciers under scrutiny" (PDF). Nature. 482 (7386): 482–483. Bibcode:2012Natur.482..482B. doi:10.1038/nature10948. PMID 22318516.

218. Heimann, Martin; Reichstein, Markus (2008). "Terrestrial ecosystem carbon dynamics and climate feedbacks" (PDF). Nature. 451 (7176): 289–292. Bibcode:2008Natur.451..289H. doi:10.1038/nature06591. PMID 18202646.

219. Davidson, Eric A.; Janssens, Ivan A. (2006). "Temperature sensitivity of soil carbon decomposition and feedbacks to climate change". Nature. 440 (7081): 165–173. Bibcode:2006Natur.440..165D. doi:10.1038/nature04514. PMID 16525463.

220. Egerton, F. N. (2001). "A history of the ecological sciences: early Greek origins" (PDF). Bulletin of the Ecological Society of America. 82 (1): 93–97.

221. Benson, Keith R. (2000). "The emergence of ecology from natural history". Endeavour. 24 (2): 59–62. doi:10.1016/S0160-9327(99)01260-0. PMID 10969480.

222. Sober, E. (1980). "Evolution, population thinking, and essentialism". Philosophy of Science. 47 (3): 350–383. doi:10.1086/288942. JSTOR 186950.

223. Hughes, J. D. (1985). "Theophrastus as ecologist". Environmental Review. 9 (4): 296–306. doi:10.2307/3984460. JSTOR 3984460.

224. Hughes, J. D. (1975). "Ecology in ancient Greece". Inquiry. 18 (2): 115–125. doi:10.1080/00201747508601756.

225. Kingsland, S. (2004). "Conveying the intellectual challenge of ecology: An historical perspective" (PDF). Frontiers in Ecology and the Environment. 2 (7): 367–374. doi:10.1890/1540-9295(2004)002[0367:CTICOE]2.0.CO;2. ISSN 1540-9295.

226. Rosenzweig, M. L. (2003). "Reconciliation ecology and the future of species diversity" (PDF). Oryx. 37 (2): 194–205. doi:10.1017/s0030605303000371.

227. Hawkins, B. A. (2001). "Ecology's oldest pattern". Endeavor. 25 (3): 133. doi:10.1016/S0160-9327(00)01369-7.

228. McIntosh, R. P. (1985). The Background of Ecology: Concept and Theory. Cambridge University Press. p. 400. ISBN 0-521-27087-1.

229. Stauffer, R. C. (1957). "Haeckel, Darwin and ecology". The Quarterly Review of Biology. 32 (2): 138–144. doi:10.1086/401754.

230. Friederichs, K. (1958). "A definition of ecology and some thoughts about basic concepts". Ecology. 39 (1): 154–159. doi:10.2307/1929981. JSTOR 1929981.

231. Hinchman, L. P.; Hinchman, S. K. (2007). "What we owe the Romantics". Environmental Values. 16 (3): 333–354. doi:10.3197/096327107X228382.

232. Goodland, R. J. (1975). "The tropical origin of ecology: Eugen Warming's jubilee". Oikos. 26 (2): 240–245. doi:10.2307/3543715. JSTOR 3543715.

233. Egerton, F. N. (2007). "A history of the ecological sciences, part 23: Linnaeus and the economy of nature". Bulletin of the Ecological Society of America. 88 (1): 72–88. doi:10.1890/0012-9623(2007)88[72:AHOTES]2.0.CO;2. ISSN 0012-9623.

234. Kormandy, E. J.; Wooster, Donald (1978). "Review: Ecology/economy of nature—synonyms?". Ecology. 59 (6): 1292–1294. doi:10.2307/1938247. JSTOR 1938247.

235. Hector, A.; Hooper, R. (2002). "Darwin and the first ecological experiment". Science. 295 (5555): 639–640. doi:10.1126/science.1064815. PMID 11809960.

236. Sinclair, G. (1826). "On cultivating a collection of grasses in pleasure-grounds or flower-gardens, and on the utility of studying the Gramineae". London Gardener's Magazine. 1. New-Street-Square: A. & R. Spottiswoode. p. 115.

237. May, R. (1999). "Unanswered questions in ecology". Philosophical Transactions

of the Royal Society B. 354 (1392): 1951–1959. doi:10.1098/rstb.1999.0534. PMC 1692702 . PMID 10670015.

238. Darwin, Charles (1859). On the Origin of Species (1st ed.). London, UK: John Murray. p. 1. ISBN 0-8014-1319-2.
239. Meysman, F. J. R.; Middelburg, Jack J.; Heip, C. H. R. (2006). "Bioturbation: A fresh look at Darwin's last idea". Trends in Ecology and Evolution. 21 (22): 688–695. doi:10.1016/j.tree.2006.08.002. PMID 16901581.
240. Acot, P. (1997). "The Lamarckian cradle of scientific ecology". Acta Biotheoretica. 45 (3–4): 185–193. doi:10.1023/A:1000631103244.
241. Forbes, S. (1887). "The lake as a microcosm" (PDF). Bulletin of the Scientific Association. Peoria, IL: 77–87.
242. Hunt, Caroline Louisa (1912). The life of Ellen H. Richards (1st ed.). Boston: Whitcomb & Barrows.
243. Clements, F. E. (1905). Research methods in ecology. Lincoln, Neb.: University Pub. Comp. ISBN 0-405-10381-6.
244. Simberloff, D. (1980). "A succession of paradigms in ecology: Essentialism to materialism and probalism". Synthese. 43: 3–39. doi:10.1007/BF00413854.
245. Gleason, H. A. (1926). "The individualistic concept of the plant association" (PDF). Bulletin of the Torrey Botanical Club. 53 (1): 7–26. doi:10.2307/2479933. JSTOR 2479933.
246. Foster, J. B.; Clark, B. (2008). "The sociology of ecology: ecological organicism versus ecosystem ecology in the social construction of ecological science, 1926–1935" (PDF). Organization & Environment. 21 (3): 311–352. doi:10.1177/1086026608321632.
247. Allee, W. C. (1932). Animal Life and Social Growth. Baltimore: The Williams & Wilkins Company and Associates.
248. Cook, R. E. (1977). "Raymond Lindeman and the trophic-dynamic concept in ecology" (PDF). Science. 198 (4312): 22–26. Bibcode:1977Sci...198...22C. doi:10.1126/science.198.4312.22. PMID 17741875.
249. Odum, E. P. (1968). "Energy flow in ecosystems: A historical review". American Zoologist. 8 (1): 11–18. doi:10.1093/icb/8.1.11. JSTOR 3881528.
250. ^ Jump up to:a b Ghilarov, A. M. (1995). "Vernadsky's biosphere concept: an historical perspective". The Quarterly Review of Biology. 70 (2): 193–203. doi:10.1086/418982. JSTOR 3036242.
251. Itô, Y. (1991). "Development of ecology in Japan, with special reference to the role of Kinji Imanishi". Journal of Ecological Research. 6 (2): 139–155. doi:10.1007/BF02347158.
252. Carson, R. (2002). Silent Spring. Houghton Mifflin Company. p. 348. ISBN 0-618-24906-0.
253. Palamar, C. R. (2008). "The justice of ecological restoration: Environmental history, health, ecology, and justice in the United States" (PDF). Human Ecology Review. 15 (1): 82–94.

254. Krebs, J. R.; Wilson, J. D.; Bradbury, R. B.; Siriwardena, G. M. (1999). "The second Silent Spring" (PDF). Nature. 400 (6745): 611–612. Bibcode:1999Natur.400..611K. doi:10.1038/23127.

CHAPTER 4

Principles of Biological Inheritance

Chapter Contents

- **Introduction**

Genetics is the study of biological inheritance. Geneticists examine a wide variety of inherited traits, from the ability to bear large numbers of fruit in trees, to eye color in mammals. This discipline is a major cornerstone of the larger field of biology. People who study it do so in a wide variety of situations, from research laboratories to pharmaceutical companies, and new breakthroughs are consistently being made.

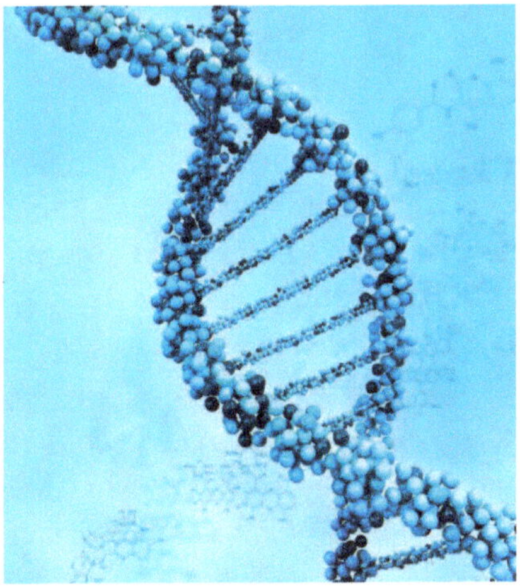

Scientists are constantly studying DNA to learn more about genetics.

INTRODUCTION

Genetics is the study of genes, genetic variation, and heredity in living organisms. It is generally considered a field of biology, but intersects frequently with many other life sciences and is strongly linked with the study of information systems.

The father of genetics is Gregor Mendel, a late 19th-century scientist and Augustinian friar. Mendel studied "trait inheritance," patterns in the way traits are handed down from parents to offspring. He observed that organisms (pea plants) inherit traits by way of discrete "units of inheritance." This term, still used today, is a somewhat ambiguous definition of what is referred to as a gene.

Trait inheritance and molecular inheritance mechanisms of genes are still primary principles of genetics in the 21st century, but modern genetics has expanded beyond inheritance to studying the function and behavior of genes. Gene structure and function, variation, and distribution are studied within the context of the cell, the organism (e.g. dominance), and within the context of a population. Genetics has given rise to a number of sub-fields, including epigenetics and population genetics. Organisms studied within the broad field span the domain of life, including bacteria, plants, animals, and humans.

Genetic processes work in combination with an organism's environment and experiences to influence development and behavior, often referred to as nature versus nurture. The intracellular or extracellular environment of a cell or organism may switch gene transcription on or off. A classic example is two seeds of genetically identical corn, one placed in a temperate climate and one in an arid climate. While the average height of the two corn stalks may be genetically determined to be equal, the one in the arid climate only grows to half the height of the one in the temperate climate due to lack of water and nutrients in its environment.

Etymology

The word genetics stems from the ancient Greek γενετικός genetikos meaning "genitive"/"generative", which in turn derives from γένεσις genesis meaning "origin"

History

The observation that living things inherit traits from their parents has been used since prehistoric times to improve crop plants and animals through selective breeding. The modern science of genetics, seeking to understand this process, began with the work of Gregor Mendel in the mid-19th century.

Prior to Mendel, Imre Festetics, a Hungarian noble, who lived in Kőszeg before Mendel, was the first who used the word "genetics." He described several rules of genetic inheritance in his work *The genetic law of the Nature* (Die genetische Gesätze der Natur, 1819). His second law is the same as what Mendel published. In his third law, he developed the basic principles of mutation (he can be considered a forerunner of Hugo de Vries).

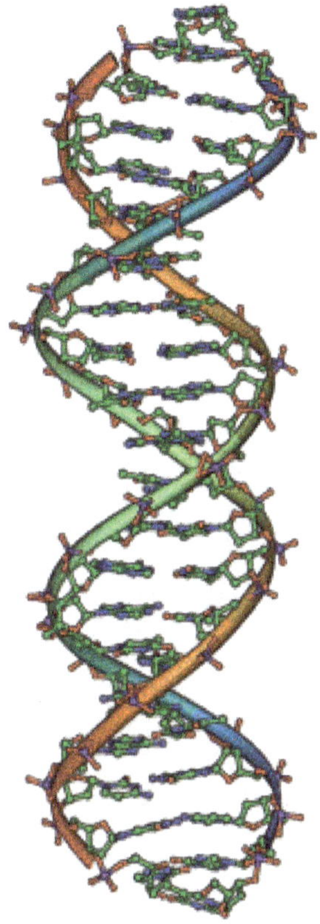

DNA, the molecular basis for biological inheritance. Each strand of DNA is a chain of nucleotides, matching each other in the center to form what look like rungs on a twisted ladder.

Other theories of inheritance preceded his work. A popular theory during Mendel›s time was the concept of blending inheritance: the idea that individuals inherit a smooth blend of traits from their parents. Mendel›s work provided examples where traits were definitely not blended after hybridization, showing that traits are produced by combinations of distinct genes rather than a continuous blend. Blending of traits in the progeny is now explained by the action of multiple genes with quantitative effects. Another theory that had some support at that time was the inheritance of acquired characteristics: the belief that individuals inherit traits strengthened by their parents. This theory (commonly associated with Jean-Baptiste Lamarck) is now known to be wrong—the experiences of individuals do not affect the genes they pass to their children, although evidence in the field of epigenetics has revived some aspects of Lamarck's theory. Other theories included the pangenesis of Charles Darwin (which had both acquired and inherited aspects) and Francis Galton›s reformulation of pangenesis as both particulate and inherited.

Mendelian and Classical Genetics

Modern genetics started with Gregor Johann Mendel, a scientist and Augustinian friar who studied the nature of inheritance in plants. In his paper "*Versuche über Pflanzenhybriden*"

("Experiments on Plant Hybridization"), presented in 1865 to the *Naturforschender Verein* (Society for Research in Nature) in Brünn, Mendel traced the inheritance patterns of certain traits in pea plants and described them mathematically. Although this pattern of inheritance could only be observed for a few traits, Mendel›s work suggested that heredity was particulate, not acquired, and that the inheritance patterns of many traits could be explained through simple rules and ratios.

The importance of Mendel›s work did not gain wide understanding until the 1890s, after his death, when other scientists working on similar problems re-discovered his research. William Bateson, a proponent of Mendel›s work, coined the word *genetics* in 1905. (The adjective *genetic*, derived from the Greek word *genesis*—γένεσις, "origin", predates the noun and was first used in a biological sense in 1860.) Bateson both acted as a mentor and was aided significantly by the work of female scientists from Newnham College at Cambridge, specifically the work of Becky Saunders, Nora Darwin Barlow, and Muriel Wheldale Onslow. Bateson popularized the usage of the word *genetics* to describe the study of inheritance in his inaugural address to the Third International Conference on Plant Hybridization in London, England, in 1906.

After the rediscovery of Mendel›s work, scientists tried to determine which molecules in the cell were responsible for inheritance. In 1911, Thomas Hunt Morgan argued that genes are on chromosomes, based on observations of a sex-linked white eye mutation in fruit flies. In 1913, his student Alfred Sturtevant used the phenomenon of genetic linkage to show that genes are arranged linearly on the chromosome.

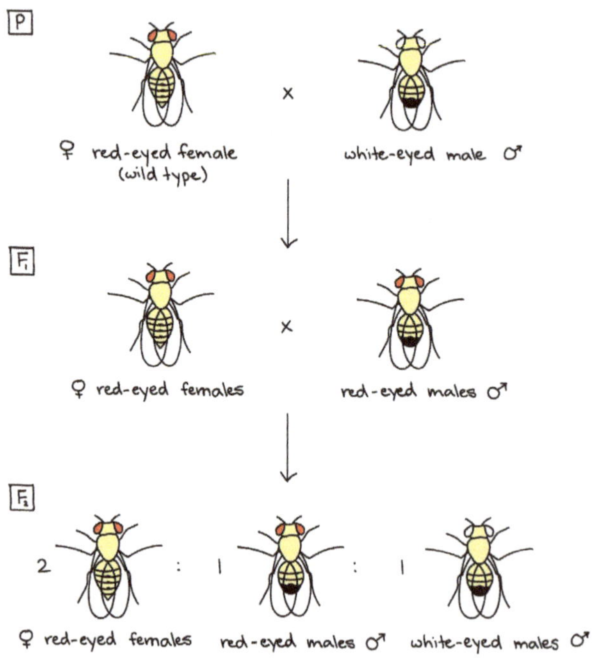

Morgan's observation of sex-linked inheritance of a mutation causing white eyes in *Drosophila* led him to the hypothesis that genes are located upon chromosomes.

Molecular Genetics

Although genes were known to exist on chromosomes, chromosomes are composed of both protein and DNA, and scientists did not know which of the two is responsible for inheritance. In 1928, Frederick Griffith discovered the phenomenon of transformation (see Griffith›s experiment): dead bacteria could transfer genetic material to "transform" other still-living bacteria. Sixteen years later, in 1944, the Avery–MacLeod–McCarty experiment identified DNA as the molecule responsible for transformation. The role of the nucleus as the repository of genetic information in eukaryotes had been established by Hämmerling in 1943 in his work on the single celled alga *Acetabularia*. The Hershey–Chase experiment in 1952 confirmed that DNA (rather than protein) is the genetic material of the viruses that infect bacteria, providing further evidence that DNA is the molecule responsible for inheritance.

James Watson and Francis Crick determined the structure of DNA in 1953, using the X-ray crystallography work of Rosalind Franklin and Maurice Wilkins that indicated DNA has a helical structure (i.e., shaped like a corkscrew). Their double-helix model had two strands of DNA with the nucleotides pointing inward, each matching a complementary nucleotide on the other strand to form what look like rungs on a twisted ladder. This structure showed that genetic information exists in the sequence of nucleotides on each strand of DNA. The structure also suggested a simple method for replication: if the strands are separated, new partner strands can be reconstructed for each based on the sequence of the old strand. This property is what gives DNA its semi-conservative nature where one strand of new DNA is from an original parent strand.

Although the structure of DNA showed how inheritance works, it was still not known how DNA influences the behavior of cells. In the following years, scientists tried to understand how DNA controls the process of protein production. It was discovered that the cell uses DNA as a template to create matching messenger RNA, molecules with nucleotides very similar to DNA. The nucleotide sequence of a messenger RNA is used to create an amino acid sequence in protein; this translation between nucleotide sequences and amino acid sequences is known as the genetic code.

With the newfound molecular understanding of inheritance came an explosion of research. A notable theory arose from Tomoko Ohta in 1973 with her amendment to the neutral theory of molecular evolution through publishing the nearly neutral theory of molecular evolution. In this theory, Ohta stressed the importance of natural selection and the environment to the rate at which genetic evolution occurs. One important development was chain-termination DNA sequencing in 1977 by Frederick Sanger. This technology allows scientists to read the nucleotide sequence of a DNA molecule. In 1983, Kary Banks Mullis developed the polymerase chain reaction, providing a quick way to isolate and amplify a specific section of DNA from a mixture. The efforts of the Human Genome Project, Department of Energy, NIH, and parallel private efforts by Celera Genomics led to the sequencing of the human genome in 2003.

Features of Inheritance

Discrete Inheritance and Mendel's Laws

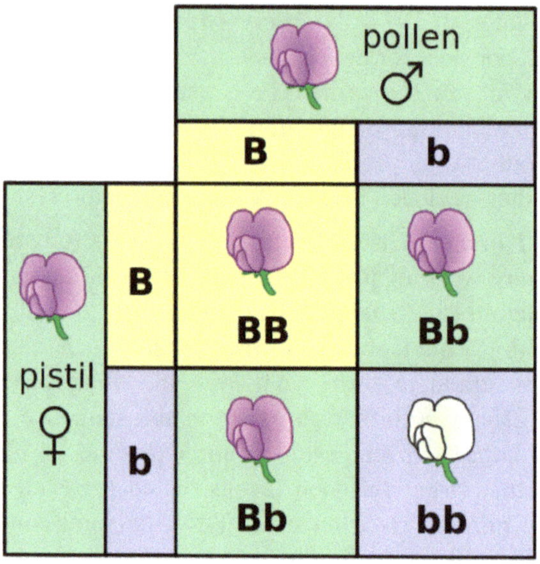

A Punnett square depicting a cross between two pea plants heterozygous for purple (B) and white (b) blossoms.

At its most fundamental level, inheritance in organisms occurs by passing discrete heritable units, called genes, from parents to offspring. This property was first observed by Gregor Mendel, who studied the segregation of heritable traits in pea plants. In his experiments studying the trait for flower color, Mendel observed that the flowers of each pea plant were either purple or white—but never an intermediate between the two colors. These different, discrete versions of the same gene are called alleles.

In the case of the pea, which is a diploid species, each individual plant has two copies of each gene, one copy inherited from each parent. Many species, including humans, have this pattern of inheritance. Diploid organisms with two copies of the same allele of a given gene are called homozygous at that gene locus, while organisms with two different alleles of a given gene are called heterozygous.

The set of alleles for a given organism is called its genotype, while the observable traits of the organism are called its phenotype. When organisms are heterozygous at a gene, often one allele is called dominant as its qualities dominate the phenotype of the organism, while the other allele is called recessive as its qualities recede and are not observed. Some alleles do not have complete dominance and instead have incomplete dominance by expressing an intermediate phenotype, or codominance by expressing both alleles at once.

When a pair of organisms reproduce sexually, their offspring randomly inherit one of the two alleles from each parent. These observations of discrete inheritance and the segregation of alleles are collectively known as Mendel›s first law or the Law of Segregation.

Notation and Diagrams

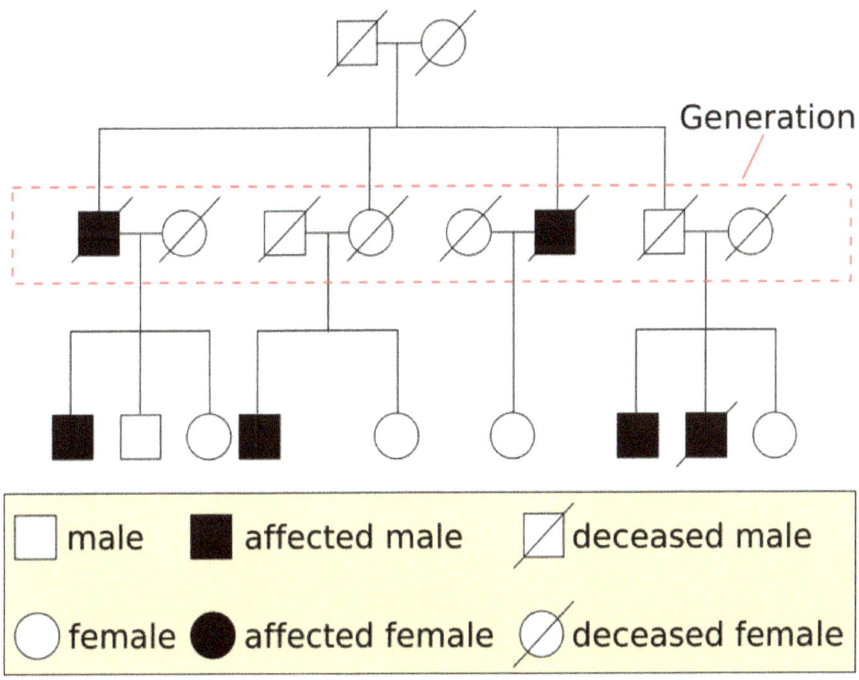

Genetic pedigree charts help track the inheritance patterns of traits.

Geneticists use diagrams and symbols to describe inheritance. A gene is represented by one or a few letters. Often a "+" symbol is used to mark the usual, non-mutant allele for a gene.

In fertilization and breeding experiments (and especially when discussing Mendel›s laws) the parents are referred to as the "P" generation and the offspring as the "F1" (first filial) generation. When the F1 offspring mate with each other, the offspring are called the "F2" (second filial) generation. One of the common diagrams used to predict the result of cross-breeding is the Punnett square.

When studying human genetic diseases, geneticists often use pedigree charts to represent the inheritance of traits. These charts map the inheritance of a trait in a family tree.

Multiple Gene Interactions

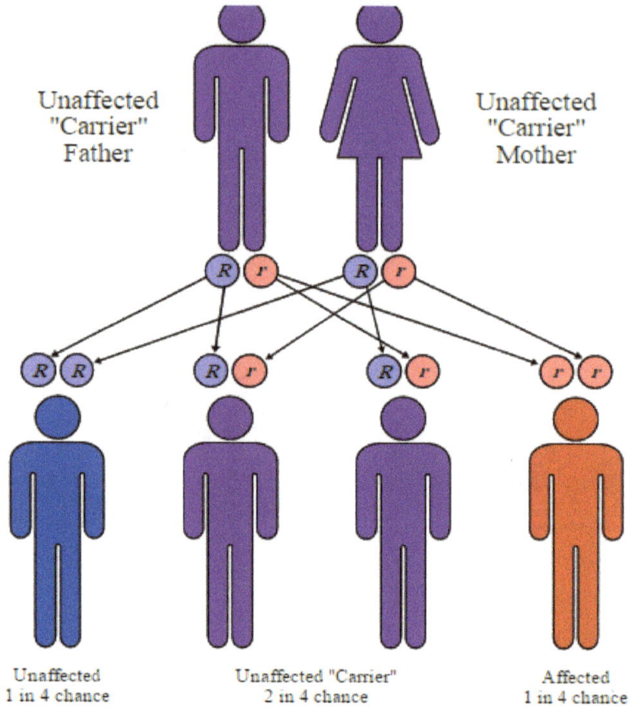

Human height is a trait with complex genetic causes. Francis Galton's data from 1889 shows the relationship between offspring height as a function of mean parent height.

Organisms have thousands of genes, and in sexually reproducing organisms these genes generally assort independently of each other. This means that the inheritance of an allele for yellow or green pea color is unrelated to the inheritance of alleles for white or purple flowers. This phenomenon, known as "Mendel's second law" or the "law of independent assortment," means that the alleles of different genes get shuffled between parents to form offspring with many different combinations. (Some genes do not assort independently, demonstrating genetic linkage, a topic discussed later in this article.)

Often different genes can interact in a way that influences the same trait. In the Blue-eyed Mary (*Omphalodes verna*), for example, there exists a gene with alleles that determine the color of flowers: blue or magenta. Another gene, however, controls whether the flowers have color at all or are white. When a plant has two copies of this white allele, its flowers are white—regardless of whether the first gene has blue or magenta alleles. This interaction between genes is called epistasis, with the second gene epistatic to the first.

Many traits are not discrete features (e.g. purple or white flowers) but are instead continuous features (e.g. human height and skin color). These complex traits are products of many genes. The influence of these genes is mediated, to varying degrees, by the environment an organism has experienced. The degree to which an organism's genes contribute to a complex trait is called heritability. Measurement of the heritability of a trait is relative—in a more variable environment, the environment has a bigger influence on the total variation of the trait. For example, human height is a trait with complex causes. It has a heritability of 89% in the United States. In Nigeria, however, where people experience a more variable access to good nutrition and health care, height has a heritability of only 62%.

Molecular Basis for Inheritance

DNA and Chromosomes

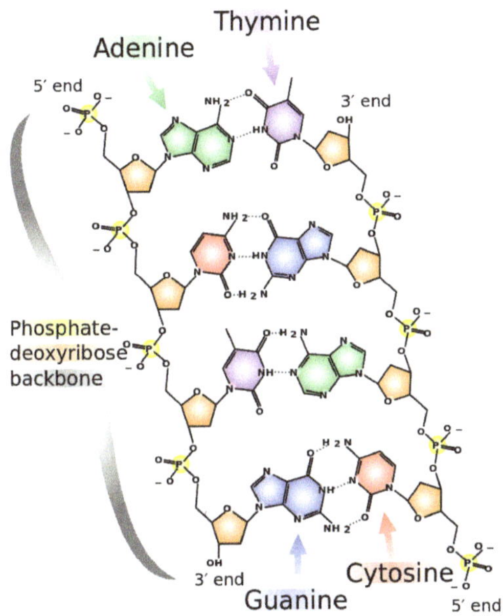

The molecular structure of DNA. Bases pair through the arrangement of hydrogen bonding between the strands.

The molecular basis for genes is deoxyribonucleic acid (DNA). DNA is composed of a chain of nucleotides, of which there are four types: adenine (A), cytosine (C), guanine (G), and thymine (T). Genetic information exists in the sequence of these nucleotides, and genes exist as stretches of sequence along the DNA chain. Viruses are the only exception to this rule—sometimes viruses use the very similar molecule RNA instead of DNA as their genetic material. Viruses cannot reproduce without a host and are unaffected by many genetic processes, so tend not to be considered living organisms.

DNA normally exists as a double-stranded molecule, coiled into the shape of a double helix. Each nucleotide in DNA preferentially pairs with its partner nucleotide on the opposite strand: A pairs with T, and C pairs with G. Thus, in its two-stranded form, each strand effectively contains all necessary information, redundant with its partner strand. This structure of DNA is the physical basis for inheritance: DNA replication duplicates the genetic information by splitting the strands and using each strand as a template for synthesis of a new partner strand.

Genes are arranged linearly along long chains of DNA base-pair sequences. In bacteria, each cell usually contains a single circular genophore, while eukaryotic organisms (such as plants and animals) have their DNA arranged in multiple linear chromosomes. These DNA strands are often extremely long; the largest human chromosome, for example, is about 247 million base pairs in length. The DNA of a chromosome is associated with structural proteins that organize, compact, and control access to the DNA, forming a material called chromatin; in eukaryotes, chromatin is usually composed of nucleosomes, segments of DNA wound around cores of histone proteins. The full set of hereditary material in an organism (usually the combined DNA sequences of all chromosomes) is called the genome.

While haploid organisms have only one copy of each chromosome, most animals and many plants are diploid, containing two of each chromosome and thus two copies of every gene. The two alleles for a gene are located on identical loci of the two homologous chromosomes, each allele inherited from a different parent.

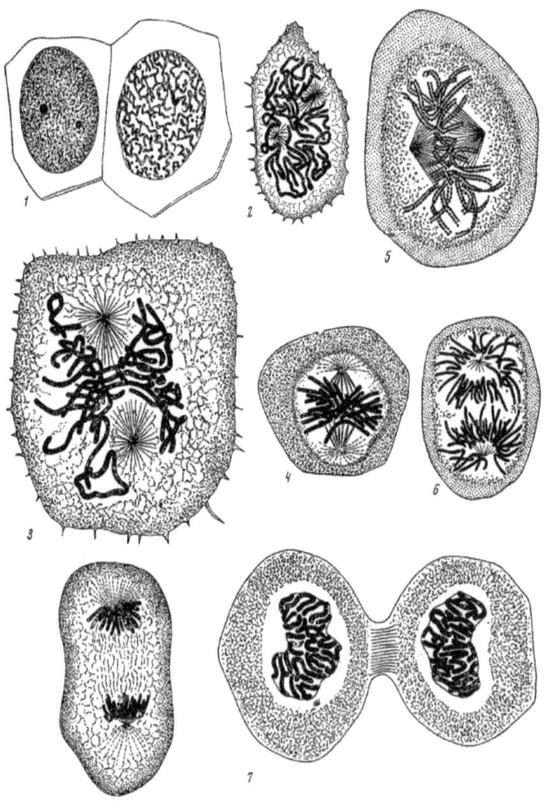

Walther Flemming's 1882 diagram of eukaryotic cell division. Chromosomes are copied, condensed, and organized. Then, as the cell divides, chromosome copies separate into the daughter cells.

Many species have so-called sex chromosomes that determine the gender of each organism. In humans and many other animals, the Y chromosome contains the gene that triggers the development of the specifically male characteristics. In evolution, this chromosome has lost most of its content and also most of its genes, while the X chromosome is similar to the other chromosomes and contains many genes. The X and Y chromosomes form a strongly heterogeneous pair.

Reproduction

When cells divide, their full genome is copied and each daughter cell inherits one copy. This process, called mitosis, is the simplest form of reproduction and is the basis for asexual reproduction. Asexual reproduction can also occur in multicellular organisms, producing offspring that inherit their genome from a single parent. Offspring that are genetically identical to their parents are called clones.

Eukaryotic organisms often use sexual reproduction to generate offspring that contain a mixture of genetic material inherited from two different parents. The process of sexual reproduction alternates between forms that contain single copies of the genome (haploid) and double copies (diploid). Haploid cells fuse and combine genetic material to create a diploid cell with paired chromosomes. Diploid organisms form haploids by dividing, without replicating their DNA, to create daughter cells that randomly inherit one of each pair of chromosomes. Most animals and many plants are diploid for most of their lifespan, with the haploid form reduced to single cell gametes such as sperm or eggs.

Although they do not use the haploid/diploid method of sexual reproduction, bacteria have many methods of acquiring new genetic information. Some bacteria can undergo conjugation, transferring a small circular piece of DNA to another bacterium. Bacteria can also take up raw DNA fragments found in the environment and integrate them into their genomes, a phenomenon known as transformation. These processes result in horizontal gene transfer, transmitting fragments of genetic information between organisms that would be otherwise unrelated.

Recombination and Genetic Linkage

The diploid nature of chromosomes allows for genes on different chromosomes to assort independently or be separated from their homologous pair during sexual reproduction wherein haploid gametes are formed. In this way new combinations of genes can occur in the offspring of a mating pair. Genes on the same chromosome would theoretically never recombine. However, they do, via the cellular process of chromosomal crossover. During crossover, chromosomes exchange stretches of DNA, effectively shuffling the gene alleles between the chromosomes. This process of chromosomal crossover generally occurs during meiosis, a series of cell divisions that creates haploid cells.

The first cytological demonstration of crossing over was performed by Harriet Creighton and Barbara McClintock in 1931. Their research and experiments on corn provided cytological evidence for the genetic theory that linked genes on paired chromosomes do in fact exchange places from one homolog to the other.

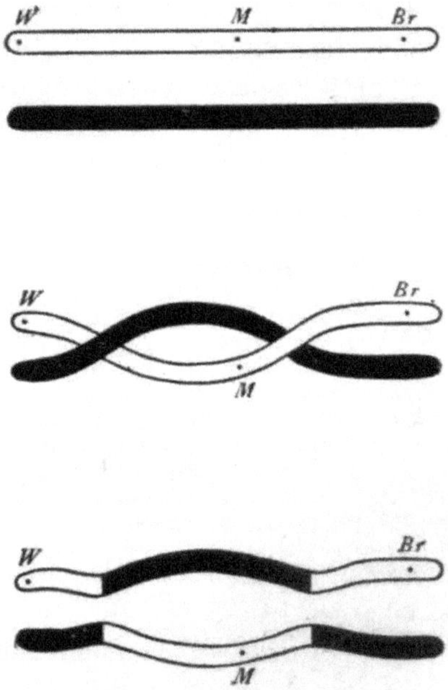

Thomas Hunt Morgan's 1916 illustration of a double crossover between chromosomes.

The probability of chromosomal crossover occurring between two given points on the chromosome is related to the distance between the points. For an arbitrarily long distance, the probability of crossover is high enough that the inheritance of the genes is effectively uncorrelated. For genes that are closer together, however, the lower probability of crossover means that the genes demonstrate genetic linkage; alleles for the two genes tend to be inherited together. The amounts of linkage between a series of genes can be combined to form a linear linkage map that roughly describes the arrangement of the genes along the chromosome.

Gene Expression

Genetic Code

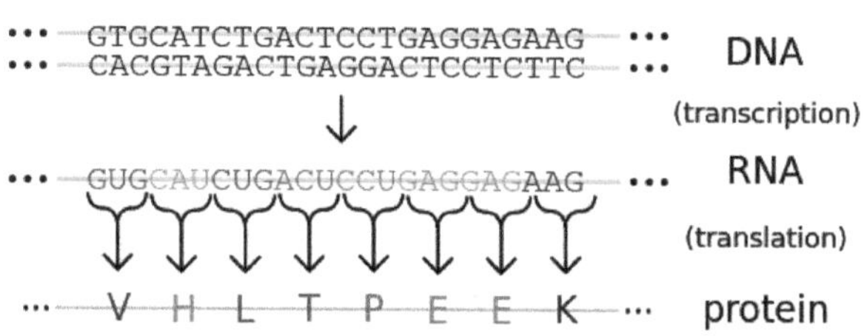

The genetic code: Using a triplet code, DNA, through a messenger RNA intermediary, specifies a protein.

Genes generally express their functional effect through the production of proteins, which are complex molecules responsible for most functions in the cell. Proteins are made up of one or more polypeptide chains, each of which is composed of a sequence of amino acids, and the DNA sequence of a gene (through an RNA intermediate) is used to produce a specific amino acid sequence. This process begins with the production of an RNA molecule with a sequence matching the gene's DNA sequence, a process called transcription.

This messenger RNA molecule is then used to produce a corresponding amino acid sequence through a process called translation. Each group of three nucleotides in the sequence, called a codon, corresponds either to one of the twenty possible amino acids in a protein or an instruction to end the amino acid sequence; this correspondence is called the genetic code. The flow of information is unidirectional: information is transferred from nucleotide sequences into the amino acid sequence of proteins, but it never transfers from protein back into the sequence of DNA—a phenomenon Francis Crick called the central dogma of molecular biology.

The specific sequence of amino acids results in a unique three-dimensional structure for that protein, and the three-dimensional structures of proteins are related to their functions. Some are simple structural molecules, like the fibers formed by the protein collagen. Proteins can bind to other proteins and simple molecules, sometimes acting as enzymes by facilitating chemical reactions within the bound molecules (without changing the structure of the protein itself). Protein structure is dynamic; the protein hemoglobin bends into slightly different forms as it facilitates the capture, transport, and release of oxygen molecules within mammalian blood.

A single nucleotide difference within DNA can cause a change in the amino acid sequence of a protein. Because protein structures are the result of their amino acid sequences, some changes can dramatically change the properties of a protein by destabilizing the structure or changing the surface of the protein in a way that changes its interaction with other proteins and molecules. For example, sickle-cell anemia is a human genetic disease that results from a single base difference within the coding region for the β-globin section of hemoglobin, causing a single amino acid change that changes hemoglobin's physical properties. Sickle-cell versions of hemoglobin stick to themselves, stacking to form fibers that distort the shape of red blood cells carrying the protein. These sickle-shaped cells no longer flow smoothly through blood vessels, having a tendency to clog or degrade, causing the medical problems associated with this disease.

Some DNA sequences are transcribed into RNA but are not translated into protein products—such RNA molecules are called non-coding RNA. In some cases, these products fold into structures which are involved in critical cell functions (e.g. ribosomal RNA and transfer RNA). RNA can also have regulatory effects through hybridization interactions with other RNA molecules (e.g. microRNA).

Nature and Nurture

Siamese cats have a temperature-sensitive pigment-production mutation.

Although genes contain all the information an organism uses to function, the environment plays an important role in determining the ultimate phenotypes an organism displays. This is the complementary relationship often referred to as “nature and nurture.” The phenotype of an organism depends on the interaction of genes and the environment. An interesting example is the coat coloration of the Siamese cat. In this case, the body temperature of the cat plays the role of the environment. The cat›s genes code for dark hair, thus the hair-producing cells in the cat make cellular proteins resulting in dark hair. But these dark hair-producing proteins are sensitive to temperature (i.e. have a mutation causing temperature-sensitivity) and denature in higher-temperature environments, failing to produce dark-hair pigment in areas where the cat has a higher body temperature. In a low-temperature environment, however, the protein’s structure is stable and produces dark-hair pigment normally. The protein remains functional in areas of skin that are colder—such as its legs, ears, tail and face—so the cat has dark-hair at its extremities.

Environment plays a major role in effects of the human genetic disease phenylketonuria. The mutation that causes phenylketonuria disrupts the ability of the body to break down the amino acid phenylalanine, causing a toxic build-up of an intermediate molecule that, in turn, causes severe symptoms of progressive intellectual disability and seizures. However, if someone with the phenylketonuria mutation follows a strict diet that avoids this amino acid, they remain normal and healthy.

A popular method for determining how genes and environment (“nature and nurture”) contribute to a phenotype involves studying identical and fraternal twins, or other siblings of multiple births. Because identical siblings come from the same zygote, they are genetically the same. Fraternal twins are as genetically different from one another as normal siblings. By comparing how often a certain disorder occurs in a pair of identical twins to how often it occurs in a pair of fraternal twins, scientists can determine whether that disorder is caused by genetic or postnatal environmental factors – whether it has "nature" or "nurture" causes. One famous example is the multiple birth study of the Genain quadruplets, who were identical quadruplets all diagnosed with schizophrenia. However such tests cannot separate genetic factors from environmental factors affecting fetal development.

Gene Regulation

The genome of a given organism contains thousands of genes, but not all these genes need to be active at any given moment. A gene is expressed when it is being transcribed into mRNA and there exist many cellular methods of controlling the expression of genes such that proteins are produced only when needed by the cell. Transcription factors are regulatory proteins that bind to DNA, either promoting or inhibiting the transcription of a gene. Within the genome of *Escherichia coli* bacteria, for example, there exists a series of genes necessary for the synthesis of the amino acid tryptophan. However, when tryptophan is already available to the cell, these genes for tryptophan synthesis are no longer needed. The presence of tryptophan directly affects the activity of the genes—tryptophan molecules bind to the tryptophan repressor (a transcription factor), changing the repressor's structure such that the repressor binds to the genes. The tryptophan repressor blocks the transcription and expression of the genes, thereby creating negative feedback regulation of the tryptophan synthesis process.

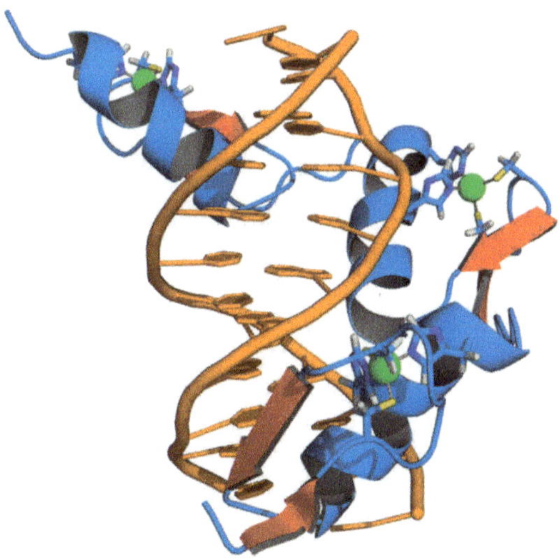

Transcription factors bind to DNA, influencing the transcription of associated genes.

Differences in gene expression are especially clear within multicellular organisms, where cells all contain the same genome but have very different structures and behaviors due to the expression of different sets of genes. All the cells in a multicellular organism derive from a single cell, differentiating into variant cell types in response to external and intercellular signals and gradually establishing different patterns of gene expression to create different behaviors. As no single gene is responsible for the development of structures within multicellular organisms, these patterns arise from the complex interactions between many cells.

Within eukaryotes, there exist structural features of chromatin that influence the transcription of genes, often in the form of modifications to DNA and chromatin that are stably inherited by daughter cells. These features are called "epigenetic" because they exist "on top" of the DNA sequence and retain inheritance from one cell generation to the next. Because of epigenetic features, different cell types grown within the same medium can retain very different properties. Although epigenetic features are generally dynamic over the course of development, some, like the phenomenon of paramutation, have multigenerational inheritance and exist as rare exceptions to the general rule of DNA as the basis for inheritance.

Genetic Change

Mutations

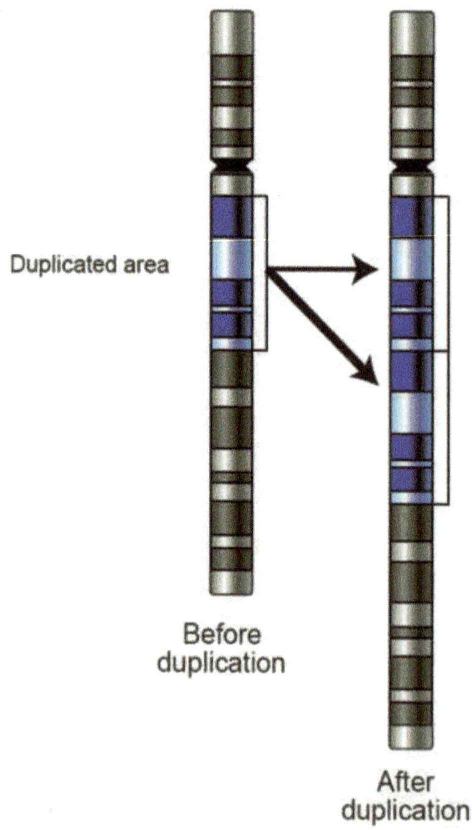

Gene duplication allows diversification by providing redundancy: one gene can mutate and lose its original function without harming the organism.

During the process of DNA replication, errors occasionally occur in the polymerization of the second strand. These errors, called mutations, can affect the phenotype of an organism, especially if they occur within the protein coding sequence of a gene. Error rates are usually very low—1 error in every 10–100 million bases—due to the "proofreading" ability of DNA polymerases. Processes that increase the rate of changes in DNA are called mutagenic: mutagenic chemicals promote errors in DNA replication, often by interfering with the structure of base-pairing, while UV radiation induces mutations by causing damage to the DNA structure. Chemical damage to DNA occurs naturally as well and cells use DNA repair mechanisms to repair mismatches and breaks. The repair does not, however, always restore the original sequence. In organisms that use chromosomal crossover to exchange DNA and recombine genes, errors in alignment during meiosis can also cause mutations. Errors in crossover are especially likely when similar sequences cause partner chromosomes to adopt a mistaken alignment; this makes some regions in genomes more prone to mutating in this way. These errors create large structural changes in DNA sequence – duplications, inversions, deletions of entire regions – or the accidental exchange of whole parts of sequences between different chromosomes (chromosomal translocation).

Natural Selection and Evolution

Mutations alter an organism's genotype and occasionally this causes different phenotypes to appear. Most mutations have little effect on an organism's phenotype, health, or reproductive fitness. Mutations that do have an effect are usually detrimental, but occasionally some can be beneficial. Studies in the fly *Drosophila melanogaster* suggest that if a mutation changes a protein produced by a gene, about 70 percent of these mutations will be harmful with the remainder being either neutral or weakly beneficial.

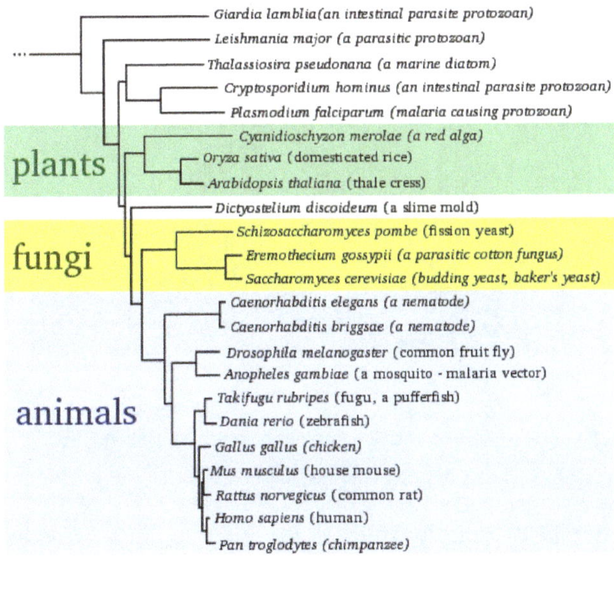

An evolutionary tree of eukaryotic organisms, constructed by the comparison of several orthologous gene sequences.

Population genetics studies the distribution of genetic differences within populations and how these distributions change over time. Changes in the frequency of an allele in a population are mainly influenced by natural selection, where a given allele provides a selective or reproductive advantage to the organism, as well as other factors such as mutation, genetic drift, genetic draft, artificial selection and migration.

Over many generations, the genomes of organisms can change significantly, resulting in evolution. In the process called adaptation, selection for beneficial mutations can cause a species to evolve into forms better able to survive in their environment. New species are formed through the process of speciation, often caused by geographical separations that prevent populations from exchanging genes with each other. The application of genetic principles to the study of population biology and evolution is known as the "modern evolutionary synthesis."

By comparing the homology between different species' genomes, it is possible to calculate the evolutionary distance between them and when they may have diverged. Genetic comparisons are generally considered a more accurate method of characterizing the relatedness between species than the comparison of phenotypic characteristics. The evolutionary distances between species can be used to form evolutionary trees; these trees represent the common descent and divergence of species over time, although they do not show the transfer of genetic material between unrelated species (known as horizontal gene transfer and most common in bacteria).

Model Organisms

The common fruit fly (*Drosophila melanogaster*) is a popular model organism in genetics research.

Although geneticists originally studied inheritance in a wide range of organisms, researchers began to specialize in studying the genetics of a particular subset of organisms. The fact that significant research already existed for a given organism would encourage new researchers to choose it for further study, and so eventually a few model organisms became the basis for most genetics research. Common research topics in model organism genetics include the study of gene regulation and the involvement of genes in development and cancer.

Organisms were chosen, in part, for convenience—short generation times and easy genetic manipulation made some organisms popular genetics research tools. Widely used model organisms include the gut bacterium *Escherichia coli*, the plant *Arabidopsis thaliana*, baker's yeast (*Saccharomyces cerevisiae*), the nematode *Caenorhabditis elegans*, the common fruit fly (*Drosophila melanogaster*), and the common house mouse (*Mus musculus*).

Medicine

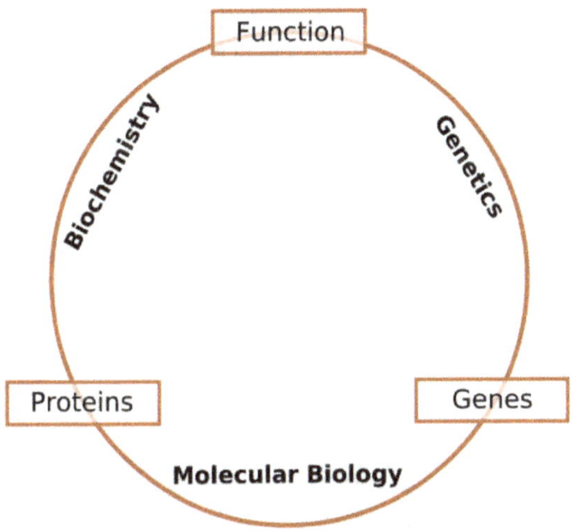

Schematic relationship between biochemistry, genetics and molecular biology.

Medical genetics seeks to understand how genetic variation relates to human health and disease.When searching for an unknown gene that may be involved in a disease, researchers commonly use genetic linkage and genetic pedigree charts to find the location on the genome associated with the disease. At the population level, researchers take advantage of Mendelian randomization to look for locations in the genome that are associated with diseases, a method especially useful for multigenic traits not clearly defined by a single gene. Once a candidate gene is found, further research is often done on the corresponding (or homologous) genes of model organisms. In addition to studying genetic diseases, the increased availability of genotyping methods has led to the field of pharmacogenetics: the study of how genotype can affect drug responses.

Individuals differ in their inherited tendency to develop cancer, and cancer is a genetic disease. The process of cancer development in the body is a combination of events. Mutations occasionally occur within cells in the body as they divide. Although these mutations will not be inherited by any offspring, they can affect the behavior of cells, sometimes causing them to grow and divide more frequently. There are biological mechanisms that attempt to stop this process; signals are given to inappropriately dividing cells that should trigger cell death, but sometimes additional mutations occur that cause cells to ignore these messages. An internal process of natural selection occurs within the body and eventually mutations accumulate within cells to promote their own growth, creating a cancerous tumor that grows and invades various tissues of the body.

Normally, a cell divides only in response to signals called growth factors and stops growing once in contact with surrounding cells and in response to growth-inhibitory signals. It usually then divides a limited number of times and dies, staying within the epithelium where it is unable to migrate to other organs. To become a cancer cell, a cell has to accumulate mutations in a number of genes (three to seven) that allow it to bypass this regulation: it no longer needs growth factors to divide, continues growing when making contact to neighbor cells, ignores inhibitory signals, keeps growing indefinitely and is immortal, escapes from the epithelium and ultimately may be able to escape from the primary tumor, cross the endothelium of a blood vessel, be transported by the bloodstream and colonize a new organ, forming deadly metastasis. Although there are some genetic predispositions in a small fraction of cancers, the major fraction is due to a set of new genetic mutations that originally appear and accumulate in one or a small number of cells that will divide to form the tumor and are not transmitted to the progeny (somatic mutations). The most frequent mutations are a loss of function of p53 protein, a tumor suppressor, or in the p53 pathway, and gain of function mutations in the Ras proteins, or in other oncogenes.

Research Methods

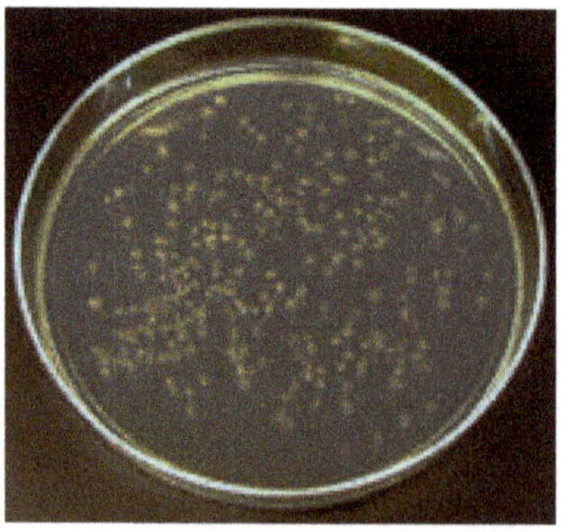

Colonies of *E. coli* produced by cellular cloning. A similar methodology is often used in molecular cloning.

DNA can be manipulated in the laboratory. Restriction enzymes are commonly used enzymes that cut DNA at specific sequences, producing predictable fragments of DNA. DNA fragments can be visualized through use of gel electrophoresis, which separates fragments according to their length.

The use of ligation enzymes allows DNA fragments to be connected. By binding ("ligating") fragments of DNA together from different sources, researchers can create recombinant DNA, the DNA often associated with genetically modified organisms.

Recombinant DNA is commonly used in the context of plasmids: short circular DNA molecules with a few genes on them. In the process known as molecular cloning, researchers can amplify the DNA fragments by inserting plasmids into bacteria and then culturing them on plates of agar (to isolate clones of bacteria cells). ("Cloning" can also refer to the various means of creating cloned ("clonal") organisms.)

DNA can also be amplified using a procedure called the polymerase chain reaction (PCR). By using specific short sequences of DNA, PCR can isolate and exponentially amplify a targeted region of DNA. Because it can amplify from extremely small amounts of DNA, PCR is also often used to detect the presence of specific DNA sequences.

DNA Sequencing and Genomics

DNA sequencing, one of the most fundamental technologies developed to study genetics, allows researchers to determine the sequence of nucleotides in DNA fragments. The technique of chain-termination sequencing, developed in 1977 by a team led by Frederick Sanger, is still routinely used to sequence DNA fragments. Using this technology, researchers have been able to study the molecular sequences associated with many human diseases.

As sequencing has become less expensive, researchers have sequenced the genomes of many organisms using a process called genome assembly, which utilizes computational tools to stitch together sequences from many different fragments. These technologies were used to sequence the human genome in the Human Genome Project completed in 2003. New high-throughput sequencing technologies are dramatically lowering the cost of DNA sequencing, with many researchers hoping to bring the cost of resequencing a human genome down to a thousand dollars.

Next-generation sequencing (or high-throughput sequencing) came about due to the ever-increasing demand for low-cost sequencing. These sequencing technologies allow the production of potentially millions of sequences concurrently. The large amount of sequence data available has created the field of genomics, research that uses computational tools to search for and analyze patterns in the full genomes of organisms. Genomics can also be considered a subfield of bioinformatics, which uses computational approaches to analyze large sets of biological data. A common problem to these fields of research is how to manage and share data that deals with human subject and personally identifiable information.

Society and Culture

On 19 March 2015, a leading group of biologists urged a worldwide ban on clinical use of methods, particularly the use of CRISPR and zinc finger, to edit the human genome in a way that can be inherited. In April 2015, Chinese researchers reported results of basic research to edit the DNA of non-viable human embryos using CRISPR.

REFERENCES

1. Griffiths, Anthony J. F.; Miller, Jeffrey H.; Suzuki, David T.; Lewontin, Richard C.; Gelbart, eds. (2000). "Genetics and the Organism: Introduction". An Introduction

to Genetic Analysis (7th ed.). New York: W. H. Freeman. ISBN 0-7167-3520-2.

2. Hartl D, Jones E (2005)
3. "Genetikos (γενετ-ικός)". Henry George Liddell, Robert Scott, A Greek-English Lexicon. Perseus Digital Library, Tufts University. Retrieved 20 February 2012.
4. "Genesis (γένεσις)". Henry George Liddell, Robert Scott, A Greek-English Lexicon. Perseus Digital Library, Tufts University. Retrieved 20 February 2012.
5. "Genetic". Online Etymology Dictionary. Retrieved 20 February 2012.
6. DK Publishing (2009). Science: The Definitive Visual Guide. Penguin. p. 362. ISBN 978-0-7566-6490-9.
7. Weiling, F (1991). "Historical study: Johann Gregor Mendel 1822–1884.". American Journal of Medical Genetics. 40 (1): 1–25; discussion 26. doi:10.1002/ajmg.1320400103. PMID 1887835.
8. Poczai P., Bell N., Hyvönen J. (2014). "Imre Festetics and the Sheep Breeders› Society of Moravia: Mendel›s Forgotten "Research Network"". PLoS Biology. 12 (1): e1001772. doi:10.1371/journal.pbio.1001772.
9. Matthew Hamilton (2011). Population Genetics. Georgetown University. p. 26. ISBN 978-1-4443-6245-9.
10. Lamarck, J-B (2008). In Encyclopædia Britannica. Retrieved from Encyclopædia Britannica Online on 16 March 2008.
11. Singer, Emily (4 February 2009). "A Comeback for Lamarckian Evolution?". Technology Review. Retrieved 14 March 2013.
12. Peter J. Bowler, The Mendelian Revolution: The Emergency of Hereditarian Concepts in Modern Science and Society (Baltimore: Johns Hopkins University Press, 1989): chapters 2 & 3.
13. Blumberg, Roger B. "Mendel›s Paper in English".
14. genetics, n., Oxford English Dictionary, 3rd ed.
15. Bateson W. "Letter from William Bateson to Alan Sedgwick in 1905". The John Innes Centre. Retrieved 15 March 2008. Note that the letter was to an Adam Sedgwick, a zoologist and "Reader in Animal Morphology" at Trinity College, Cambridge
16. genetic, adj., Oxford English Dictionary, 3rd ed.
17. Richmond, Marsha L. (November 2007). "Opportunities for women in early genetics". Nature Reviews Genetics. 8 (11): 897–902. doi:10.1038/nrg2200. PMID 17893692. Retrieved April 23, 2015.
18. Bateson, W (1907). "The Progress of Genetic Research". In Wilks, W. Report of the Third 1906 International Conference on Genetics: Hybridization (the cross-breeding of genera or species), the cross-breeding of varieties, and general plant breeding. London: Royal Horticultural Society.
19. Initially titled the "International Conference on Hybridisation and Plant Breeding", the title was changed as a result of Bateson›s speech. See: Cock AG, Forsdyke DR (2008). Treasure your exceptions: the science and life of William Bateson.

Springer. p. 248. ISBN 978-0-387-75687-5.

20. Moore, John A. (1983). "Thomas Hunt Morgan—The Geneticist". Integrative and Comparative Biology. 23 (4): 855–865. doi:10.1093/icb/23.4.855.
21. Sturtevant AH (1913). "The linear arrangement of six sex-linked factors in Drosophila, as shown by their mode of association" (PDF). Journal of Experimental Biology. 14: 43–59. doi:10.1002/jez.1400140104.
22. Avery, OT; MacLeod, CM; McCarty, M (1944). "Studies on the Chemical Nature of the Substance Inducing Transformation of Pneumococcal Types: Induction of Transformation by a Desoxyribonucleic Acid Fraction Isolated from Pneumococcus Type III". The Journal of Experimental Medicine. 79 (2): 137–58. doi:10.1084/jem.79.2.137. PMC 2135445 . PMID 19871359.Reprint: Avery, OT; MacLeod, CM; McCarty, M (1979). "Studies on the chemical nature of the substance inducing transformation of pneumococcal types. Inductions of transformation by a desoxyribonucleic acid fraction isolated from pneumococcus type III". The Journal of Experimental Medicine. 149 (2): 297–326. doi:10.1084/jem.149.2.297. PMC 2184805 . PMID 33226.
23. "Cell and Molecular Biology", Pragya Khanna. I. K. International Pvt Ltd, 2008. p. 221. ISBN 81-89866-59-1, ISBN 978-81-89866-59-4
24. Hershey, AD; Chase, M (1952). "Independent functions of viral protein and nucleic acid in growth of bacteriophage". The Journal of General Physiology. 36 (1): 39–56. doi:10.1085/jgp.36.1.39. PMC 2147348 . PMID 12981234.
25. Judson, Horace (1979). The Eighth Day of Creation: Makers of the Revolution in Biology. Cold Spring Harbor Laboratory Press. pp. 51–169. ISBN 0-87969-477-7.
26. Watson, J. D.; Crick, FH (1953). "Molecular Structure of Nucleic Acids: A Structure for Deoxyribose Nucleic Acid" (PDF). Nature. 171 (4356): 737–8. Bibcode:1953Natur.171..737W. doi:10.1038/171737a0. PMID 13054692.
27. Watson, J. D.; Crick, FH (1953). "Genetical Implications of the Structure of Deoxyribonucleic Acid" (PDF). Nature. 171 (4361): 964–7. Bibcode:1953Natur.171..964W. doi:10.1038/171964b0. PMID 13063483.
28. Stratmann, S. A. (1 Nov 2013). "DNA replication at the single molecule level". Chemical Society Reviews. 43 (4): 1201–20. doi:10.1039/c3cs60391a. PMID 24395040.
29. Frederick Betz (2010). Managing Science: Methodology and Organization of Research. Springer. p. 76. ISBN 978-1-4419-7488-4.
30. Stanley A. Rice (2009). Encyclopedia of Evolution. Infobase Publishing. p. 134. ISBN 978-1-4381-1005-9.
31. Sahotra Sarkar (1998). Genetics and Reductionism. Cambridge University Press. p. 140. ISBN 978-0-521-63713-8.
32. Ohta, Tomoko (1973). "Slightly Deleterious Mutant Substitutions in Evolution". Nature. 246 (5428): 96–98. Bibcode:1973Natur.246...96O. doi:10.1038/246096a0. PMID 4585855.

33. Sanger, F; Nicklen, S; Coulson, AR (1977). "DNA sequencing with chain-terminating inhibitors". Proceedings of the National Academy of Sciences of the United States of America. 74 (12): 5463–7. Bibcode:1977PNAS...74.5463S. doi:10.1073/pnas.74.12.5463. PMC 431765 . PMID 271968.
34. Saiki, RK; Scharf, S; Faloona, F; Mullis, KB; Horn, GT; Erlich, HA; Arnheim, N (1985). "Enzymatic amplification of beta-globin genomic sequences and restriction site analysis for diagnosis of sickle cell anemia". Science. 230 (4732): 1350–4. Bibcode:1985Sci...230.1350S. doi:10.1126/science.2999980. PMID 2999980.
35. "Human Genome Project Information". Human Genome Project. Retrieved 15 March 2008.
36. "The sequence of the human genome". Science. 291.
37. Griffiths, Anthony J. F.; Miller, Jeffrey H.; Suzuki, David T.; Lewontin, Richard C.; Gelbart, eds. (2000). "Patterns of Inheritance: Introduction". An Introduction to Genetic Analysis (7th ed.). New York: W. H. Freeman. ISBN 0-7167-3520-2.
38. Griffiths, Anthony J. F.; Miller, Jeffrey H.; Suzuki, David T.; Lewontin, Richard C.; Gelbart, eds. (2000). "Mendel›s experiments". An Introduction to Genetic Analysis (7th ed.). New York: W. H. Freeman. ISBN 0-7167-3520-2.
39. Griffiths, Anthony J. F.; Miller, Jeffrey H.; Suzuki, David T.; Lewontin, Richard C.; Gelbart, eds. (2000). "Mendelian genetics in eukaryotic life cycles". An Introduction to Genetic Analysis (7th ed.). New York: W. H. Freeman. ISBN 0-7167-3520-2.
40. Griffiths, Anthony J. F.; Miller, Jeffrey H.; Suzuki, David T.; Lewontin, Richard C.; Gelbart, eds. (2000). "Interactions between the alleles of one gene". An Introduction to Genetic Analysis (7th ed.). New York: W. H. Freeman. ISBN 0-7167-3520-2.
41. Cheney, Richard W. "Genetic Notation". Christopher Newport University. Archived from the original on 3 January 2008. Retrieved 18 March 2008.
42. Griffiths, Anthony J. F.; Miller, Jeffrey H.; Suzuki, David T.; Lewontin, Richard C.; Gelbart, eds. (2000). "Human Genetics". An Introduction to Genetic Analysis (7th ed.). New York: W. H. Freeman. ISBN 0-7167-3520-2.
43. Griffiths, Anthony J. F.; Miller, Jeffrey H.; Suzuki, David T.; Lewontin, Richard C.; Gelbart, eds. (2000). "Gene interaction and modified dihybrid ratios". An Introduction to Genetic Analysis (7th ed.). New York: W. H. Freeman. ISBN 0-7167-3520-2.
44. Mayeux, R (2005). "Mapping the new frontier: complex genetic disorders". The Journal of Clinical Investigation. 115 (6): 1404–7. doi:10.1172/JCI25421. PMC 1137013 . PMID 15931374.
45. Griffiths, Anthony J. F.; Miller, Jeffrey H.; Suzuki, David T.; Lewontin, Richard C.; Gelbart, eds. (2000). "Quantifying heritability". An Introduction to Genetic Analysis (7th ed.). New York: W. H. Freeman. ISBN 0-7167-3520-2.
46. Luke, A; Guo, X; Adeyemo, AA; Wilks, R; Forrester, T; Lowe Jr, W; Comuzzie, AG; Martin, LJ; Zhu, X; Rotimi, CN; Cooper, RS (2001). "Heritability of obesity-

related traits among Nigerians, Jamaicans and US black people". International Journal of Obesity and Related Metabolic Disorders. 25 (7): 1034–41. doi:10.1038/sj.ijo.0801650. PMID 11443503.

47. Pearson, H (2006). "Genetics: what is a gene?". Nature. 441 (7092): 398–401. Bibcode:2006Natur.441..398P. doi:10.1038/441398a. PMID 16724031.
48. Prescott, L (1993). Microbiology. Wm. C. Brown Publishers. ISBN 0-697-01372-3.
49. Griffiths, Anthony J. F.; Miller, Jeffrey H.; Suzuki, David T.; Lewontin, Richard C.; Gelbart, eds. (2000). "Mechanism of DNA Replication". An Introduction to Genetic Analysis (7th ed.). New York: W. H. Freeman. ISBN 0-7167-3520-2.
50. Gregory, SG; Barlow, KF; McLay, KE; Kaul, R; Swarbreck, D; Dunham, A; Scott, CE; Howe, KL; et al. (2006). "The DNA sequence and biological annotation of human chromosome 1". Nature. 441 (7091): 315–21. Bibcode:2006Natur.441..315G. doi:10.1038/nature04727. PMID 16710414.
51. Alberts et al. (2002), II.4. DNA and chromosomes: Chromosomal DNA and Its Packaging in the Chromatin Fiber
52. Griffiths, Anthony J. F.; Miller, Jeffrey H.; Suzuki, David T.; Lewontin, Richard C.; Gelbart, eds. (2000). "Sex chromosomes and sex-linked inheritance". An Introduction to Genetic Analysis (7th ed.). New York: W. H. Freeman. ISBN 0-7167-3520-2.
53. Griffiths, Anthony J. F.; Miller, Jeffrey H.; Suzuki, David T.; Lewontin, Richard C.; Gelbart, eds. (2000). "Bacterial conjugation". An Introduction to Genetic Analysis (7th ed.). New York: W. H. Freeman. ISBN 0-7167-3520-2.
54. Griffiths, Anthony J. F.; Miller, Jeffrey H.; Suzuki, David T.; Lewontin, Richard C.; Gelbart, eds. (2000). "Bacterial transformation". An Introduction to Genetic Analysis (7th ed.). New York: W. H. Freeman. ISBN 0-7167-3520-2.
55. Griffiths, Anthony J. F.; Miller, Jeffrey H.; Suzuki, David T.; Lewontin, Richard C.; Gelbart, eds. (2000). "Nature of crossing-over". An Introduction to Genetic Analysis (7th ed.). New York: W. H. Freeman. ISBN 0-7167-3520-2.
56. "A Correlation of Cytological and Genetical Crossing-Over in Zea Mays". Prod Natl Acad Sci USA. 17.
57. Jack E. Staub (1994). Crossover: Concepts and Applications in Genetics, Evolution, and Breeding. University of Wisconsin Press. p. 55. ISBN 978-0-299-13564-5.
58. Griffiths, Anthony J. F.; Miller, Jeffrey H.; Suzuki, David T.; Lewontin, Richard C.; Gelbart, eds. (2000). "Linkage maps". An Introduction to Genetic Analysis (7th ed.). New York: W. H. Freeman. ISBN 0-7167-3520-2.
59. Berg JM, Tymoczko JL, Stryer L, Clarke ND (2002). "I. 5. DNA, RNA, and the Flow of Genetic Information: Amino Acids Are Encoded by Groups of Three Bases Starting from a Fixed Point". Biochemistry (5th ed.). New York: W. H. Freeman and Company.
60. Crick, F (1970). "Central dogma of molecular biology" (PDF). Nature. 227 (5258): 561–3. Bibcode:1970Natur.227..561C. doi:10.1038/227561a0. PMID 4913914.
61. Alberts et al. (2002), I.3. Proteins: The Shape and Structure of Proteins

62. Alberts et al. (2002), I.3. Proteins: Protein Function

63. "How Does Sickle Cell Cause Disease?". Brigham and Women›s Hospital: Information Center for Sickle Cell and Thalassemic Disorders. 11 April 2002. Retrieved 23 July 2007.

64. Imes, DL; Geary, LA; Grahn, RA; Lyons, LA (2006). "Albinism in the domestic cat (Felis catus) is associated with a tyrosinase (TYR) mutation". Animal Genetics. 37 (2): 175–8. doi:10.1111/j.1365-2052.2005.01409.x. PMC 1464423 . PMID 16573534.

65. "MedlinePlus: Phenylketonuria". NIH: National Library of Medicine. Retrieved 15 March 2008.

66. For example, Ridley, M. (2003). Nature via Nurture: Genes, Experience and What Makes Us Human. Fourth Estate. p. 73. ISBN 978-1-84115-745-0.

67. Rosenthal, David (1964). The Genain Quadruplets: A Case Study and Theoretical Analysis of Heredity and Environment in Schizophrenia. New York: Basic Books. doi:10.1002/bs.3830090407.

68. Brivanlou, AH; Darnell Jr, JE (2002). "Signal transduction and the control of gene expression". Science. 295 (5556): 813–8. Bibcode:2002Sci...295..813B. doi:10.1126/science.1066355. PMID 11823631.

69. Alberts et al. (2002), II.3. Control of Gene Expression – The Tryptophan Repressor is a Simple Switch That Turns Genes On and Off in Bacteria

70. Jaenisch, R; Bird, A (2003). "Epigenetic regulation of gene expression: how the genome integrates intrinsic and environmental signals". Nature Genetics. 33 Suppl (3s): 245–54. doi:10.1038/ng1089. PMID 12610534.

71. Chandler, VL (2007). "Paramutation: from maize to mice". Cell. 128 (4): 641–5. doi:10.1016/j.cell.2007.02.007. PMID 17320501.

72. Griffiths, Anthony J. F.; Miller, Jeffrey H.; Suzuki, David T.; Lewontin, Richard C.; Gelbart, eds. (2000). "Spontaneous mutations". An Introduction to Genetic Analysis (7th ed.). New York: W. H. Freeman. ISBN 0-7167-3520-2.

73. Freisinger, E; Grollman, AP; Miller, H; Kisker, C (2004). "Lesion (in)tolerance reveals insights into DNA replication fidelity". The EMBO Journal. 23 (7): 1494–505. doi:10.1038/sj.emboj.7600158. PMC 391067 . PMID 15057282.

74. Griffiths, Anthony J. F.; Miller, Jeffrey H.; Suzuki, David T.; Lewontin, Richard C.; Gelbart, eds. (2000). "Induced mutations". An Introduction to Genetic Analysis (7th ed.). New York: W. H. Freeman. ISBN 0-7167-3520-2.

75. Griffiths, Anthony J. F.; Miller, Jeffrey H.; Suzuki, David T.; Lewontin, Richard C.; Gelbart, eds. (2000). "Chromosome Mutation I: Changes in Chromosome Structure: Introduction". An Introduction to Genetic Analysis (7th ed.). New York: W. H. Freeman. ISBN 0-7167-3520-2.

76. Moselio Schaechter (2009). Encyclopedia of Microbiology. Academic Press. p. 551. ISBN 978-0-12-373944-5.

77. Mike Calver; Alan Lymbery; Jennifer McComb; Mike Bamford (2009).

Environmental Biology. Cambridge University Press. p. 118. ISBN 978-0-521-67982-4.

78. Sawyer, SA; Parsch, J; Zhang, Z; Hartl, DL (2007). "Prevalence of positive selection among nearly neutral amino acid replacements in Drosophila". Proceedings of the National Academy of Sciences of the United States of America. 104 (16): 6504–10. Bibcode:2007PNAS..104.6504S. doi:10.1073/pnas.0701572104. PMC 1871816 . PMID 17409186.

79. Griffiths, Anthony J. F.; Miller, Jeffrey H.; Suzuki, David T.; Lewontin, Richard C.; Gelbart, eds. (2000). "Variation and its modulation". An Introduction to Genetic Analysis (7th ed.). New York: W. H. Freeman. ISBN 0-7167-3520-2.

80. Griffiths, Anthony J. F.; Miller, Jeffrey H.; Suzuki, David T.; Lewontin, Richard C.; Gelbart, eds. (2000). "Selection". An Introduction to Genetic Analysis (7th ed.). New York: W. H. Freeman. ISBN 0-7167-3520-2.

81. Gillespie, John H. (2001). "Is the population size of a species relevant to its evolution?". Evolution. 55 (11): 2161–2169. doi:10.1111/j.0014-3820.2001.tb00732.x. PMID 11794777.

82. Griffiths, Anthony J. F.; Miller, Jeffrey H.; Suzuki, David T.; Lewontin, Richard C.; Gelbart, eds. (2000). "Random events". An Introduction to Genetic Analysis (7th ed.). New York: W. H. Freeman. ISBN 0-7167-3520-2.

83. Darwin, Charles (1859). On the Origin of Species (1st ed.). London: John Murray. p. 1. ISBN 0-8014-1319-2.

84. Earlier related ideas were acknowledged in Darwin, Charles (1861). On the Origin of Species (3rd ed.). London: John Murray. xiii. ISBN 0-8014-1319-2.

85. Gavrilets, S (2003). "Perspective: Models of Speciation: What Have We Learned in 40 Years?". Evolution. 57 (10): 2197–215. doi:10.1554/02-727. PMID 14628909.

86. Wolf, YI; Rogozin, IB; Grishin, NV; Koonin, EV (2002). "Genome trees and the tree of life". Trends in Genetics. 18 (9): 472–9. doi:10.1016/S0168-9525(02)02744-0. PMID 12175808.

87. "The Use of Model Organisms in Instruction". University of Wisconsin: Wisconsin Outreach Research Modules. Retrieved 15 March 2008.

88. "NCBI: Genes and Disease". NIH: National Center for Biotechnology Information. Retrieved 15 March 2008.

89. Davey Smith, G; Ebrahim, S (2003). "›Mendelian randomization›: Can genetic epidemiology contribute to understanding environmental determinants of disease?". International Journal of Epidemiology. 32 (1): 1–22. doi:10.1093/ije/dyg070. PMID 12689998.

90. "Pharmacogenetics Fact Sheet". NIH: National Institute of General Medical Sciences. Retrieved 15 March 2008.

91. Frank, SA (2004). "Genetic predisposition to cancer – insights from population genetics". Nature Reviews Genetics. 5 (10): 764–72. doi:10.1038/nrg1450. PMID 15510167.

92. Strachan T, Read AP (1999). Human Molecular Genetics 2 (second ed.). John

Wiley & Sons Inc. Chapter 18: Cancer Genetics

93. Lodish et al. (2000), Chapter 7: 7.1. DNA Cloning with Plasmid Vectors
94. Lodish et al. (2000), Chapter 7: 7.7. Polymerase Chain Reaction: An Alternative to Cloning
95. Brown TA (2002). "Section 2, Chapter 6: 6.1. The Methodology for DNA Sequencing". Genomes 2 (2nd ed.). Oxford: Bios. ISBN 1-85996-228-9.
96. Brown (2002), Section 2, Chapter 6: 6.2. Assembly of a Contiguous DNA Sequence
97. Service, RF (2006). "The race for the $1000 genome". Science. 311 (5767): 1544–6. doi:10.1126/science.311.5767.1544. PMID 16543431.
98. Hall, Nell (May 2007). "Advanced sequencing technologies and their wider impact in microbiology". J. Exp. Biol. 209 (Pt 9): 1518–1525. doi:10.1242/jeb.001370. PMID 17449817.
99. Church, George M. (January 2006). "Genomes for all". Sci. Am. 294 (1): 46–54. doi:10.1038/scientificamerican0106-46. PMID 16468433.(subscription required)
100. Wade, Nicholas (19 March 2015). "Scientists Seek Ban on Method of Editing the Human Genome". New York Times. Retrieved 20 March 2015.
101. Pollack, Andrew (3 March 2015). "A Powerful New Way to Edit DNA". New York Times. Retrieved 20 March 2015.
102. Baltimore, David; Berg, Paul; Botchan, Dana; Charo, R. Alta; Church, George; Corn, Jacob E.; Daley, George Q.; Doudna, Jennifer A.; Fenner, Marsha; Greely, Henry T.; Jinek, Martin; Martin, G. Steven; Penhoet, Edward; Puck, Jennifer; Sternberg, Samuel H.; Weissman, Jonathan S.; Yamamoto, Keith R. (19 March 2015). "A prudent path forward for genomic engineering and germline gene modification". Science. 348: 36–8. Bibcode:2015Sci...348...36B. doi:10.1126/science.aab1028. PMC 4394183 . PMID 25791083. Retrieved 20 March 2015.
103. Lanphier, Edward; Urnov, Fyodor; Haecker, Sarah Ehlen; Werner, Michael; Smolenski, Joanna (26 March 2015). "Don›t edit the human germ line". Nature. 519: 410–411. Bibcode:2015Natur.519..410L. doi:10.1038/519410a. PMID 25810189. Retrieved 20 March 2015.
104. Kolata, Gina (23 April 2015). "Chinese Scientists Edit Genes of Human Embryos, Raising Concerns". New York Times. Retrieved 24 April 2015.
105. Liang, Puping; et al. (18 April 2015). "CRISPR/Cas9-mediated gene editing in human tripronuclear zygotes". Protein & Cell. 6: 363–72. doi:10.1007/s13238-015-0153-5. PMC 4417674 . PMID 25894090. Retrieved 24 April 2015.

CHAPTER 5

Genetic Differences Within and Between Populations

Chapter Contents

- **Introduction**

Darwinian population genetics, or simply population genetics, is a central feature, if not the central feature, of the modern evolutionary synthesis, or neo-Darwinism. The modern evolutionary synthesis is a combination of Darwin's theory of the evolution of species by natural selection, Mendel's theory of genetics as the basis for biological inheritance, and mathematical population genetics. Put together by dozens of scientists throughout the 1930s and 1940s, Darwinian population genetics is our best model of the process that incrementally created all life on Earth, evolution and natural selection.

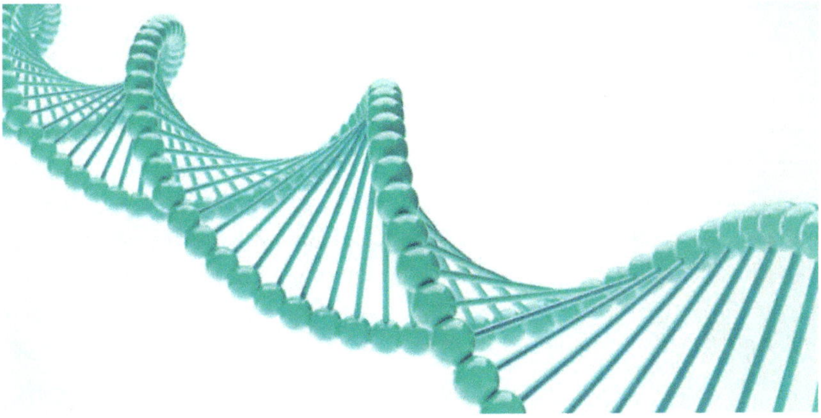

INTRODUCTION

Population genetics is a subfield of genetics that deals with genetic differences within and between populations, and is a part of evolutionary biology. Studies in this branch of biology examine such phenomena as adaptation, speciation, and population structure.

Population genetics was a vital ingredient in the emergence of the modern evolutionary synthesis. Its primary founders were Sewall Wright, J. B. S. Haldane and Ronald Fisher, who also laid the foundations for the related discipline of quantitative genetics. Traditionally a highly mathematical discipline, modern population genetics encompasses theoretical, lab, and field work. Population genetic models are used both for statistical inference from DNA sequence data and for proof/disproof of concept.

What sets population genetics apart today from newer, more phenotypic approaches to modelling evolution, such as evolutionary game theory and adaptive dynamics, is its emphasis on genetic phenomena as dominance, epistasis, and the degree to which genetic recombination breaks up linkage disequilibrium. This makes it appropriate for comparison to population genomics data.

History

Ronald Fisher

Population genetics began as a reconciliation of Mendelian inheritance and biostatistics models. Natural selection will only cause evolution if there is enough genetic variation in a population. Before the discovery of Mendelian genetics, one common hypothesis was blending inheritance. But with blending inheritance, genetic variance would be rapidly lost, making evolution by natural or sexual selection implausible. The Hardy–Weinberg principle provides the solution to how variation is maintained in a population with Mendelian inheritance. According to this principle, the frequencies of alleles (variations in a gene) will remain constant in the absence of selection, mutation, migration and genetic drift.

The next key step was the work of the British biologist and statistician Ronald Fisher. In a series of papers starting in 1918 and culminating in his 1930 book *The Genetical Theory of Natural Selection*, Fisher showed that the continuous variation measured by the biometricians could be produced by the combined action of many discrete genes, and that natural selection could change allele frequencies in a population, resulting in evolution. In a series of papers beginning in 1924, another British geneticist, J.B.S. Haldane worked out the mathematics of allele frequency change at a single gene locus under a broad range of conditions. Haldane also applied statistical analysis to real-world examples of natural selection, such as the Peppered moth evolution and industrial melanism, and showed that selection coefficients could be larger than Fisher assumed, leading to more rapid adaptive evolution as a camouflage strategy following increased pollution.

J.B.S.Haldane

The American biologist Sewall Wright, who had a background in animal breeding experiments, focused on combinations of interacting genes, and the effects of inbreeding on small, relatively isolated populations that exhibited genetic drift. In 1932, Wright introduced the concept of an adaptive landscape and argued that genetic drift and inbreeding could drive a small, isolated sub-population away from an adaptive peak, allowing natural selection to drive it towards different adaptive peaks.

The work of Fisher, Haldane and Wright founded the discipline of population genetics. This integrated natural selection with Mendelian genetics, which was the critical first step in developing a unified theory of how evolution worked. John Maynard Smith was Haldane's pupil, whilst W.D. Hamilton was heavily influenced by the writings of Fisher. The American George R. Price worked with both Hamilton and Maynard Smith. American Richard Lewontin and Japanese Motoo Kimura were heavily influenced by Wright.

Modern Evolutionary Synthesis

The mathematics of population genetics were originally developed as the beginning of the modern evolutionary synthesis. According to Beatty,population genetics defines the core of the modern synthesis. In the first few decades of the 20th century, most field naturalists continued to believe that Lamarckian and orthogenic mechanisms of evolution provided the best explanation for the complexity they observed in the living world. However, as the field of genetics continued to develop, those views became less tenable. During the modern evolutionary synthesis, these ideas were purged, and only evolutionary causes that could be expressed in the mathematical framework of population genetics were retained. Consensus was reached as to which evolutionary factors might influence evolution, but not as to the relative importance of the various factors.

Theodosius Dobzhansky, a postdoctoral worker in T. H. Morgan›s lab, had been influenced by the work on genetic diversity by Russian geneticists such as Sergei Chetverikov. He helped to bridge the divide between the foundations of microevolution developed by the population geneticists and the patterns of macroevolution observed by field biologists, with his 1937 book *Genetics and the Origin of Species*. Dobzhansky examined the genetic diversity of wild populations and showed that, contrary to the assumptions of the population geneticists, these populations had large amounts of genetic diversity, with marked differences between sub-populations. The book also took the highly mathematical work of the population geneticists and put it into a more accessible form. Many more biologists were influenced by population genetics via Dobzhansky than were able to read the highly mathematical works in the original.

In Great Britain E.B. Ford, the pioneer of ecological genetics, continued throughout the 1930s and 1940s to empirically demonstrate the power of selection due to ecological factors including the ability to maintain genetic diversity through genetic polymorphisms such as human blood types. Ford›s work, in collaboration with Fisher, contributed to a shift in emphasis during the course of the modern synthesis towards natural selection as the dominant force. During the modern evolutionary synthesis, these ideas were purged, and only evolutionary causes that could be expressed in the mathematical framework of population genetics were retained.

Neutral Theory and Origin-Fixation Dynamics

The original, modern synthesis view of population genetics assumes that mutations provide ample raw material, and focuses only on the change in frequency of alleles within populations. The main processes influencing allele frequencies are natural selection, genetic drift, gene flow and recurrent mutation. Fisher and Wright had some fundamental disagreements about the relative roles of selection and drift.

The availability of molecular data on all genetic differences led to the neutral theory of molecular evolution. In this view, many mutations are deleterious and so never observed, and most of the remainder are neutral, i.e. are not under selection. With the fate of each neutral mutation left to chance (genetic drift), the direction of evolutionary change is driven by which mutations occur, and so cannot be captured by models of change in the frequency of (existing) alleles alone.

The origin-fixation view of population genetics generalizes this approach beyond strictly neutral mutations, and sees the rate at which a particular change happens as the product of the mutation rate and the fixation probability.

Four Processes

Selection

Natural selection, which includes sexual selection, is the fact that some traits make it more likely for an organism to survive and reproduce. Population genetics describes natural selection by defining fitness as a propensity or probability of survival and reproduction in a particular environment. The fitness is normally given by the symbol **w**=1-**s** where **s** is the selection coefficient. Natural selection acts on phenotypes, so population genetic models assume relatively simple relationships to predict the phenotype and hence fitness from the allele at one or a small number of loci. In this way, natural selection converts differences in the fitness of individuals with different phenotypes into changes in allele frequency in a population over successive generations.

Before the advent of population genetics, many biologists doubted that small differences in fitness were sufficient to make a large difference to evolution. Population geneticists addressed this concern in part by comparing selection to genetic drift. Selection can overcome genetic drift when **s** is greater than 1 divided by the effective population size. When this criterion is met, the probability that a new advantageous mutant becomes fixed is approximately equal to **2s**. The time until fixation of such an allele depends little on genetic drift, and is approximately proportional to log(sN)/s.

Dominance

Dominance means that the phenotypic and/or fitness effect of one allele at a locus depends on which allele is present in the second copy for that locus. Consider three genotypes at one locus, with the following fitness values

Genotype:	A_1A_1	A_1A_2	A_2A_2
Relative fitness:	1	1-hs	1-s

s is the selection coefficient and h is the dominance coefficient. The value of h yields the following information:

h=0	A_1 dominant, A_2 recessive
h=1	A_2 dominant, A_1 recessive
0<h<1	incomplete dominance
h<0	overdominance
h>1	Underdominance

Epistasis

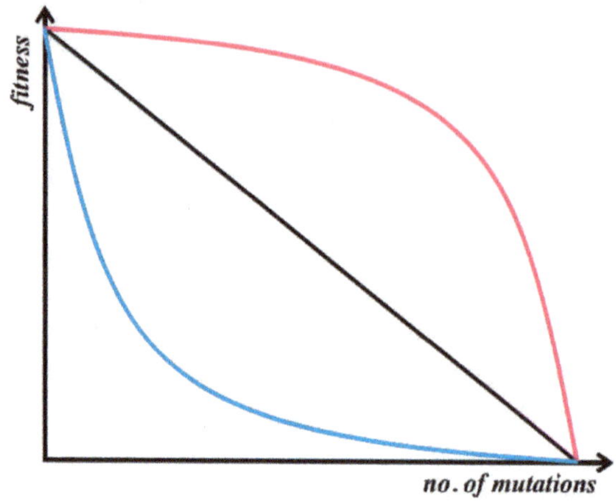

The logarithm of fitness as a function of the number of deleterious mutations. Synergistic epistasis is represented by the red line - each subsequent deleterious mutation has a larger proportionate effect on the organism's fitness. Antagonistic epistasis is in blue. The black line shows the non-epistatic case, where fitness is the product of the contributions from each of its loci.

Epistasis means that the phenotypic and/or fitness effect of an allele at one locus depends on which alleles are present at other loci. Selection does not act on a single locus, but on a phenotype that arises through development from a complete genotype. Many population genetics models of sexual species are "single locus" models, where the fitness of an individual is calculated as the product of the contributions from each of its loci - this assumes that there is no epistasis.

In fact, the genotype to fitness landscape is more complex. Population genetics must either model this complexity in detail, or capture it by some simpler average rule. Empirically, beneficial mutations tend to have a smaller fitness benefit when added to a genetic background that already has high fitness: this is known as diminishing returns epistasis. When deleterious mutations also have a smaller fitness effect on high fitness backgrounds, this is known as synergistic epistasis. However, the effect of deleterious mutations tends on average to be very close to multiplicative, or can even show the opposite pattern, known as antagonistic epistasis.

Synergistic epistasis is central to some theories of the purging of mutation load and to the evolution of sexual reproduction.

Mutation

Mutation is the ultimate source of genetic variation in the form of new alleles. In addition, mutation may influence the direction of evolution when there is mutation bias, i.e. different probabilities for different mutations to occur. For example, recurrent mutation that tends to be in the opposite direction to selection can lead to mutation-selection balance. At the molecular level, if mutation from G to A happens more often than mutation from A to G, then genotypes with A will tend to evolve. Different insertion vs. deletion mutation biases in different taxa can lead to the evolution of different genome sizes.Developmental or mutational biases have also been observed in morphological evolution. For example, according to the phenotype-first theory of evolution, mutations can eventually cause the genetic assimilation of traits that were previously induced by the environment.

Mutation bias effects are superimposed on other processes. If selection would favor either one out of two mutations, but there is no extra advantage to having both, then the mutation that occurs the most frequently is the one that is most likely to become fixed in a population.

Mutation can have no effect, alter the product of a gene, or prevent the gene from functioning. Studies in the fly *Drosophila melanogaster* suggest that if a mutation changes a protein produced by a gene, this will probably be harmful, with about 70 percent of these mutations having damaging effects, and the remainder being either neutral or weakly beneficial. Most loss of function mutations are selected against. But when selection is weak, mutation bias towards loss of function can affect evolution. For example, pigments are no longer useful when animals live in the darkness of caves, and tend to be lost. This kind of loss of function can occur because of mutation bias, and/or because the function had a cost, and once the benefit of the function disappeared, natural selection leads to the loss. Loss of sporulation ability in a bacterium during laboratory evolution appears to have been caused by mutation bias, rather than natural selection against the cost of maintaining sporulation ability. When there is no selection for loss of function, the speed at which loss evolves depends more on the mutation rate than it does on the effective population size, indicating that it is driven more by mutation bias than by genetic drift.

Mutations can involve large sections of DNA becoming duplicated, usually through genetic recombination. This leads to copy-number variation within a population. Duplications are a major source of raw material for evolving new genes. Other types of mutation occasionally create new genes from previously noncoding DNA.

Genetic Drift

Genetic drift is a change in allele frequencies caused by random sampling. That is, the alleles in the offspring are a random sample of those in the parents. Genetic drift may cause gene variants to disappear completely, and thereby reduce genetic variability. In contrast to natural selection, which makes gene variants more common or less common depending on their reproductive success, the changes due to genetic drift are not driven by environmental or adaptive pressures, and are equally likely to make an allele more common as less common.

The effect of genetic drift is larger for alleles present in few copies than when an allele is present in many copies. The population genetics of genetic drift are described using either branching processes or a diffusion equation describing changes in allele frequency. These approaches are usually applied to the Wright-Fisher and Moran models of population genetics. Assuming genetic drift is the only evolutionary force acting on an allele, after t generations in many replicated populations, starting with allele frequencies of p and q, the variance in allele frequency across those populations is

$$V_t \approx pq\left(1 - \exp\left\{-\frac{t}{2N_e}\right\}\right)$$

Ronald Fisher held the view that genetic drift plays at the most a minor role in evolution, and this remained the dominant view for several decades. No population genetics perspective have ever given genetic drift a central role by itself, but some have made genetic drift important in combination with another non-selective force. The shifting balance theory of Sewall Wright held that the combination of population structure and genetic drift was important. Motoo Kimura›s neutral theory of molecular evolution claims that most genetic differences within and between populations are caused by the combination of neutral mutations and genetic drift.

The role of genetic drift by means of sampling error in evolution has been criticized by John H Gillespie and Will Provine, who argue that selection on linked sites is a more important stochastic force, doing the work traditionally ascribed to genetic drift by means of sampling error. The mathematical properties of genetic draft are different from those of genetic drift. The direction of the random change in allele frequency is autocorrelated across generations.

Gene Flow

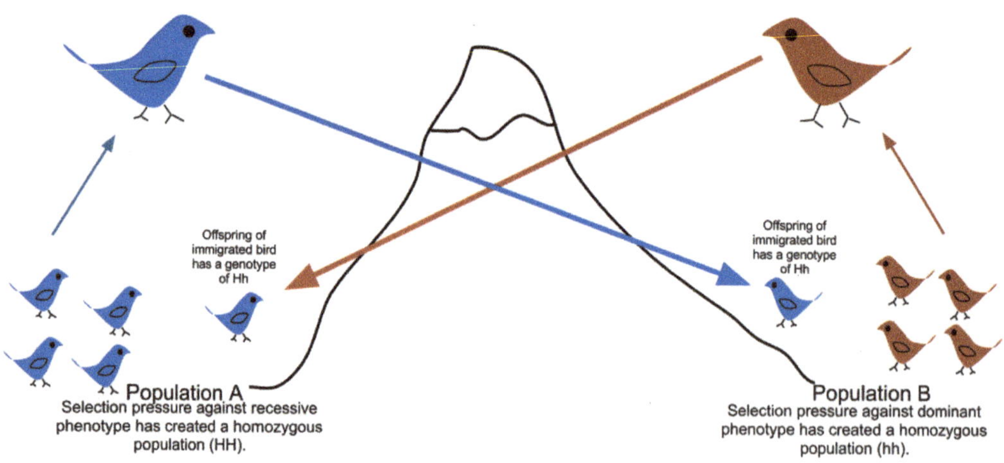

Gene flow is the transfer of alleles from one population to another population through immigration of individuals. In this example, one of the birds from population A immigrates to population B, which has fewer of the dominant alleles, and through mating incorporates its alleles into the other population.

The Great Wall in 1907

Because of physical barriers to migration, along with limited tendency for individuals to move or spread (vagility), and tendency to remain or come back to natal place (philopatry), natural populations rarely all interbreed as convenient in theoretical random models (panmixy). There is usually a geographic range within which individuals are more closely related to one another than those randomly selected from the general population. This is described as the extent to which a population is genetically structured.Genetic structuring can be caused by migration due to historical climate change, species range expansion or current availability of habitat. Gene flow is hindered by mountain ranges, oceans and deserts or even man-made structures such as the Great Wall of China, which has hindered the flow of plant genes.

Gene flow is the exchange of genes between populations or species, breaking down the structure. Examples of gene flow within a species include the migration and then breeding of organisms, or the exchange of pollen. Gene transfer between species includes the formation of hybrid organisms and horizontal gene transfer. Population genetic models can be used to identify which populations show significant genetic isolation from one another, and to reconstruct their history.

Subjecting a population to isolation leads to inbreeding depression. Migration into a population can introduce new genetic variants. If a significant proportion of individuals or gametes migrate, it can also change allele frequencies, e.g. giving rise to migration load.

In the presence of gene flow, other barriers to hybridization between two diverging populations of an outcrossing species are required for the populations to become new species.

Horizontal Gene Transfer

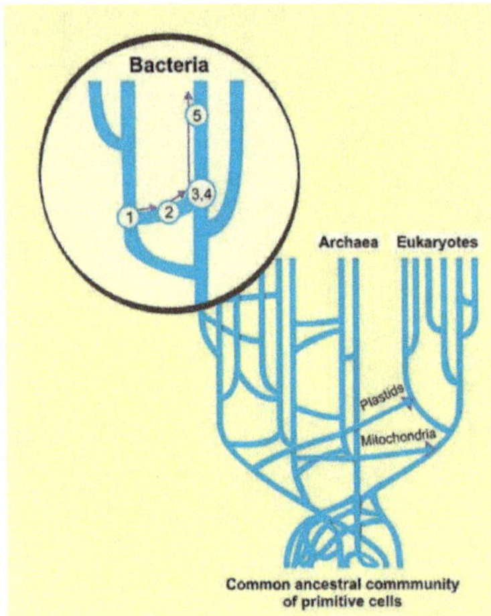

Current tree of life showing vertical and horizontal gene transfers.

Horizontal gene transfer is the transfer of genetic material from one organism to another organism that is not its offspring; this is most common among prokaryotes. In medicine, this contributes to the spread of antibiotic resistance, as when one bacteria acquires resistance genes it can rapidly transfer them to other species. Horizontal transfer of genes from bacteria to eukaryotes such as the yeast *Saccharomyces cerevisiae* and the adzuki bean beetle *Callosobruchus chinensis* may also have occurred. An example of larger-scale transfers are the eukaryotic bdelloid rotifers, which appear to have received a range of genes from bacteria, fungi, and plants. Viruses can also carry DNA between organisms, allowing transfer of genes even across biological domains. Large-scale gene transfer has also occurred between the ancestors of eukaryotic cells and prokaryotes, during the acquisition of chloroplasts and mitochondria.

Linkage

If all genes are in linkage equilibrium, the effect of an allele at one locus can be averaged across the gene pool at other loci. In reality, one allele is frequently found in linkage disequilibrium with genes at other loci, especially with genes located nearby on the same chromosome. Recombination breaks up this linkage disequilibrium too slowly to avoid genetic hitchhiking, where an allele at one locus rises to high frequency because it is linked to an allele under selection at a nearby locus. Linkage also slows down the rate of adaptation, even in sexual populations. The effect of linkage disequilibrium in slowing down the rate of adaptive evolution arises from a combination of the Hill-Robertson effect (delays in bringing beneficial mutations together) and background selection (delays in separating beneficial mutations from deleterious hitchhikers).

Linkage is a problem for population genetic models that treat one gene locus at a time. It can, however, be exploited as a method for detecting the action of natural selection via selective sweeps.

In the extreme case of an asexual population, linkage is complete, and population genetic equations can be derived and solved in terms of a travelling wave of genotype frequencies along a simple fitness landscape. Most microbes, such as bacteria, are asexual. The population genetics of their adaptation have two contrasting regimes. When the product of the beneficial mutation rate and population size is small, asexual populations follow a "successional regime" of origin-fixation dynamics, with adaptation rate strongly dependent on this product. When the product is much larger, asexual populations follow a concurrent mutations regime with adaptation rate less dependent on the product, characterized by clonal interference and the appearance of a new beneficial mutation before the last one has fixed.

Applications

Explaining Levels of Genetic Variation

Neutral theory predicts that the level of genetic diversity in a population will be proportional to the product of the population size and the neutral mutation rate. The fact that levels of genetic diversity vary much less than population sizes do is known as the "paradox of variation". While high levels of genetic diversity were one of the original arguments in favor of neutral theory, the paradox of variation has been one of the strongest arguments against neutral theory.

It is clear that levels of genetic diversity vary greatly within a species as a function of local recombination rate, due to both genetic hitchhiking and background selection. Most current solutions to the paradox of variation invoke some level of selection at linked sites. For example, one analysis suggests that larger populations have more selective sweeps, which remove more neutral genetic diversity. A negative correlation between mutation rate and population size may also contribute.

Life history affects genetic diversity more than population history does, e.g. r-strategists have more genetic diversity.

Detecting Selection

Population genetics models are used to infer which genes are undergoing selection. One common approach is to look for regions of high linkage disequilibrium and low genetic variance along the chromosome, to detect recent selective sweeps.

A second common approach is the McDonald–Kreitman test. The McDonald–Kreitman test compares the amount of variation within a species (polymorphism) to the divergence between species (substitutions) at two types of sites, one assumed to be neutral. Typically, synonymous sites are assumed to be neutral. Genes undergoing positive selection have an excess of divergent sites relative to polymorphic sites. The test can also be use to obtain a genome-wide estimate of the proportion of substitutions that are fixed by positive selection, α. According to the neutral theory of molecular evolution, this number should be near zero. High numbers have therefore been interpreted as a genome-wide falsification of neutral theory.

Demographic Inference

The simplest test for population structure in a sexually reproducing, diploid species, is to see whether genotype frequencies follow Hardy-Weinberg proportions as a function of allele frequencies. For example, in the simplest case of a single locus with two alleles denoted **A** and **a** at frequencies p and q, random mating predicts freq(**AA**) = p^2 for the **AA** homozygotes , freq(**aa**) = q^2 for the **aa** homozygotes, and freq(**Aa**) = $2pq$ for the heterozygotes. In the absence of population structure, Hardy-Weinberg proportions are reached within 1-2 generations of random mating. More typically, there is an excess of homozygotes, indicative of population structure. The extent of this excess can be quantified as the inbreeding coefficient, F. When individuals can be assigned to different subpopulations, the degree of population structure is usually calculated using F_{ST}, which is a measure of the proportion of genetic variance that can be explained by population structure.

Coalescent theory relates genetic diversity in a sample to demographic history of the population from which it was taken. It normally assumes neutrality, and so sequences from more neutrally-evolving portions of genomes are therefore selected for such analyses. It can be used to infer both the relationships between species (phylogenetics) and the population structure, demographic history (e.g. population bottlenecks or population growth), and introgression within a species.

Another approach to demographic inference relies on the allele frequency spectrum.

Evolution of Genetic Systems

By assuming that there are loci that control the genetic system itself, population genetic models are created to describe the evolution of dominance and other forms of robustness, the evolution of sexual reproduction and recombination rates, the evolution of mutation rates, the evolution of evolutionary capacitors, the evolution of costly signalling traits, the evolution of ageing, and the evolution of co-operation. For example, most mutations are deleterious, so the optimal mutation rate for a species may be a trade-off between the damage from a high deleterious mutation rate and the metabolic costs of maintaining systems to reduce the mutation rate, such as DNA repair enzymes.

One important aspect of such models is that selection is only strong enough to purge deleterious mutations and hence overpower mutational bias towards degradation if the selection coefficient s is greater than the inverse of the effective population size. This is known as the drift barrier and is related to the nearly neutral theory of molecular evolution. Drift barrier theory predicts that species with large effective population sizes will have highly streamlined, efficient genetic systems, while those with small population sizes will have bloated and complex genomes containing for example introns and transposable elements. However, somewhat paradoxically, species with large population sizes might be so tolerant to the consequences of certain types of errors that they evolve higher error rates, e.g. in transcription and translation, than small populations.

REFERENCES

1. Servedio, Maria R.; Brandvain, Yaniv; Dhole, Sumit; Fitzpatrick, Courtney L.; Goldberg, Emma E.; Stern, Caitlin A.; Van Cleve, Jeremy; Yeh, D. Justin (9 December 2014). "Not Just a Theory—The Utility of Mathematical Models in Evolutionary Biology". PLoS Biology. 12 (12): e1002017. doi:10.1371/journal.pbio.1002017.
2. Ewens W.J. (2004). Mathematical Population Genetics (2nd Edition). Springer-Verlag, New York. ISBN 0-387-20191-2.
3. Bowler, Peter J. (2003). Evolution : the history of an idea (3rd ed.). Berkeley: University of California Press. pp. 325–339. ISBN 978-0-520-23693-6.
4. Larson, Edward J. (2004). Evolution : the remarkable history of a scientific theory (Modern Library ed.). New York: Modern Library. pp. 221–243. ISBN 978-0-679-64288-6.
5. Beatty, John. "The Synthesis and the Synthetic Theory". Integrating Scientific Disciplines. Springer Netherlands. pp. 125–135. ISBN 9789024733422.
6. Mayr, edited by Ernst; Mayer, William B. Provine ; with a new preface by Ernst (1998). The Evolutionary synthesis : perspectives on the unification of biology ([New ed]. ed.). Cambridge, Mass.: Harvard University Press. pp. 295–298. ISBN 9780674272262.
7. Provine, W. B. (1988). "Progress in evolution and meaning in life". Evolutionary progress. University of Chicago Press. pp. 49–79.
8. Provine, William B. (1978). "The role of mathematical population geneticists in the evolutionary synthesis of the 1930s and 1940s". Studies of the History of Biology. 2: 167–192.
9. Mayr, Ernst (1988). Toward a New Philosophy of Biology: Observations of an Evolutionist. Cambridge, MA: Belknap Press of Harvard University Press. p. 402. ISBN 0-674-89665-3.
10. Mayr, edited by Ernst; Mayer, William B. Provine ; with a new preface by Ernst (1998). The Evolutionary synthesis : perspectives on the unification of biology ([New ed]. ed.). Cambridge, Mass.: Harvard University Press. pp. 338–341. ISBN 9780674272262.
11. McCandlish, David M.; Stoltzfus, Arlin (September 2014). "Modeling Evolution Using the Probability of Fixation: History and Implications". The Quarterly Review of Biology. 89 (3): 225–252. doi:10.1086/677571.
12. Wright and Fisher on Inbreeding and Random Drift by James F. Crow, published in Genetics Published 2010
13. JBS Haldane (1927). "A Mathematical Theory of Natural and Artificial Selection, Part V: Selection and Mutation". Mathematical Proceedings of the Cambridge Philosophical Society. 23 (7): 838–844. Bibcode:1927PCPS...23..838H. doi:10.1017/S0305004100015644.

14. Orr, H. A. (2010). "The population genetics of beneficial mutations". Philosophical Transactions of the Royal Society B: Biological Sciences. 365 (1544): 1195–1201. doi:10.1098/rstb.2009.0282.
15. Hermisson J, Pennings PS; Pennings (2005). "Soft sweeps: molecular population genetics of adaptation from standing genetic variation". Genetics. 169 (4): 2335–2352. doi:10.1534/genetics.104.036947. PMC 1449620 . PMID 15716498.
16. Gillespie, John (2004). Population Genetics: A Concise Guide, Second Edition. Johns Hopkins University Press. ISBN 0-8018-8008-4.
17. Berger, D.; Postma, E. (13 October 2014). "Biased Estimates of Diminishing-Returns Epistasis? Empirical Evidence Revisited". Genetics. 198 (4): 1417–1420. doi:10.1534/genetics.114.169870.
18. Kouyos, Roger D.; Silander, Olin K.; Bonhoeffer, Sebastian (June 2007). "Epistasis between deleterious mutations and the evolution of recombination". Trends in Ecology & Evolution. 22 (6): 308–315. doi:10.1016/j.tree.2007.02.014.
19. Crow, JF (5 August 1997). "The high spontaneous mutation rate: is it a health risk?". Proceedings of the National Academy of Sciences of the United States of America. 94 (16): 8380–6. doi:10.1073/pnas.94.16.8380. PMC 33757 . PMID 9237985.
20. Smith N.G.C., Webster M.T., Ellegren, H.; Webster; Ellegren (2002). "Deterministic Mutation Rate Variation in the Human Genome". Genome Research. 12 (9): 1350–1356. doi:10.1101/gr.220502. PMC 186654 . PMID 12213772.
21. Petrov DA, Sangster TA, Johnston JS, Hartl DL, Shaw KL; Sangster; Johnston; Hartl; Shaw (2000). "Evidence for DNA loss as a determinant of genome size". Science. 287 (5455): 1060–1062. Bibcode:2000Sci...287.1060P. doi:10.1126/science.287.5455.1060. PMID 10669421.
22. Petrov DA (2002). "DNA loss and evolution of genome size in Drosophila". Genetica. 115 (1): 81–91. doi:10.1023/A:1016076215168. PMID 12188050.
23. Kiontke K, Barriere A , Kolotuev I, Podbilewicz B , Sommer R, Fitch DHA , Felix MA; Barrière; Kolotuev; Podbilewicz; Sommer; Fitch; Félix (2007). "Trends, stasis, and drift in the evolution of nematode vulva development". Current Biology. 17 (22): 1925–1937. doi:10.1016/j.cub.2007.10.061. PMID 18024125.
24. Braendle C, Baer CF, Felix MA; Baer; Félix (2010). Barsh, Gregory S, ed. "Bias and Evolution of the Mutationally Accessible Phenotypic Space in a Developmental System". PLoS Genetics. 6 (3): e1000877. doi:10.1371/journal.pgen.1000877. PMC 2837400 . PMID 20300655.
25. Palmer, RA (2004). "Symmetry breaking and the evolution of development". Science. 306 (5697): 828–833. Bibcode:2004Sci...306..828P. doi:10.1126/science.1103707. PMID 15514148.
26. West-Eberhard, M-J. (2003). Developmental plasticity and evolution. New York: Oxford University Press. ISBN 978-0-19-512235-0.
27. Stoltzfus, A & Yampolsky, L.Y. (2009). "Climbing Mount Probable: Mutation

as a Cause of Nonrandomness in Evolution". J Hered. 100 (5): 637–647. doi:10.1093/jhered/esp048. PMID 19625453.

28. Yampolsky, L.Y. & Stoltzfus, A (2001). "Bias in the introduction of variation as an orienting factor in evolution". Evol Dev. 3 (2): 73–83. doi:10.1046/j.1525-142x.2001.003002073.x. PMID 11341676.
29. Sawyer SA, Parsch J, Zhang Z, Hartl DL; Parsch; Zhang; Hartl (2007). "Prevalence of positive selection among nearly neutral amino acid replacements in Drosophila". Proc. Natl. Acad. Sci. U.S.A. 104 (16): 6504–10. Bibcode:2007PNAS..104.6504S. doi:10.1073/pnas.0701572104. PMC 1871816 . PMID 17409186.
30. Haldane, JBS (1933). "The Part Played by Recurrent Mutation in Evolution". American Naturalist. 67 (708): 5–19. doi:10.1086/280465. JSTOR 2457127.
31. Protas, Meredith; Conrad, M; Gross, JB; Tabin, C; Borowsky, R (2007). "Regressive evolution in the Mexican cave tetra, Astyanax mexicanus". Current Biology. 17 (5): 452–454. doi:10.1016/j.cub.2007.01.051. PMC 2570642 . PMID 17306543.
32. Maughan H, Masel J, Birky WC, Nicholson WL; Masel; Birky Jr; Nicholson (2007). "The roles of mutation accumulation and selection in loss of sporulation in experimental populations of Bacillus subtilis". Genetics. 177 (2): 937–948. doi:10.1534/genetics.107.075663. PMC 2034656 . PMID 17720926.
33. Masel J, King OD, Maughan H; King; Maughan (2007). "The loss of adaptive plasticity during long periods of environmental stasis". American Naturalist. 169 (1): 38–46. doi:10.1086/510212. PMC 1766558 . PMID 17206583.
34. Hastings, P J; Lupski, JR; Rosenberg, SM; Ira, G (2009). "Mechanisms of change in gene copy number". Nature Reviews Genetics. 10 (8): 551–564. doi:10.1038/nrg2593. PMC 2864001 . PMID 19597530.
35. Long M, Betrán E, Thornton K, Wang W; Betrán; Thornton; Wang (November 2003). "The origin of new genes: glimpses from the young and old". Nat. Rev. Genet. 4 (11): 865–75. doi:10.1038/nrg1204. PMID 14634634.
36. Liu N, Okamura K, Tyler DM; Okamura; Tyler; Phillips; Chung; Lai (2008). "The evolution and functional diversification of animal microRNA genes". Cell Res. 18 (10): 985–96. doi:10.1038/cr.2008.278. PMC 2712117 . PMID 18711447.
37. McLysaght, Aoife; Hurst, Laurence D. (25 July 2016). "Open questions in the study of de novo genes: what, how and why". Nature Reviews Genetics. 17 (9): 567–578. doi:10.1038/nrg.2016.78.
38. Masel J (2011). "Genetic drift". Current Biology. 21 (20): R837–R838. doi:10.1016/j.cub.2011.08.007. PMID 22032182.
39. Futuyma, Douglas (1998). Evolutionary Biology. Sinauer Associates. p. Glossary. ISBN 0-87893-189-9.
40. Avers, Charlotte (1989). "Process and Pattern in Evolution". Oxford University Press.
41. Wahl L.M. (2011). "Fixation when N and s Vary: Classic Approaches Give Elegant New Results". Genetics. 188 (4): 783–785. doi:10.1534/genetics.111.131748.

PMC 3176088 . PMID 21828279.

42. Nicholas H. Barton; Derek E. G. Briggs; Jonathan A. Eisen; David B. Goldstein; Nipam H. Patel (2007). Evolution. Cold Spring Harbor Laboratory Press. p. 417. ISBN 0-87969-684-2.
43. Futuyma, Douglas (1998). Evolutionary Biology. Sinauer Associates. p. 320. ISBN 0-87893-189-9.
44. Gillespie JH (2000). "Genetic Drift in an Infinite Population: The Pseudohitchhiking Model". Genetics. 155 (2): 909–919. PMC 1461093 . PMID 10835409.
45. Provine, William B. The "Random Genetic Drift" Fallacy. CreateSpace.
46. Neher, Richard A.; Shraiman, Boris I. (August 2011). "Genetic Draft and Quasi-Neutrality in Large Facultatively Sexual Populations". Genetics. Bethesda, MD: Genetics Society of America. 188 (4): 975–996. doi:10.1534/genetics.111.128876. ISSN 0016-6731. PMC 3176096 . PMID 21625002.
47. Buston, PM; Pilkington, J. G.; et al. (2007). "Are clownfish groups composed of close relatives? An analysis of microsatellite DNA vraiation in Amphiprion percula". Molecular Ecology. 12 (3): 733–742. doi:10.1046/j.1365-294X.2003.01762.x. PMID 12675828.
48. Repaci, V; Stow AJ, Briscoe DA (2007). "Fine-scale genetic structure, co-founding and multiple mating in the Australian allodapine bee (Ramphocinclus brachyurus)". Journal of Zoology. 270 (4): 687–691. doi:10.1111/j.1469-7998.2006.00191.x.
49. Su H, Qu L, He K, Zhang Z, Wang J, Chen Z, Gu H; Qu; He; Zhang; Wang; Chen; Gu (2003). "The Great Wall of China: a physical barrier to gene flow?". Heredity. 90 (3): 212–9. doi:10.1038/sj.hdy.6800237. PMID 12634804.
50. Gravel, S., S. (2012). "Population Genetics Models of Local Ancestry". Genetics. 1202 (2): 4811. arXiv:1202.4811 . Bibcode:2012arXiv1202.4811G. doi:10.1534/genetics.112.139808.
51. Morjan C, Rieseberg L; Rieseberg (2004). "How species evolve collectively: implications of gene flow and selection for the spread of advantageous alleles". Mol. Ecol. 13 (6): 1341–56. doi:10.1111/j.1365-294X.2004.02164.x. PMC 2600545 . PMID 15140081.
52. Bolnick, Daniel I.; Nosil, Patrik (September 2007). "NATURAL SELECTION IN POPULATIONS SUBJECT TO A MIGRATION LOAD". Evolution. 61 (9): 2229–2243. doi:10.1111/j.1558-5646.2007.00179.x.
53. Boucher Y, Douady CJ, Papke RT, Walsh DA, Boudreau ME, Nesbo CL, Case RJ, Doolittle WF; Douady; Papke; Walsh; Boudreau; Nesbø; Case; Doolittle (2003). "Lateral gene transfer and the origins of prokaryotic groups". Annu Rev Genet. 37: 283–328. doi:10.1146/annurev.genet.37.050503.084247. PMID 14616063.
54. Walsh T (2006). "Combinatorial genetic evolution of multiresistance". Curr. Opin. Microbiol. 9 (5): 476–82. doi:10.1016/j.mib.2006.08.009. PMID 16942901.
55. Kondo N, Nikoh N, Ijichi N, Shimada M, Fukatsu T; Nikoh; Ijichi; Shimada; Fukatsu (2002). "Genome fragment of Wolbachia endosymbiont transferred to

X chromosome of host insect". Proc. Natl. Acad. Sci. U.S.A. 99 (22): 14280–5. Bibcode:2002PNAS...9914280K. doi:10.1073/pnas.222228199. PMC 137875 . PMID 12386340.

56. Sprague G (1991). "Genetic exchange between kingdoms". Curr. Opin. Genet. Dev. 1 (4): 530–3. doi:10.1016/S0959-437X(05)80203-5. PMID 1822285.
57. Gladyshev EA, Meselson M, Arkhipova IR; Meselson; Arkhipova (May 2008). "Massive horizontal gene transfer in bdelloid rotifers". Science. 320 (5880): 1210–3. Bibcode:2008Sci...320.1210G. doi:10.1126/science.1156407. PMID 18511688.
58. Baldo A, McClure M; McClure (1 September 1999). "Evolution and horizontal transfer of dUTPase-encoding genes in viruses and their hosts". J. Virol. 73 (9): 7710–21. PMC 104298 . PMID 10438861.
59. Poole A, Penny D; Penny (2007). "Evaluating hypotheses for the origin of eukaryotes". BioEssays. 29 (1): 74–84. doi:10.1002/bies.20516. PMID 17187354.
60. Weissman, D. B.; Hallatschek, O. (15 January 2014). "The Rate of Adaptation in Large Sexual Populations with Linear Chromosomes". Genetics. 196 (4): 1167–1183. doi:10.1534/genetics.113.160705.
61. Weissman, Daniel B.; Barton, Nicholas H.; McVean, Gil (7 June 2012). "Limits to the Rate of Adaptive Substitution in Sexual Populations". PLoS Genetics. 8 (6): e1002740. doi:10.1371/journal.pgen.1002740.
62. Neher, R. A.; Shraiman, B. I.; Fisher, D. S. (30 November 2009). "Rate of Adaptation in Large Sexual Populations". Genetics. 184 (2): 467–481. doi:10.1534/genetics.109.109009.
63. Michael M. Desai, Daniel S. Fisher; Fisher (2007). "Beneficial Mutation Selection Balance and the Effect of Linkage on Positive Selection". Genetics. 176 (3): 1759–1798. doi:10.1534/genetics.106.067678. PMC 1931526 . PMID 17483432.
64. Lewontin, [by] R. C. (1973). The genetic basis of evolutionary change ([4th printing.] ed.). New York: Columbia University Press. ISBN 978-0231033923.
65. Ellegren, Hans; Galtier, Nicolas (6 June 2016). "Determinants of genetic diversity". Nature Reviews Genetics. 17 (7): 422–433. doi:10.1038/nrg.2016.58.
66. Corbett-Detig, Russell B.; Hartl, Daniel L.; Sackton, Timothy B.; Barton, Nick H. (10 April 2015). "Natural Selection Constrains Neutral Diversity across A Wide Range of Species". PLOS Biology. 13 (4): e1002112. doi:10.1371/journal.pbio.1002112.
67. Sung, W.; Ackerman, M. S.; Miller, S. F.; Doak, T. G.; Lynch, M. (17 October 2012). "Drift-barrier hypothesis and mutation-rate evolution". Proceedings of the National Academy of Sciences. 109 (45): 18488–18492. doi:10.1073/pnas.1216223109.
68. Charlesworth, J. Eyre-Walker (2008). "The McDonald–Kreitman Test and Slightly Deleterious Mutations". Molecular Biology and Evolution. 25 (6): 1007–1015. doi:10.1093/molbev/msn005.

69. Eyre-Walker, A (2006). "The genomic rate of adaptive evolution"(PDF). Trends in Ecology and Evolution. 21 (10): 569–575.
70. Smith, N. G. C.; Eyre-Walker, A. (2002). "Adaptive protein evolution in Drosophila". Nature. 415 (6875): 1022–1024. doi:10.1038/4151022a. PMID 11875568.
71. Hahn, M.W. (2008). "Toward a selection theory of molecular evolution". Evolution. 62: 255–265. doi:10.1111/j.1558-5646.2007.00308.x. PMID 18302709.
72. Gutenkunst, Ryan N.; Hernandez, Ryan D.; Williamson, Scott H.; Bustamante, Carlos D.; McVean, Gil (23 October 2009). "Inferring the Joint Demographic History of Multiple Populations from Multidimensional SNP Frequency Data". PLoS Genetics. 5 (10): e1000695. doi:10.1371/journal.pgen.1000695.
73. Sniegowski P, Gerrish P, Johnson T, Shaver A; Gerrish; Johnson; Shaver (2000). "The evolution of mutation rates: separating causes from consequences". BioEssays. 22 (12): 1057–66. doi:10.1002/1521-1878(200012)22:12<1057::AID-BIES3>3.0.CO;2-W. PMID 11084621.
74. Lynch, Michael, John S. Conery; Conery (2003). "The origins of genome complexity". Science. 302 (5649): 1401–1404. Bibcode:2003Sci...302.1401L. doi:10.1126/science.1089370. PMID 14631042.
75. Rajon, E.; Masel, J. (3 January 2011). "Evolution of molecular error rates and the consequences for evolvability". Proceedings of the National Academy of Sciences. 108 (3): 1082–1087. doi:10.1073/pnas.1012918108. PMC 3024668 . PMID 21199946.

CHAPTER 6

Human Nutrition

Chapter Contents

- **Introduction**

In the Human Nutrition concentration students integrate chemistry, biology, physiology, psychology, sociology and nutrition to understand the impact of diet and lifestyles on human health. Students enter human services professions and health professional schools such as medical, dental and physical therapy.

INTRODUCTION

Human nutrition refers to the provision of essential nutrients necessary to support human life and health. Generally, people can survive up to 40 days without food, a period largely depending on the amount of water consumed, stored body fat, muscle mass and genetic factors.

Example of nutrients (in this case for magnesium)

Poor nutrition is a chronic problem often linked to poverty, poor nutrition understanding and practices, and deficient sanitation and food security. Lack of proper nutrition contributes to lower academic performance, lower test scores, and eventually less successful students and a less productive and competitive economy. Malnutrition and its consequences are immense contributors to deaths and disabilities worldwide. Promoting good nutrition helps children grow, promotes human development and advances economic growth and eradication of poverty.

Overview

The human body contains chemical compounds, such as water, carbohydrates (sugar, starch, and fiber), amino acids (in proteins), fatty acids (in lipids), and nucleic acids (DNA and RNA). These compounds consist of elements such as carbon, hydrogen, oxygen, nitrogen, phosphorus, calcium, iron, zinc, magnesium, manganese, and so on. All the chemical compounds and elements contained in the human body occur in various forms and combinations such as hormones, vitamins, phospholipids and hydroxyapatite. These compounds are found in the human body and in the different types of organisms that humans eat.

Any study done to determine nutritional status must take into account the state of the body before and after experiments, as well as the chemical composition of the whole diet and of all the materials excreted and eliminated from the body (including urine and feces). Comparing food to waste material can help determine the specific compounds and elements absorbed and metabolized by the body. The effects of nutrients may only be discernible over an extended period of time, during which all food and waste must be analyzed. The number of variables involved in such experiments is high, making nutritional studies time-consuming and expensive, which explains why the science of human nutrition is still slowly evolving.

Nutrients

The seven major classes of nutrients are: carbohydrates, fats, fiber, minerals, proteins, vitamins, and water. These nutrient classes are categorized as either macronutrients (needed in relatively large amounts) or micronutrients (needed in smaller quantities). The macronutrients are carbohydrates, fats, fiber, proteins, and water. The micronutrients are minerals and vitamins.

The macronutrients (excluding fiber and water) provide structural material (amino acids from which proteins are built, and lipids from which cell membranes and some signaling molecules are built), and energy. Some of the structural material can be used to generate energy internally, and in either case it is measured in Joules or kilocalories (often called "Calories" and written with a capital 'C' to distinguish them from little 'c' calories). Carbohydrates and proteins provide 17 kJ approximately (4 kcal) of energy per gram, while fats provide 37 kJ (9 kcal) per gram, though the net energy from either depends on such factors as absorption and digestive effort, which vary substantially from instance to instance.

Vitamins, minerals, fiber, and water do not provide energy, but are required for other reasons. A third class of dietary material, fiber (i.e., nondigestible material such as cellulose), seems also to be required, for both mechanical and biochemical reasons, though the exact reasons remain unclear. For all age groups, males need to consume higher amounts of macronutrients than females. In general, intakes increase with age until the second or third decade of life.

Molecules of carbohydrates and fats consist of carbon, hydrogen, and oxygen atoms. Carbohydrates range from simple monosaccharides (glucose, fructose, galactose) to complex polysaccharides (starch). Fats are triglycerides, made of assorted fatty acid monomers bound to a glycerol backbone. Some fatty acids, but not all, are essential in the diet: they cannot be synthesized in the body. Protein molecules contain nitrogen atoms in addition to carbon, oxygen, and hydrogen. The fundamental components of protein are nitrogen-containing amino acids, some of which are essential in the sense that humans cannot make them internally. Some of the amino acids are convertible (with the expenditure of energy) to glucose and can be used for energy production just as ordinary glucose. By breaking down existing protein, some glucose can be produced internally; the remaining amino acids are discarded, primarily as urea in urine. This occurs naturally when atrophy takes place, or during periods of starvation.

Carbohydrates

Grain products: rich sources of complex and simple carbohydrates

Carbohydrates may be classified as monosaccharides, disaccharides or polysaccharides depending on the number of monomer (sugar) units they contain. They are a diverse group of substances, with a range of chemical, physical and physiological properties. They make up a large part of foods such as rice, noodles, bread, and other grain-based products.

Monosaccharides contain one sugar unit, disaccharides two, and polysaccharides three or more. Monosaccharides include glucose, fructose and galactose. Disaccharides include sucrose, lactose, and maltose; purified sucrose, for instance, is used as table sugar. Polysaccharides, which include starch and glycogen, are often referred to as ‹complex› carbohydrates because they are typically long multiple-branched chains of sugar units. The difference is that complex carbohydrates take longer to digest and absorb since their sugar units must be separated from the chain before absorption. The spike in blood glucose levels after ingestion of simple sugars is thought to be related to some of the heart and vascular diseases, which have become more common in recent times. Simple sugars form a greater part of modern diets than in the past, perhaps leading to more cardiovascular disease. The degree of causation is still not clear.

Simple carbohydrates are absorbed quickly, and therefore raise blood-sugar levels more rapidly than other nutrients. However, the most important plant carbohydrate nutrient, starch, varies in its absorption. Gelatinized starch (starch heated for a few minutes in the presence of water) is far more digestible than plain starch, and starch which has been divided into fine particles is also more absorbable during digestion. The increased effort and decreased availability reduces the available energy from starchy foods substantially and can be seen experimentally in rats and anecdotally in humans. Additionally, up to a third of dietary starch may be unavailable due to mechanical or chemical difficulty.

Fat

A molecule of dietary fat typically consists of several fatty acids (containing long chains of carbon and hydrogen atoms), bonded to a glycerol. They are typically found as triglycerides (three fatty acids attached to one glycerol backbone). Fats may be classified as saturated or unsaturated depending on the detailed structure of the fatty acids involved. Saturated fats have all of the carbon atoms in their fatty acid chains bonded to hydrogen atoms, whereas unsaturated fats have some of these carbon atoms double-bonded, so their molecules have relatively fewer hydrogen atoms than a saturated fatty acid of the same length. Unsaturated fats may be further classified as monounsaturated (one double-bond) or polyunsaturated (many double-bonds). Furthermore, depending on the location of the double-bond in the fatty acid chain, unsaturated fatty acids are classified as omega-3 or omega-6 fatty acids. Trans fats are a type of unsaturated fat with *trans*-isomer bonds; these are rare in nature and in foods from natural sources; they are typically created in an industrial process called (partial) hydrogenation.

Many studies have shown that consumption of unsaturated fats, particularly monounsaturated fats, is associated with better health in humans. Saturated fats, typically from animal sources, are next in order of preference, while trans fats are associated with a variety of disease and should be avoided. Saturated and some trans fats are typically solid at room temperature (such as butter or lard), while unsaturated fats are typically liquids (such as olive oil or flaxseed oil). Trans fats are very rare in nature, but have properties useful in the food processing industry, such as rancidity resistance.

Most fatty acids are not essential, meaning the body can produce them as needed, generally from other fatty acids and always by expending energy to do so. However, in humans, at least two fatty acids are essential and must be included in the diet. An appropriate balance of essential fatty acids – omega-3 and omega-6 fatty acids – seems also important for health, though definitive experimental demonstration has been elusive. Both of these omega long-chain polyunsaturated fatty acids are substrates for a class of eicosanoids known as prostaglandins, which have roles throughout the human body. They are hormones, in some respects. The omega-3 eicosapentaenoic acid (EPA), which can be made in the human body from the omega-3 essential fatty acid alpha-linolenic acid (LNA), or taken in through marine food sources, serves as a building block for series 3 prostaglandins (e.g. weakly inflammatory PGE3). The omega-6 dihomo-gamma-linolenic acid (DGLA) serves as a building block for series 1 prostaglandins (e.g. anti-inflammatory PGE1), whereas arachidonic acid (AA) serves as a building block for series 2 prostaglandins (e.g., pro-inflammatory PGE 2). Both DGLA and AA can be made from the omega-6 linoleic acid (LA) in the human body, or can be taken in directly through food. An appropriately balanced intake of omega-3 and omega-6 partly determines the relative production of different prostaglandins: one reason a balance between omega-3 and omega-6 is believed important for cardiovascular health. In industrialized societies, people typically consume large amounts of processed vegetable oils, which have reduced amounts of the essential fatty acids along with too much of omega-6 fatty acids relative to omega-3 fatty acids.

Fiber

Dietary fiber is a carbohydrate, specifically a polysaccharide, which is incompletely absorbed in humans and in some animals. Like all carbohydrates, when it is metabolized, it can produce four Calories (kilocalories) of energy per gram, but in most circumstances, it accounts for less than that because of its limited absorption and digestibility. The two subcategories are insoluble and soluble fiber. Insoluble dietary fiber consists mainly of cellulose, a large carbohydrate polymer that is indigestible by humans, because humans do not have the required enzymes to break it down, and the human digestive system does not harbor enough of the types of microbes that can do so. Soluble dietary fiber comprises a variety of oligosaccharides, waxes, esters, resistant starches, and other carbohydrates that dissolve or gelatinize in water. Many of these soluble fibers can be fermented or partially fermented by microbes in the human digestive system to produce short-chain fatty acids which are absorbed and therefore introduce some caloric content.

Whole grains, beans and other legumes, fruits (especially plums, prunes, and figs), and vegetables are good sources of dietary fiber. Fiber is important to digestive health and is thought to reduce the risk of colon cancer. For mechanical reasons, fiber can help in alleviating both constipation and diarrhea. Fiber provides bulk to the intestinal contents, and insoluble fiber especially stimulates peristalsis – the rhythmic muscular contractions of the intestines which move digesta along the digestive tract. Some soluble fibers produce a solution of high viscosity; this is essentially a gel, which slows the movement of food through the intestines. Additionally, fiber, perhaps especially that from whole grains, may help lessen insulin spikes and reduce the risk of type 2 diabetes.

Protein

Proteins are the basis of many animal body structures (e.g. muscles, skin, and hair) and form the enzymes which catalyse chemical reactions throughout the body. Each protein molecule is composed of amino acids which contain nitrogen and sometimes sulphur (these components are responsible for the distinctive smell of burning protein, such as the keratin in hair). The body requires amino acids to produce new proteins (protein retention) and to replace damaged proteins (maintenance). Amino acids are soluble in the digestive juices within the small intestine, where they are absorbed into the blood. Once absorbed, they cannot be stored in the body, so they are either metabolized as required or excreted in the urine.

Proteins consist of amino acids in different proportions. The most important aspect and defining characteristic of protein from a nutritional standpoint is its amino acid composition. Amino acids which an animal cannot synthesize on its own from smaller molecules are deemed essential. The synthesis of some amino acids can be limited under special pathophysiological conditions, such as prematurity in the infant or individuals in severe catabolic distress, and they are called conditionally essential.

A vegetarian diet can adequately supply protein, support pregnancy, childhood and athletic endeavors, and lower the risk of cardiovascular disease and cancer.

Minerals

Dietary minerals are the chemical elements required by living organisms, other than the four elements carbon, hydrogen, nitrogen, and oxygen that are present in nearly all organic molecules. The term mineral is archaic, since the intent is to describe simply the less common elements in the diet. Some are heavier than the four just mentioned – including several metals, which often occur as ions in the body. Some dietitians recommend that these be supplied from foods in which they occur naturally, or at least as complex compounds, or sometimes even from natural inorganic sources (such as calcium carbonate from ground oyster shells). Some are absorbed much more readily in the ionic forms found in such sources. On the other hand, minerals are often artificially added to the diet as supplements; the most famous is likely iodine in iodized salt which prevents goiter.

Essential Dietary Minerals

Include the following:

- Chlorine as chloride ions; very common electrolyte; see sodium, below
- Magnesium, required for processing ATP and related reactions (builds bone, causes strong peristalsis, increases flexibility, increases alkalinity). Approximately 50% is in bone, the remaining 50% is almost all inside body cells, with only about 1% located in extracellular fluid. Food sources include oats, buckwheat, tofu, nuts, caviar, green leafy vegetables, legumes, and chocolate.
- Phosphorus, required component of bones; essential for energy processing Approximately 80% is found in inorganic portion of bones and teeth. Phosphorus is a component of every cell, as well as important metabolites, including DNA, RNA, ATP, and phospholipids. Also important in pH regulation. Food sources include cheese, egg yolk, milk, meat, fish, poultry, whole-grain cereals, and many others.
- Potassium, a very common electrolyte (heart and nerve health). With sodium, potassium is involved in maintaining normal water balance, osmotic equilibrium, and acid-base balance. In addition to calcium, it is important in the regulation of neuromuscular activity. Food sources include bananas, avocados, vegetables, potatoes, legumes, fish, and mushrooms.
- Sodium, a very common electrolyte; not generally found in dietary supplements, despite being needed in large quantities, because the ion is very common in food: typically as sodium chloride, or common salt

Trace Minerals

Many elements are required in smaller amounts (microgram quantities), usually because they play a catalytic role in enzymes. Some trace mineral elements (RDA < 200 mg/ day) are, in alphabetical order:

- Cobalt required for biosynthesis of vitamin B_{12} family of coenzymes
- Copper required component of many redox enzymes, including cytochrome c oxidase

- Chromium required for sugar metabolism
- Iodine required not only for the biosynthesis of thyroxin, but probably, for other important organs as breast, stomach, salivary glands, thymus etc. (see Iodine deficiency); for this reason iodine is needed in larger quantities than others in this list, and sometimes classified with the macrominerals; It can be found in ionized salt
- Iron required for many enzymes, and for hemoglobin and some other proteins
- Manganese (processing of oxygen)
- Molybdenum required for xanthine oxidase and related oxidases
- Nickel present in urease
- Selenium required for peroxidase (antioxidant proteins)
- Zinc required for several enzymes such as carboxypeptidase, liver alcohol dehydrogenase, carbonic anhydrase

Vitamins

As with the minerals discussed above, some vitamins are recognized as essential nutrients, necessary in the diet for good health. (Vitamin D is the exception: it can alternatively be synthesized in the skin, in the presence of UVB radiation.) Certain vitamin-like compounds that are recommended in the diet, such as carnitine, are thought useful for survival and health, but these are not essential dietary nutrients because the human body has some capacity to produce them from other compounds. Moreover, thousands of different phytochemicals have recently been discovered in food (particularly in fresh vegetables), which may have desirable properties including antioxidant activity (see below); experimental demonstration has been suggestive but inconclusive. Other essential nutrients not classed as vitamins include essential amino acids (see above), essential fatty acids (see above), and the minerals discussed in the preceding section.

Vitamin deficiencies may result in disease conditions: goiter, scurvy, osteoporosis, impaired immune system, disorders of cell metabolism, certain forms of cancer, symptoms of premature aging, and poor psychological health (including eating disorders), among many others.

Malnutrition

Malnutrition refers to insufficient, excessive, or imbalanced consumption of nutrients. In developed countries, the diseases of malnutrition are most often associated with nutritional imbalances or excessive consumption. Although there are more people in the world who are malnourished due to excessive consumption, according to the United Nations World Health Organization, the greatest challenge in developing nations today is not starvation, but insufficient nutrition – the lack of nutrients necessary for the growth and maintenance of vital functions. The causes of malnutrition are directly linked to inadequate macronutrient consumption and disease, and are indirectly linked to factors like "household food security, maternal and child care, health services, and the environment."

Illnesses

Nutrients	Deficiency	Excess
Food Energy	starvation, marasmus	obesity, diabetes mellitus, cardiovascular disease
Simple carbohydrates	None.	diabetes mellitus, obesity
Complex carbohydrates	none	obesity
Saturated fat	low sex hormone levels	cardiovascular disease
Trans fat	none	cardiovascular disease
Unsaturated fat	none	obesity
Fat	malabsorption of fat-soluble vitamins, rabbit starvation (if protein intake is high), during development: stunted brain development and reduced brain weight.	cardiovascular disease
Omega-3 fats	cardiovascular disease	bleeding, hemorrhages
Omega-6 fats	none	cardiovascular disease, cancer
Cholesterol	during development: deficiencies in myelinization of the brain.	cardiovascular disease
Protein	kwashiorkor	
Sodium	hyponatremia	hypernatremia, hypertension
Iron	anemia	cirrhosis, cardiovascular disease
Iodine	goiter, hypothyroidism	Iodine toxicity (goiter, hypothyroidism)
Vitamin A	xerophthalmia and night blindness, low testosterone levels	hypervitaminosis A (cirrhosis, hair loss)
Vitamin B_1	beriberi	
Vitamin B_2	cracking of skin and corneal unclearation	
Niacin	pellagra	dyspepsia, cardiac arrhythmias, birth defects
Vitamin B_{12}	pernicious anemia	
Vitamin C	scurvy	diarrhea causing dehydration
Vitamin D	rickets, osteoporosis, balance, immune system, inflammation	hypervitaminosis D (dehydration, vomiting, constipation)
Vitamin E	nervous disorders	hypervitaminosis E (anticoagulant: excessive bleeding)
Vitamin K	hemorrhage	
Calcium	osteoporosis, tetany, carpopedal spasm, laryngospasm, cardiac arrhythmias	fatigue, depression, confusion, anorexia, nausea, vomiting, constipation, pancreatitis, increased urination
Magnesium	hypertension	weakness, nausea, vomiting, impaired breathing, and hypotension
Potassium	hypokalemia, cardiac arrhythmias	hyperkalemia, palpitations

Mental Agility

Research indicates that improving the awareness of nutritious meal choices and establishing long-term habits of healthy eating has a positive effect on a cognitive and spatial memory capacity, potentially increasing a student's potential to process and retain academic information. Some organizations have begun working with teachers, policymakers, and managed foodservice contractors to mandate improved nutritional content and increased nutritional resources in school cafeterias from primary to university level institutions. Health and nutrition have been proven to have close links with overall educational success. Currently less than 10% of American college students report that they eat the recommended five servings of fruit and vegetables daily. Better nutrition has been shown to affect both cognitive and spatial memory performance; a study showed those with higher blood sugar levels performed better on certain memory tests. In another study, those who consumed yogurt performed better on thinking tasks when compared to those who consumed caffeine free diet soda or confections. Nutritional deficiencies have been shown to have a negative effect on learning behavior in mice as far back as 1951."Better learning performance is associated with diet induced effects on learning and memory ability".

- The "nutrition-learning nexus" demonstrates the correlation between diet and learning and has application in a higher education setting..
- We find that better nourished children perform significantly better in school, partly because they enter school earlier and thus have more time to learn but mostly because of greater learning productivity per year of schooling."
- 91% of college students feel that they are in good health while only 7% eat their recommended daily allowance of fruits and vegetables.
- Nutritional education is an effective and workable model in a higher education setting.
- More "engaged" learning models that encompass nutrition is an idea that is picking up steam at all levels of the learning cycle.

Mental Disorders

Nutritional supplement treatment may be appropriate for major depression, bipolar disorder, schizophrenia, and obsessive compulsive disorder, the four most common mental disorders in developed countries. It is because Lakhan and Vieira mentioned that the supplements possess amino acids that may change into neurotransmitters and improve mental disorders. Supplements that have been studied most for mood elevation and stabilization include eicosapentaenoic acid and docosahexaenoic acid (each of which are an omega-3 fatty acid contained in fish oil, but not in flaxseed oil), vitamin B_{12}, folic acid, and inositol.

Cancer

Cancer has become common in developing countries. According to a study by the International Agency for Research on Cancer, "In the developing world, cancers of the liver, stomach and esophagus were more common, often linked to consumption of carcinogenic preserved foods, such as smoked or salted food, and parasitic infections that attack organs." Lung cancer rates are rising rapidly in poorer nations because of increased use of tobacco. Developed countries "tended to have cancers linked to affluence or a 'Western lifestyle' – cancers of the colon, rectum, breast and prostate – that can be caused by obesity, lack of exercise, diet and age."

A comprehensive worldwide report, "Food, Nutrition, Physical Activity and the Prevention of Cancer: a Global Perspective", compiled by the World Cancer Research Fund and the American Institute for Cancer Research, reports that there is a significant relation between lifestyle (including food consumption) and cancer prevention. The same report recommends eating mostly foods of plant origin and aiming to meet nutritional needs through diet alone, while limiting consumption of energy-dense foods, red meat, alcoholic drinks and salt and avoiding sugary drinks, processed meat and moldy cereals (grains) or pulses (legumes). Protein consumption leads to an increase in IGF-1, which plays a role in cancer development.

Metabolic Syndrome and Obesity

Several lines of evidence indicate lifestyle-induced hyperinsulinemia and reduced insulin function (i.e. insulin resistance) as decisive factors in many disease states. For example, hyperinsulinemia and insulin resistance are strongly linked to chronic inflammation, which in turn is strongly linked to a variety of adverse developments such as arterial microinjuries and clot formation (i.e. heart disease) and exaggerated cell division (i.e. cancer). Hyperinsulinemia and insulin resistance (the so-called metabolic syndrome) are characterized by a combination of abdominal obesity, elevated blood sugar, elevated blood pressure, elevated blood triglycerides, and reduced HDL cholesterol.

Obesity can unfavourably alter hormonal and metabolic status via resistance to the hormone leptin, and a vicious cycle may occur in which insulin/leptin resistance and obesity aggravate one another. The vicious cycle is putatively fuelled by continuously high insulin/leptin stimulation and fat storage, as a result of high intake of strongly insulin/leptin stimulating foods and energy. Both insulin and leptin normally function as satiety signals to the hypothalamus in the brain; however, insulin/leptin resistance may reduce this signal and therefore allow continued overfeeding despite large body fat stores.

There is a debate about how and to what extent different dietary factors – such as intake of processed carbohydrates, total protein, fat, and carbohydrate intake, intake of saturated and trans fatty acids, and low intake of vitamins/minerals – contribute to the development of insulin and leptin resistance. Evidence indicates that diets possibly protective against metabolic syndrome include low saturated and trans fat intake and foods rich in dietary fiber, such as high consumption of fruits and vegetables and moderate intake of low-fat dairy products.

Hyponatremia

Excess water intake, without replenishment of sodium and potassium salts, leads to hyponatremia, which can further lead to water intoxication at more dangerous levels. A well-publicized case occurred in 2007, when Jennifer Strange died while participating in a water-drinking contest. More usually, the condition occurs in long-distance endurance events (such as marathon or triathlon competition and training) and causes gradual mental dulling, headache, drowsiness, weakness, and confusion; extreme cases may result in coma, convulsions, and death. The primary damage comes from swelling of the brain, caused by increased osmosis as blood salinity decreases. Effective fluid replacement techniques include Water aid stations during running/cycling races, trainers providing water during team games such as Soccer and devices such as Camel Baks which can provide water for a person without making it too hard to drink the water.

Global Nutrition Challenges

The challenges facing global nutrition are disease, child malnutrition, obesity, and vitamin deficiency.

Disease

The most common non-infectious diseases worldwide, that contribute most to the global mortality rate, are cardiovascular diseases, various cancers, diabetes, and chronic respiratory problems, all of which are linked to poor nutrition. Nutrition and diet are closely associated with the leading causes of death, including cardiovascular disease and cancer. Obesity and high sodium intake can contribute to ischemic heart disease, while consumption of fruits and vegetables can decrease the risk of developing cancer.

Foodborne and infectious diseases can result in malnutrition, and malnutrition exacerbates infectious disease. Poor nutrition leaves children and adults more susceptible to contracting life-threatening diseases such as diarrheal infections and respiratory infections. According to the WHO, in 2011, 6.9 million children died of infectious diseases like pneumonia, diarrhea, malaria, and neonatal conditions, of which at least one third were associated with undernutrition.

Child Malnutrition

According to UNICEF, in 2011, 101 million children across the globe were underweight and one in four children, 165 million, were stunted in growth. Simultaneously, there are 43 million children under five who are overweight or obese. Nearly 20 million children under 5 suffer from severe acute malnutrition, a life-threatening condition requiring urgent treatment.According to estimations at UNICEF, hunger will be responsible for 5.6 million deaths of children under the age of five this year. These all represent significant public health emergencies. This is because proper maternal and child nutrition has immense consequences for survival, acute and chronic disease incidence, normal growth, and economic productivity of individuals.

Childhood malnutrition is common and contributes to the global burden of disease. Childhood is a particularly important time to achieve good nutrition status, because poor nutrition has the capability to lock a child in a vicious cycle of disease susceptibility and recurring sickness, which threatens cognitive and social development. Undernutrition and bias in access to food and health services leaves children less likely to attend or perform well in school.

Under Nutrition

UNICEF defines under nutrition "as the outcome of insufficient food intake (hunger) and repeated infectious diseases. Under nutrition includes being underweight for one's age, too short for one's age (stunted), dangerously thin (wasted), and deficient in vitamins and minerals (micronutrient malnutrient). Under nutrition causes 53% of deaths of children under five across the world. It has been estimated that undernutrition is the underlying cause for 35% of child deaths. The Maternal and Child Nutrition Study Group estimate that under nutrition, "including fetal growth restriction, stunting, wasting, deficiencies of vitamin A and zinc along with suboptimum breastfeeding- is a cause of 3.1 million child deaths and infant mortality, or 45% of all child deaths in 2011".

When humans are undernourished, they no longer maintain normal bodily functions, such as growth, resistance to infection, or have satisfactory performance in school or work. Major causes of under nutrition in young children include lack of proper breast feeding for infants and illnesses such as diarrhea, pneumonia, malaria, and HIV/AIDS. According to UNICEF 146 million children across the globe, that one out of four under the age of five, are underweight. The amount of underweight children has decreased since 1990, from 33 percent to 28 percent between 1990 and 2004. Underweight and stunted children are more susceptible to infection, more likely to fall behind in school, more likely to become overweight and develop non-infectious diseases, and ultimately earn less than their non-stunted coworkers. Therefore, undernutrition can accumulate deficiencies in health which results in less productive individuals and societies

Many children are born with the inherent disadvantage of low birth weight, often caused by intrauterine growth restriction and poor maternal nutrition, which results in worse growth, development, and health throughout the course of their lifetime. Children born at low birthweight (less than 5.5 pounds), are less likely to be healthy and are more susceptible to disease and early death. Those born at low birthweight also are likely to have a depressed immune system, which can increase their chances of heart disease and diabetes later on in life. Because 96% of low birthweight occurs in the developing world, low birthweight is associated with being born to a mother in poverty with poor nutritional status that has had to perform demanding labor.

Stunting and other forms of undernutrition reduces a child's chance of survival and hinders their optimal growth and health. Stunting has demonstrated association with poor brain development, which reduces cognitive ability, academic performance, and eventually earning potential. Important determinants of stunting include the quality and frequency of infant and child feeding, infectious disease susceptibility, and the mother's nutrition and health status. Undernourished mothers are more likely to birth stunted children, perpetuating a cycle of undernutrition and poverty. Stunted children are more likely to develop obesity and chronic diseases upon reaching adulthood. Therefore, malnutrition resulting in stunting can further worsen the obesity epidemic, especially in low and middle income countries. This creates even new economic and social challenges for vulnerable impoverished groups.

Data on global and regional food supply shows that consumption rose from 2011-2012 in all regions. Diets became more diverse, with a decrease in consumption of cereals and roots and an increase in fruits, vegetables, and meat products. However, this increase masks the discrepancies between nations, where Africa, in particular, saw a decrease in food consumption over the same years. This information is derived from food balance sheets that reflect national food supplies, however, this does not necessarily reflect the distribution of micro and macronutrients. Often inequality in food access leaves distribution which uneven, resulting in undernourishment for some and obesity for others.

Undernourishment, or hunger, according to the FAO, is dietary intake below the minimum daily energy requirement. The amount of undernourishment is calculated utilizing the average amount of food available for consumption, the size of the population, the relative disparities in access to the food, and the minimum calories required for each individual. According to FAO, 868 million people (12% of the global population) were undernourished in 2012. This has decreased across the world since 1990, in all regions except for Africa, where undernourishment has steadily increased. However, the rates of decrease are not sufficient to meet the first Millennium Development Goal of halving hunger between 1990 and 2015. The global financial, economic, and food price crisis in 2008 drove many people to hunger, especially women and children. The spike in food prices prevented many people from escaping poverty, because the poor spend a larger proportion of their income on food and farmers are net consumers of food. High food prices cause consumers to have less purchasing power and to substitute more-nutritious foods with low-cost alternatives.

Adult Overweight and Obesity

Malnutrition in industrialized nations is primarily due to excess calories and non-nutritious carbohydrates, which has contributed to the obesity epidemic affecting both developed and some developing nations. In 2008, 35% of adults above the age of 20 years were overweight (BMI 25 kg/m), a prevalence that has doubled worldwide between 1980 and 2008. Also 10% of men and 14% of women were obese, with an BMI greater than 30. Rates of overweight and obesity vary across the globe, with the highest prevalence in the Americas, followed by European nations, where over 50% of the population is overweight or obese.

Obesity is more prevalent amongst high income and higher middle income groups than lower divisions of income. Women are more likely than men to be obese, where the rate of obesity in women doubled from 8% to 14% between 1980 and 2008. Being overweight as a child has become an increasingly important indicator for later development of obesity and non-infectious diseases such as heart disease. In several western European nations, the prevalence of overweight and obese children rose by 10% from 1980 to 1990, a rate that has begun to accelerate recently.

Vitamin and Mineral Malnutrition

Vitamins and minerals are essential to the proper functioning and maintenance of the human body. Globally, particularly in developing nations, deficiencies in Iodine, Iron and Zinc amog others are said to impair human health when these minerals are not ingested in an adequate quantity. There are 20 trace elements and minerals that are essential in small quantities to body function and overall human health.

Iron deficiency is the most common inadequate nutrient worldwide, affecting approximately 2 billion people. Globally, anemia affects 1.6 billion people, and represents a public health emergency in children under five and mothers. The World Health Organization estimates that there exists 469 million women of reproductive age and approximately 600 million preschool and school-age children worldwide who are anemic. Anemia, especially iron-deficient anemia, is a critical problem for cognitive developments in children, and its presence leads to maternal deaths and poor brain and motor development in children. The development of anemia affects mothers and children more because infants and children have higher iron requirements for growth. Health consequences for iron deficiency in young children include increased perinatal mortality, delayed mental and physical development, negative behavioral consequences, reduced auditory and visual function, and impaired physical performance. The harm caused by iron deficiency during child development cannot be reversed and result in reduced academic performance, poor physical work capacity, and decreased productivity in adulthood. Mothers are also very susceptible to iron-deficient anemia because women lose iron during menstruation, and rarely supplement it in their diet. Maternal iron deficiency anemia increases the chances of maternal mortality, contributing to at least 18% of maternal deaths in low and middle income countries.

Vitamin A plays an essential role in developing the immune system in children, therefore, it is considered an essential micronutrient that can greatly affect health. However, because of the expense of testing for deficiencies, many developing nations have not been able to fully detect and address vitamin A deficiency, leaving vitamin A deficiency considered a silent hunger. According to estimates, subclinical vitamin A deficiency, characterized by low retinol levels, affects 190 million pre-school children and 19 million mothers worldwide. The WHO estimates that 5.2 million of these children under 5 are affected by night blindness, which is considered clinical vitamin A deficiency. Severe vitamin A deficiency (VAD) for developing children can result in visual impairments, anemia and weakened immunity, and increase their risk of morbidity and mortality from infectious disease. This also presents a problem for women, with WHO estimating that 9.8 million women are affected by night blindness. Clinical vitamin A deficiency is particularly common among pregnant women, with prevalence rates as high as 9.8% in South-East Asia.

Estimates say that 28.5% of the global population is iodine deficient, representing 1.88 billion individuals. Although salt iodization programs have reduced the prevalence of iodine deficiency, this is still a public health concern in 32 nations. Moderate deficiencies are common in Europe and Africa, and over consumption is common in the Americas. Iodine-deficient diets can interfere with adequate thyroid hormone production, which is responsible for normal growth in the brain and nervous system. This ultimately leads to poor school performance and impaired intellectual capabilities.

Infant and Young Child Feeding

Improvement of breast feeding practices, like early initiation and exclusive breast feeding for the first two years of life, could save the lives of 1.5 million children annually. Nutrition interventions targeted at infants aged 0–5 months first encourages early initiation of breastfeeding. Though the relationship between early initiation of breast feeding and improved health outcomes has not been formally established, a recent study in Ghana suggests a causal relationship between early initiation and reduced infection-caused neonatal deaths. Also, experts promote exclusive breastfeeding, rather than using formula, which has shown to promote optimal growth, development, and health of infants. Exclusive breasfeeding often indicates nutritional status because infants that consume breast milk are more likely to receive all adequate nourishment and nutrients that will aid their developing body and immune system. This leaves children less likely to contract diarrheal diseases and respiratory infections.

Besides the quality and frequency of breastfeeding, the nutritional status of mothers affects infant health. When mothers do not receive proper nutrition, it threatens the wellness and potential of their children. Well-nourished women are less likely to experience risks of birth and are more likely to deliver children who will develop well physically and mentally.Maternal undernutrition increases the chances of low-birth weight, which can increase the risk of infections and asphyxia in fetuses, increasing the probability of neonatal deaths.Growth failure during intrauterine conditions, associated with improper mother nutrition, can contribute to lifelong health complications. Approximately 13 million children are born with intrauterine growth restriction annually.

International Food Insecurity and Malnutrition

According to UNICEF, South Asia has the highest levels of underweight children under five, followed by sub-Saharan Africans nations, with Industrialized countries and Latin nations having the lowest rates.

United States

In the United States, 2% of children are underweight, with under 1% stunted and 6% are wasting.

In the US, dietitians are registered (RD) or licensed (LD) with the Commission for Dietetic Registration and the American Dietetic Association, and are only able to use the title dietitian, as described by the business and professions codes of each respective state, when they have met specific educational and experiential prerequisites and passed a national registration or licensure examination, respectively. In California, registered dietitians must abide by the *"Business and Professions Code of Section 2585-2586.8"*. Anyone may call themselves a nutritionist, including unqualified dietitians, as this term is unregulated. Some states, such as the State of Florida, have begun to include the title "nutritionist" in state licensure requirements. Most governments provide guidance on nutrition, and some also impose mandatory disclosure/labeling requirements for processed food manufacturers and restaurants to assist consumers in complying with such guidance..

In the US, nutritional standards and recommendations are established jointly by the US Department of Agriculture and US Department of Health and Human Services. Dietary and physical activity guidelines from the USDA are presented in the concept of a plate of food which in 2011 superseded the food pyramid that had replaced the Four Food Groups. The Senate committee currently responsible for oversight of the USDA is the *Agriculture, Nutrition and Forestry Committee*. Committee hearings are often televised on C-SPAN. The U.S. Department of Health and Human Services provides a sample week-long menu which fulfills the nutritional recommendations of the government. Canada›s Food Guide is another governmental recommendation..

Industrialized Countries

According to UNICEF, the Commonwealth of Independent States has the lowest rates of stunting and wasting, at 14 percent and 3 percent. The nations of Estonia, Finland,

Iceland, Lithuania and Sweden have the lowest prevalence of low birthweight children in the world- at 4%. Proper prenatal nutrition is responsible for this small prevalence of low birthweight infants. However, low birthweight rates are increasing, due to the use of fertility drugs, resulting in multiple births, women bearing children at an older age, and the advancement of technology allowing more pre-term infants to survive. Industrialized nations more often face malnutrition in the form of over-nutrition from excess calories and non-nutritious carbohydrates, which has contributed greatly to the public health epidemic of obesity. Disparities, according to gender, geographic location and socio-economic position, both within and between countries, represent the biggest threat to child nutrition in industrialized countries. These disparities are a direct product of social inequalities and social inequalities are rising throughout the industrialized world, particularly in Europe.

South Asia

South Asia has the highest percentage and number of underweight children under five in the world, at approximately 78 million children. Patterns of stunting and wasting are similar, where 44% have not reached optimal height and 15% are wasted, rates much higher than any other regions. This region of the world has extremely high rates of child underweight- 46% of its child population under five is underweight. India, Bangladesh, and Pakistan alone account for half the globe's underweight child population. South Asian nations have made progress towards the MDGs, considering the rate has decreased from 53% since 1990, however, a 1.7% decrease of underweight prevalence per year will not be sufficient to meet the 2015 goal. Some nations, such as Afghanistan, Bangladesh, and Sri Lanka, on the other hand, have made significant improvements, all decreasing their prevalence by half in ten years. While India and Pakistan have made modest improvements, Nepal has made no significant improvement in underweight child prevalence. Other forms of undernutrition have continued to persist with high resistance to improvement, such as the prevalence of stunting and wasting, which has not changed significantly in the past 10 years. Causes of this poor nutrition include energy-insufficient diets, poor sanitation conditions, and the gender disparities in educational and social status. Girls and women face discrimination especially in nutrition status, where South Asia is the only region in the world where girls are more likely to be underweight than boys. In South Asia, 60% of children in the lowest quintile are underweight, compared to only 26% in the highest quintile, and the rate of reduction of underweight is slower amongst the poorest.

Eastern/South Africa

The Eastern and Southern African nations have shown no improvement since 1990 in the rate of underweight children under five. They have also made no progress in halving hunger by 2015, the most prevalent Millennium Development Goal. This is due primarily to the prevalence of famine, declined agricultural productivity, food emergencies, drought, conflict, and increased poverty. This, along with HIV/AIDS, has inhibited the nutrition development of nations such as Lesotho, Malawi, Mozambique, Swaziland, Zambia and Zimbabwe. Botswana has made remarkable achievements in reducing underweight prevalence, dropping 4% in 4 years, despite its place as the second leader in HIV prevalence amongst adults in the globe.South Africa, the wealthiest nation in this region, has the second lowest proportion of underweight children at 12%, but has been steadily increasing in underweight prevalence since 1995. Almost half of Ethiopian children are underweight, and along with Nigeria, they account for almost one-third of the underweight under five in all of Sub-Saharan Africa.

West/Central Africa

West/Central Africa has the highest rate of children under five underweight in the world. Of the countries in this region, the Congo has the lowest rate at 14%, while the nations of Democratic Republic of the Congo, Ghana, Guinea, Mali, Nigeria, Senegal and Togo are improving slowly. In Gambia, rates decreased from 26% to 17% in four years, and their coverage of vitamin A supplementation reaches 91% of vulnerable populations. This region has the next highest proportion of wasted children, with 10% of the population under five not at optimal weight. Little improvement has been made between the years of 1990 and 2004 in reducing the rates of underweight children under five, whose rate stayed approximately the same. Sierra Leone has the highest child under five mortality rate in the world, due predominantly to its extreme infant mortality rate, at 238 deaths per 1000 live births. Other contributing factors include the high rate of low birthweight children (23%) and low levels of exclusive breast feeding (4%). Anemia is prevalent in these nations, with unacceptable rates of iron deficient anemia. The nutritional status of children is further indicated by its high rate of child wasting - 10%. Wasting is a significant problem in Sahelian countries – Burkina Faso, Chad, Mali, Mauritania and Niger – where rates fall between 11% and 19% of under fives, affecting more than 1 million children.

Middle East/North Africa

Six countries in the Middle East and North Africa region are on target to meet goals for reducing underweight children by 2015, and 12 countries have prevalence rates below 10%.However, the nutrition of children in the region as a whole has degraded for the past ten years due to the increasing portion of underweight children in three populous nations – Iraq, Sudan, and Yemen. Forty six percent of all children in Yemen are underweight, a percentage that has worsened by 4% since 1990. In Yemen, 53% of children under five are stunted and 32% are born at low birth weight. Sudan has an underweight prevalence of 41%, and the highest proportion of wasted children in the region at 16%. One percent of households in Sudan consume iodized salt. Iraq has also seen an increase in child underweight since 1990. Djibouti, Jordan, the Occupied Palestinian Territory (OPT), Oman, the Syrian Arab Republic and Tunisia are all projected to meet minimum nutrition goals, with OPT, Syrian AR, and Tunisia the fastest improving regions. This region demonstrates that undernutrition does not always improve with economic prosperity, where the United Arab Emirates, for example, despite being a wealthy nation, has similar child death rates due to malnutrition to those seen in Yemen.

East Asia/Pacific

The East Asia/Pacific region has reached its goals on nutrition, in part due to the improvements contributed by China, the region's most populous country. China has reduced its underweight prevalence from 19 percent to 8 percent between 1990 and 2002. China played the largest role in the world in decreasing the rate of children under five underweight between 1990 and 2004, halving the prevalence. This reduction of underweight prevalence has aided in the lowering of the under 5 mortality rate from 49 to 31 of 1000. They also have a low birthweight rate at 4%, a rate comparable to industrialized countries, and over 90% of households receive adequate iodized salts. However, large disparities exist between children in rural and urban areas, where 5 provinces in China leave 1.5 million children iodine deficient and susceptible to diseases. Singapore, Vietnam, Malaysia, and Indonesia are all projected to reach nutrition MDGs. Singapore has the lowest under five mortality rate of any nation, besides Iceland, in the world, at 3%. Cambodia has the highest rate of child mortality in the region (141 per 1,000 live births), while still its proportion of underweight children increased by 5 percent to 45% in 2000. Further nutrient indicators show that only 12 per cent of Cambodian babies are exclusively breastfed and only 14 per cent of households consume iodized salt.

Latin America/Caribbean

This region has undergone the fastest progress in decreasing poor nutrition status of children in the world. The Latin American region has reduced underweight children prevalence by 3.8% every year between 1990 and 2004, with a current rate of 7% underweight. They also have the lowest rate of child mortality in the developing world, with only 31 per 1000 deaths, and the highest iodine consumption. Cuba has seen improvement from 9 to 4 percent underweight under 5 between 1996 and 2004. The prevalence has also decreased in the Dominican Republic, Jamaica, Peru, and Chile. Chile has a rate of underweight under 5, at merely 1%. The most populous nations, Brazil and Mexico, mostly have relatively low rates of underweight under 5, with only 6% and 8%. Guatemala has the highest percentage of underweight and stunted children in the region, with rates above 45%. There are disparities amongst different populations in this region. For example, children in rural areas have twice the prevalence of underweight at 13%, compared to urban areas at 5%.

Nutrition Access Disparities

Occurring throughout the world, lack of proper nutrition is both a consequence and cause of poverty. Impoverished individuals are less likely to have access to nutritious food and to escape from poverty than those who have healthy diets. Disparities in socioeconomic status, both between and within nations, provide the largest threat to child nutrition in industrialized nations, where social inequality is on the rise. According to UNICEF, children living in the poorest households are twice as likely to be underweight as those in the richest. Those in the lowest wealth quintile and whose mothers have the least education demonstrate the highest rates of child mortality and stunting. Throughout the developing world, socioeconomic inequality in childhood malnutrition is more severe than in upper income brackets, regardless of the general rate of malnutrition. Concurrently, the greatest increase in childhood obesity has been seen in the lower middle income bracket.

According to UNICEF, children in rural locations are more than twice as likely to be underweight as compared to children under five in urban areas. In Latin American/ Caribbean nations, "Children living in rural areas in Bolivia, Honduras, Mexico and Nicaragua are more than twice as likely to be underweight as children living in urban areas. That likelihood doubles to four times in Peru." In the United States, the incidence of low birthweight is on the rise among all populations, but particularly among minorities.

According to UNICEF, boys and girls have almost identical rates as underweight children under age 5 across the world, except in South Asia.

Nutrition Policy

Nutrition Interventions

Nutrition directly influences progress towards meeting the Millennium Goals of eradicating hunger and poverty through health and education. Therefore, nutrition interventions take a multi-faceted approach to improve the nutrition status of various populations. Policy and programming must target both individual behavioral changes and policy approaches to public health. While most nutrition interventions focus on delivery through the health-sector, non-health sector interventions targeting agriculture, water and sanitation, and education are important as well. Global nutrition micro-nutrient deficiencies often receive large-scale solution approaches by deploying large governmental and non-governmental organizations. For example, in 1990, iodine deficiency was particularly prevalent, with one in five households, or 1.7 billion people, not consuming adequate iodine, leaving them at risk to develop associated diseases. Therefore, a global campaign to iodize salt to eliminate iodine deficiency successfully boosted the rate to 69% of households in the world consuming adequate amounts of iodine.

Emergencies and crises often exacerbate undernutrition, due to the aftermath of crises that include food insecurity, poor health resources, unhealthy environments, and poor healthcare practices. Therefore, the repercussions of natural disasters and other emergencies can exponentially increase the rates of macro and micronutrient deficiencies in populations. Disaster relief interventions often take a multi-faceted public health approach. UNICEF's programming targeting nutrition services amongst disaster settings include nutrition assessments, measles immunization, vitamin A supplementation, provision of fortified foods and micronutrient supplements, support for breastfeeding and complementary feeding for infants and young children, and therapeutic and supplementary feeding. For example, during Nigeria's food crisis of 2005, 300,000 children received therapeutic nutrition feeding programs through the collaboration of UNICEF, the Niger government, the World Food Programme, and 24 NGOs utilizing community and facility based feeding schemes.

Interventions aimed at pregnant women, infants, and children take a behavioral and program-based approach. Behavioral intervention objectives include promoting proper breast-feeding, the immediate initiation of breastfeeding, and its continuation through 2 years and beyond. UNICEF recognizes that to promote these behaviors, healthful environments must be established conducive to promoting these behaviors, like healthy hospital environments, skilled health workers, support in the public and workplace, and removing negative influences.Finally, other interventions include provisions of adequate micro and macro nutrients such as iron, anemia, and vitamin A supplements and vitamin-fortified foods and ready-to-use products. Programs addressing micro-nutrient deficiencies, such as those aimed at anemia, have attempted to provide iron supplementation to pregnant and lactating women. However, because supplementation often occurs too late, these programs have had little effect. Interventions such as women's nutrition, early and exclusive breastfeeding, appropriate complementary food and micronutrient supplementation have proven to reduce stunting and other manifestations of undernutrition. A Cochrane review of community-based maternal health packages showed that this community-based approach improved the initiation of breastfeeding within one hour of birth. Some programs have had adverse effects. One example is the "Formula for Oil" relief program in Iraq, which resulted in the replacement of breastfeeding for formula, which has negatively affected infant nutrition.

Implementation and Delivery Platforms

In April 2010, the World Bank and the IMF released a policy briefing entitled "Scaling up Nutrition (SUN): A Framework for action" that represented a partnered effort to address the Lancet's Series on under nutrition, and the goals it set out for improving under nutrition. They emphasized the 1000 days after birth as the prime window for effective nutrition intervention, encouraging programming that was cost-effective and showed significant cognitive improvement in populations, as well as enhanced productivity and economic growth.This document was labeled the SUN framework, and was launched by the UN General Assembly in 2010 as a road map encouraging the coherence of stakeholders like governments, academia, UN system organizations and foundations in working towards reducing under nutrition. The SUN framework has initiated a transformation in global nutrition- calling for country-based nutrition programs, increasing evidence based and cost–effective interventions, and "integrating nutrition within national strategies for gender equality, agriculture, food security, social protection, education, water supply, sanitation, and health care". Government often plays a role in implementing nutrition programs through policy. For instance, several East Asian nations have enacted legislation to increase iodization of salt to increase household consumption. Political commitment in the form of evidence-based effective national policies and programs, trained skilled community nutrition workers, and effective communication and advocacy can all work to decrease malnutrition. Market and industrial production can play a role as well. For example, in the Philippines, improved production and market availability of iodized salt increased household consumption. While most nutrition interventions are delivered directly through governments and health services, other sectors, such as agriculture, water and sanitation, and education, are vital for nutrition promotion as well.

Nutrition Education

Nutrition is taught in schools in many countries. In England and Wales the Personal and Social Education and Food Technology curricula include nutrition, stressing the importance of a balanced diet and teaching how to read nutrition labels on packaging. In many schools a Nutrition class will fall within the Family and Consumer Science or Health departments. In some American schools, students are required to take a certain number of FCS or Health related classes. Nutrition is offered at many schools, and if it is not a class of its own, nutrition is included in other FCS or Health classes such as: Life Skills, Independent Living, Single Survival, Freshmen Connection, Health etc. In many Nutrition classes, students learn about the food groups, the food pyramid, Daily Recommended Allowances, calories, vitamins, minerals, malnutrition, physical activity, healthy food choices and how to live a healthy life.

A 1985 US National Research Council report entitled *Nutrition Education in US Medical Schools* concluded that nutrition education in medical schools was inadequate. Only 20% of the schools surveyed taught nutrition as a separate, required course. A 2006 survey found that this number had risen to 30%.

Nutrition for Special Populations

Sports Nutrition

Individuals with highly active lifestyles require more nutrients. Furthermore, recent studies declared that increasing dietary intake of carbohydrates, proteins and Vitamin D may be part of improving athlete's level of performance

Protein

Protein milkshakes, made from protein powder (center) and milk (left), are a common bodybuilding supplement.

Protein is an important component of every cell in the body. Hair and nails are mostly made of protein. The body uses protein to build and repair tissues. Also protein is used to make enzymes, hormones, and other body chemicals. Protein is an important building block of bones, muscles, cartilage, skin, and blood.

The protein requirement for each individual differs, as do opinions about whether and to what extent physically active people require more protein. The 2005 Recommended Dietary Allowances (RDA), aimed at the general healthy adult population, provide for an intake of 0.8 – 1 grams of protein per kilogram of body weight (according to the BMI formula), with the review panel stating that no additional dietary protein is suggested for healthy adults undertaking resistance or endurance exercise.

Carbohydrates

The main fuel used by the body during exercise is carbohydrates, which is stored in muscle as glycogen – a form of sugar. During exercise, muscle glycogen reserves can be used up, especially when activities last longer than 90 min. Because the amount of glycogen stored in the body is limited, it is important for athletes to replace glycogen by consuming a diet high in carbohydrates. Meeting energy needs can help improve performance during the sport, as well as improve overall strength and endurance..

There are different kinds of carbohydrates: simple (for example from fruits) and complex (for example from grains such as wheat). Simple sugars can be from an unrefined natural source, or may be refined and added to processed food. A typical American consumes about 50% of their carbohydrates as refined sugars. Over the course of a year, the average American consumes 204 litres (54 US gallons @ 3.781 per gallon) of soft drinks, which contain the highest amount of added sugars. Even though carbohydrates are necessary for humans to function, they are not all equally healthful. When machinery has been used to remove bits of high fiber, the carbohydrates are refined. These are the carbohydrates found in white bread and fast food.

Maternal Nutrition

Studies have shown that nutritional status of women pre-conception (specifically body weight) has an influence on foetal growth. However during pregnancy, nutritional intake by the mother is believed not to have any effect during the first and second trimesters, but maternal nutritional intake in the last trimester is widely believed to have some effect on the foetal growth, and therefore maternal nutrition plays a vital role in the development and growth of the foetus.

Paediatric Nutrition

Adequate nutrition is essential for the growth of children from infancy right through until adolescence. Some nutrients are specifically required for growth on top of nutrients required for normal body maintenance, in particular calcium and iron.

Elderly Nutrition

Chwang (2012) reports that malnutrition is higher among the elderly. This article highlights the uneven distribution of nutrients between developed and undeveloped countries. For example, most European countries has a high fat intake; while, the Malaysian elderly were found to have insufficient intake of calcium, thiamine and riboflavin.

From Antiquity to 1900

Anaxagoras

Around 3000 BC the Vedic texts made mention of scientific research on nutrition. The Bible's Book of Daniel recounts first recorded nutritional experiment. During an invasion of Judah, King Nebuchadnezzar of Babylon captured Daniel and his friends. Selected as court servants, they were to share in the king's fine foods and wine. But they objected, preferring vegetables (pulses) and water in accordance with their Jewish dietary restrictions. The king's chief steward reluctantly agreed to a trial. Daniel and his friends received their diet for 10 days. On comparison with the king's men, they appeared healthier, and were allowed to continue with their diet. Around 475 BC, Anaxagoras stated that food is absorbed by the human body and therefore contained "homeomerics (generative components), suggesting the existence of nutrients. Around 400 BC, Hippocrates said: "Let food be your medicine and medicine be your food."

The 16th-century scientist and artist Leonardo da Vinci (1452–1519) compared metabolism to a burning candle. In 1747 Dr. James Lind, a physician in the British navy, performed the first attested scientific nutrition experiment, discovering that lime juice saved sailors who had been at sea for years from scurvy, a deadly and painful bleeding disorder. The discovery was ignored for forty years, but after about 1850 British sailors became known as limeys. (Scientists would not identify the essential vitamin C within lime juice until the 1930s.)

Around 1770 Antoine Lavoisier, the Father of Nutrition and Chemistry, discovered the details of metabolism, demonstrating that the oxidation of food is the source of body heat. In 1790 George Fordyce recognized calcium as necessary for fowl survival. In the early 19th century, the elements carbon, nitrogen, hydrogen and oxygen were recognized as the primary components of food, and methods to measure their proportions were developed.

In 1816 François Magendie discovered that dogs fed only carbohydrates and fat lost their body protein and died in a few weeks, but dogs also fed protein survived, identifying protein as an essential dietary component. In 1840, Justus Liebig discovered the chemical makeup of carbohydrates (sugars), fats (fatty acids) and proteins (amino acids). In the 1860s Claude Bernard discovered that body fat can be synthesized from carbohydrate and protein, showing that the energy in blood glucose can be stored as fat or as glycogen.In the early 1880s Kanehiro Takaki observed that Japanese sailors (whose diets consisted almost entirely of white rice) developed beriberi (or endemic neuritis, a disease causing heart problems and paralysis), but British sailors and Japanese naval officers did not. Adding various types of vegetables and meats to the diets of Japanese sailors prevented the disease.

In 1896 Eugen Baumann observed iodine in thyroid glands. In 1897, Christiaan Eijkman worked with natives of Java, who also suffered from beriberi. Eijkman observed that chickens fed the native diet of white rice developed the symptoms of beriberi, but remained healthy when fed unprocessed brown rice with the outer bran intact. Eijkman cured the natives by feeding them brown rice, demonstrating that food can cure disease. Over two decades later, nutritionists learned that the outer rice bran contains vitamin B

From 1900 to the Present

In the early 20th century Carl von Voit and Max Rubner independently measured caloric energy expenditure in different species of animals, applying principles of physics in nutrition. In 1906, Wilcock and Hopkins showed that the amino acid tryptophan was necessary for the survival of rats. He fed them a special mixture of food containing all the nutrients he believed were essential for survival, but the rats died. A second group of rats to which he also fed an amount of milk containing vitamins. Gowland Hopkins recognized "accessory food factors" other than calories, protein and minerals, as organic materials essential to health but which the body cannot synthesize. In 1907 Stephen M. Babcock and Edwin B. Hart conducted the single-grain experiment. This experiment ran through 1911.

In 1912 Casimir Funk coined the term vitamin to label a vital factor in the diet: from the words "vital" and "amine," because these unknown substances preventing scurvy, beriberi, and pellagra, were thought then to derive from ammonia. The vitamins were studied in the first half of the 20th century. In 1913 Elmer McCollum discovered the first vitamins, fat-soluble vitamin A and water-soluble vitamin B (in 1915; later identified as a complex of several water-soluble vitamins) and named vitamin C as the then-unknown substance preventing scurvy. Lafayette Mendel (1872-1935) and Thomas Osborne (1859–1929) also performed pioneering work on vitamins A and B. In 1919 Sir Edward Mellanby incorrectly identified rickets as a vitamin A deficiency, because he could cure it in dogs with cod-liver oil. In 1922 McCollum destroyed the vitamin A in cod liver oil but found it still cured rickets, thus identifying vitamin D. Also in 1922, H.M. Evans and L.S. Bishop discovered vitamin E as essential for rat pregnancy, and originally called it "food factor X" until 1925.

In 1925 Hart discovered that iron absorption requires trace amounts of copper. In 1927 Adolf Otto Reinhold Windaus synthesized vitamin D, for which he won the Nobel Prize in Chemistry in 1928. In 1928 Albert Szent-Györgyi isolated ascorbic acid, and in

1932 proved that it is vitamin C by preventing scurvy. In 1935 he synthesized it, and in 1937 won a Nobel Prize for his efforts. Szent-Györgyi concurrently elucidated much of the citric acid cycle. In the 1930s William Cumming Rose identified essential amino acids, necessary protein components which the body cannot synthesize. In 1935 Eric Underwood and Hedley Marston independently discovered the necessity of cobalt. In 1936 Eugene Floyd Dubois showed that work and school performance relate to caloric intake. In 1938 Erhard Fernholz discovered the chemical structure of vitamin E. It was synthesised by Paul Karrer (1889–1971).

From 1940 rationing in the United Kingdom – during and after World War II – took place according to nutritional principles drawn up by Elsie Widdowson and others. In 1941 the National Research Council established the first Recommended Dietary Allowances (RDAs). In 1992 the U.S. Department of Agriculture introduced the Food Guide Pyramid. In 2002 a Natural Justice study showed a relation between nutrition and violent behavior. In 2005 a study found that in addition to bad nutrition, adenovirus may cause obesity.

REFERENCES

1. Lieberson AD (8 November 2004). "How long can a person survive without food?". Scientific American. Retrieved 17 December 2014.
2. Janiszewski S (13 May 2011). "The Science of Starvation: How long can humans survive without food or water?". The Public Library of Science, Science Blog Network. Retrieved 17 December 2014.
3. Progress for Children: A Report Card on Nutrition (No. 4), UNICEF, May 2006, ISBN 978-92-806-3988-9. http://www.unicef.org/nutrition/index_33685.html
4. World Health Organization. (2013). Essential Nutrition Actions: improving maternal, newborn, infant and young child health and nutrition. Washington, DC: WHO. http://www.who.int/nutrition/publications/infantfeeding/essential_nutrition_actions/en/index.html
5. Lean, Michael E.J. (2015). "Principles of Human Nutrition". Medicine. 43 (2): 61–65. doi:10.1016/j.mpmed.2014.11.009. Retrieved March 31, 2015.
6. World Health Organization, Food and Agricultural Organization of the United Nations (2004). Vitamin and mineral requirements in human nutrition (2. ed.). Geneva [u.a.]: World Health Organization. ISBN 9241546123.
7. Berg J, Tymoczko JL, Stryer L (2002). Biochemistry (5th ed.). San Francisco: W.H. Freeman. p. 603. ISBN 0-7167-4684-0.
8. "National Nutrition Survey: Nutrient Intakes and Physical Measurements". Australian Bureau of Statistics. 1995.
9. Cummings, J H; Stephen, A M (2007). "Carbohydrate terminology and classification" (PDF). European Journal of Clinical Nutrition. 61: S5–S18. doi:10.1038/sj.ejcn.1602936. PMID 17992187.
10. "Grams of Carbohydrates in White Bread - Carb Counter". www.carb-counter.net. Retrieved 2016-03-18.
11. "American Rice, Inc.". www.amrice.com. Retrieved 2016-03-18.

12. "Carbohydrates That Contain Monosaccharides". Healthy eating.
13. Lean, Michael E.J. (2015). "Principles of human nutrition". Medicine. 43 (2): 61–65. doi:10.1016/j.mpmed.2014.11.009.
14. ^ Jump up to:a b Dietary Reference Intakes: The Essential Guide to Nutrient Requirements, published by the Institute of Medicine's Food and Nutrition Board, currently available online at http://fnic.nal.usda.gov/dietary-guidance/dietary-reference-intakes/dri-reports
15. Craig, W. J.; Mangels, AR (2009). "Position of the American Dietetic Association: Vegetarian diets". Journal of the American Dietetic Association. 109 (7): 1266–82. PMID 19562864.
16. Dinu, M; Abbate, R; Gensini, G. F.; Casini, A; Sofi, F (2016). "Vegetarian, vegan diets and multiple health outcomes: A systematic review with meta-analysis of observational studies". Critical Reviews in Food Science and Nutrition: 0. doi :10.1080/10408398.2016.1138447. PMID 26853923.
17. L. Kathleen Mahan; Janice L. Raymond; Sylvia Escott-Stump (2012). Krausw's Food and the Nutrition Care Process (13th ed.). St. Louis: Elsevier. ISBN 978-1-4377-2233-8.
18. USDA National Nutrient Database for Standard Reference, SR26, 2013
19. D. E. C. Corbridge (1995). Phosphorus: An Outline of its Chemistry, Biochemistry, and Technology (5th ed.). Amsterdam: Elsevier. ISBN 0-444-89307-5.
20. Lippard, S. J.; Berg, J. M. (1994). Principles of Bioinorganic Chemistry. Mill Valley, CA: University Science Books. ISBN 0-935702-73-3.
21. Kapil, Umesh (2016-11-22). "Health Consequences of Iodine Deficiency". Sultan Qaboos University Medical Journal. 7 (3): 267–272. ISSN 2075-051X. PMC 3074887 . PMID 21748117.
22. Shils; et al. (2005). Modern Nutrition in Health and Disease. Lippincott Williams and Wilkins. ISBN 0-7817-4133-5.
23. Berardi, John. "The Big T: How Your Lifestyle Influences Your Testosterone Levels". Deepfitness.com. Archived from the original on 25 November 2013. Retrieved 8 October 2013.
24. European Society of Cardiology; Atar; Borch-Johnsen; Boysen; Burell; Cifkova; Dallongeville; De Backer; Ebrahim; Gjelsvik; Herrmann-Lingen; Hoes; Humphries; Knapton; Perk; Priori; Pyorala; Reiner; Ruilope; Sans-Menendez; Scholte Op Reimer; Weissberg; Wood; Yarnell; Zamorano; Walma; Fitzgerald; Cooney; Dudina; European Society of Cardiology (ESC) Committee for Practice Guidelines (CPG) (2007). "European guidelines on cardiovascular disease prevention in clinical practice: executive summary" (PDF). European Heart Journal. 28 (19): 2375–2414. doi:10.1093/eurheartj/ehm316. PMID 17726041.
25. L, Winkler; Schlag B; Kessner C; Maess J; Dargel R. (1980). "Development of brain lipids in rats receiving a fat-free diet with respect to myelinization". Acta Biol Med Ger. 39 (11–12): 1197–203. PMID 7245988.
26. Haque, ZU; Mozaffar Z. (1992). "Importance of dietary cholesterol for the

maturation of mouse brain myelin". Biosci Biotechnol Biochem. 56 (8): 1351–4. doi:10.1271/bbb.56.1351. PMID 1369207.

27. Behrman, J. R. (1996). "The Impact of Health and Nutrition on Education". The World Bank Research Observer. 11 (1): 23–37. doi:10.1093/wbro/11.1.23. JSTOR 3986477.
28. American College Health Association (2007). "American College Health Association National College Health Assessment Spring 2006 Reference Group Data Report (Abridged): The American College Health Association". Journal of American College Health. 55 (4): 195–206. doi:10.3200/JACH.55.4.195-206. PMID 17319325.
29. Benton, David; Sargent, Julia (1992). "Breakfast, blood glucose and memory". Biological Psychology. 33 (2–3): 207–10. doi:10.1016/0301-0511(92)90032-P. PMID 1525295.
30. Kanarek, Robin B.; Swinney, David (1990). "Effects of food snacks on cognitive performance in male college students". Appetite. 14 (1): 15–27. doi:10.1016/0195-6663(90)90051-9. PMID 2310175.
31. Whitley JR, O'Dell BL, Hogan AG; O'Dell; Hogan (September 1951). "Effect of diet on maze learning in second generation rats; folic acid deficiency". J. Nutr. 45 (1): 153–60. PMID 14880969.
32. Umezawa M, Kogishi K, Tojo H, et al. (February 1999). "High-linoleate and high-alpha-linolenate diets affect learning ability and natural behavior in SAMR1 mice". J. Nutr. 129 (2): 431–7. PMID 10024623.
33. Glewwe, Paul; Jacoby, Hanan G; King, Elizabeth M (2001). "Early childhood nutrition and academic achievement: A longitudinal analysis". Journal of Public Economics. 81 (3): 345–68. doi:10.1016/S0047-2727(00)00118-3.
34. Guernsey L (1993). "Many colleges clear their tables of steak, substitute fruit and pasta". Chronicle of Higher Education. 39 (26): A30.
35. Duster T, Waters A (2006). "Engaged learning across the curriculum: The vertical integration of food for thought". Liberal Education. 92 (2): 42.
36. Lakhan SE, Vieira KF; Vieira (2008). "Nutritional therapies for mental disorders". Nutr J. 7 (1): 2. doi:10.1186/1475-2891-7-2. PMC 2248201 . PMID 18208598.
37. Coren, Michael (2005-03-10). "Study: Cancer no longer rare in poorer countries". CNN. Retrieved 2007-01-01.
38. Fernández-García JC; et al. (2013). "Inflammation, oxidative stress and metabolic syndrome: dietary modulation". Curr Vasc Pharmacol. 11 (October): 1570–1611. PMID 24168441.
39. Feldeisen SE, Tucker KL; Tucker (2007). "Nutritional strategies in the prevention and treatment of metabolic syndrome". Applied Physiology, Nutrition, and Metabolism. National Research Council of Canada Research Press. 32 (1): 46–60. doi:10.1139/h06-101. PMID 17332784. Retrieved November 3, 2013.
40. "Why is too much water dangerous?". BBC News. 2007-01-15. Retrieved

2008-11-09.

41. WHO (2013). Global Nutrition Policy. Report of a WHO Expert Committee. Geneva, World Health Organization. http://apps.who.int/iris/bitstream/10665/84408/1/9789241505529_eng.pdf
42. "UNICEF 2012 Progress Report: Rapid progress in child survival". World Health Organization. 18 September 2012. Retrieved 28 May 2016.
43. WHO. World health statistics 2013: a wealth of information on global public health. Geneva, WHO, 2013.[page needed]
44. Liu, Li; Johnson, Hope L; Cousens, Simon; Perin, Jamie; Scott, Susana; Lawn, Joy E; Rudan, Igor; Campbell, Harry; Cibulskis, Richard; Li, Mengying; Mathers, Colin; Black, Robert E; Child Health Epidemiology Reference Group of WHO UNICEF (2012). "Global, regional, and national causes of child mortality: An updated systematic analysis for 2010 with time trends since 2000". The Lancet. 379 (9832): 2151–61. doi:10.1016/S0140-6736(12)60560-1. PMID 22579125.
45. UNICEF, WHO, World Bank. UNICEF-WHO-World Bank Joint child malnutrition estimates. New York, Geneva & Washington DC, UNICEF, WHO & World Bank, 2012 (http://www.who.int/nutgrowthdb/estimates/en/index.html, accessed 27 March 2013)
46. Black, Robert E; Victora, Cesar G; Walker, Susan P; Bhutta, Zulfiqar A; Christian, Parul; De Onis, Mercedes; Ezzati, Majid; Grantham-Mcgregor, Sally; Katz, Joanne; Martorell, Reynaldo; Uauy, Ricardo; Maternal Child Nutrition Study Group (2013). "Maternal and child undernutrition and overweight in low-income and middle-income countries". The Lancet. 382 (9890): 427–51. doi:10.1016/S0140-6736(13)60937-X. PMID 23746772.
47. Murray C, Lopez A; Lopez (1997). "Global mortality, disability, and the contribution of risk factors: Global Burden of Disease Study". Lancet. 349 (9063): 1436–42. doi:10.1016/S0140-6736(96)07495-8. PMID 9164317.
48. Black RE; et al. (2008). "Maternal and child undernutrition: global and regional exposures and health consequences". Lancet. 371 (9608): 243–260. doi:10.1016/s0140-6736(07)61690-0. PMID 18207566.
49. IMPROVING CHILD NUTRITION > UNICEF. (April 2013). IMPROVING CHILD NUTRITION: The achievable imperative for global progress. http://www.unicef.org/publications/index_68661.html
50. FAO (2012). The state of food insecurity in the world 2012: Economic growth is necessary but not sufficient to accelerate reduction of hunger and malnutrition. Rome, Food and Agricultural Organization of the United Nations. http://www.fao.org/publications/sofi/en/ (Accessed 7 December 2012.).
51. UNSCN (2009). Global financial and economic crisis – The most vulnerable are at increased risk of hunger and malnutrition. United Nations Standing Committee on Nutrition. http://www.unscn.org/en/publications/nutrition_briefs/#Nutrition_impacts_of_global_food_and_financial_crises.
52. IBRD, World Bank (2012). Global Monitoring Report 2012: Food prices, nutrition,

and the Millennium Development Goals. International Bank for Reconstruction and Development (IBRD)/World Bank, Washington, DC..

53. Darnton-Hill, Ian, C. Nishida and W.P.T. James, 'A life course approach to diet, nutrition and the prevention of chronic diseases', Public Health Nutrition, vol. 7, no. 1A, 2004, pp. 101–121.
54. Finucane, Mariel M; Stevens, Gretchen A; Cowan, Melanie J; Danaei, Goodarz; Lin, John K; Paciorek, Christopher J; Singh, Gitanjali M; Gutierrez, Hialy R; Lu, Yuan; Bahalim, Adil N; Farzadfar, Farshad; Riley, Leanne M; Ezzati, Majid; Global Burden of Metabolic Risk Factors of Chronic Diseases Collaborating Group (Body Mass Index) (2011). "National, regional, and global trends in body-mass index since 1980: Systematic analysis of health examination surveys and epidemiological studies with 960 country-years and 9·1 million participants". The Lancet. 377 (9765): 557–67. doi:10.1016/S0140-6736(10)62037-5. PMID 21295846.
55. "WHO (2011a). Global status report on noncommunicable diseases 2010". Geneva, World Health Organization.
56. Stein, A. J. (2010). "Global Impacts of Human Mineral Malnutrition". Plant and Soil. 335 (1/2): 133–154. doi:10.1007/s11104-009-0228-2.
57. WHO. Iron deficiency anemia: assessment, prevention, and control. A guide for program managers. Geneva, WHO, 2001
58. WHO (2001). Iron deficiency anemia: Assessment, prevention, and control. A guide for program managers. Geneva, World Health Organization.
59. WHO, Centers for Disease Control. Worldwide prevalence of anemia 1993–2005: WHO global database of anemia. Geneva, WHO, 2008.
60. W Institute of Medicine. Dietary reference intakes for vitamin A, vitamin K, arsenic, boron, chromium, copper, iodine, iron, manganese, molybdenum, nickel, silicon, vanadium, and zinc. Washington DC, National Academy Press, 2001
61. Algarín, Cecilia; Peirano, Patricio; Garrido, Marcelo; Pizarro, Felipe; Lozoff, Betsy (2003). "Iron Deficiency Anemia in Infancy: Long-Lasting Effects on Auditory and Visual System Functioning". Pediatric Research. 53 (2): 217–23. doi:10.1203/01.PDR.0000047657.23156.55. PMID 12538778.
62. WHO. Mortality and burden of disease attributable to selected major risks. Geneva, WHO, 2009 (http://www.who.int/healthinfo/global_burden_disease/GlobalHealthRisks_ report_full.pdf, accessed 17 May 2013).[dead link][page needed]
63. WHO (2009). Global prevalence of vitamin A deficiency in populations at risk 1995–2005. WHO Global Database on Vitamin A Deficiency. Geneva, World Health Organization. http://www.edu-lib.us/whqlibdoc.who.int/publications/2009/9789241598019_eng.pdf
64. WHO. Global prevalence of vitamin A deficiency in populations at risk 1995–2005: WHO Global database of vitamin A deficiency. Geneva, WHO, 2009.
65. Sommer A, West KP Jr. Vitamin A deficiency: health, survival, and vision. New

York, Oxford University Press, 1996[page needed]

66. Lozoff, Betsy; Jimenez, Elias; Wolf, Abraham W. (1991). "Long-Term Developmental Outcome of Infants with Iron Deficiency". New England Journal of Medicine. 325 (10): 687–94. doi:10.1056/NEJM199109053251004. PMID 1870641.
67. Andersson M, Karumbunathan V, Zimmermann MB; Karumbunathan; Zimmermann (2012). "Global iodine status in 2011 and trends over the past decade". Journal of Nutrition. 142 (4): 744–750. doi:10.3945/jn.111.149393. PMID 22378324.
68. Jones, Gareth; Steketee, Richard W; Black, Robert E; Bhutta, Zulfiqar A; Morris, Saul S; Bellagio Child Survival Study Group (2003). "How many child deaths can we prevent this year?". The Lancet. 362 (9377): 65–71. doi:10.1016/S0140-6736(03)13811-1. PMID 12853204.
69. WHO. Report of the expert consultation on the optimal duration of exclusive breastfeeding. Geneva, WHO, 2001.[page needed]
70. Ramakrishnan, U; Yip, R (2002). "Experiences and challenges in industrialized countries: Control of iron deficiency in industrialized countries". The Journal of Nutrition. 132 (4 Suppl): 820S–4S. PMID 11925488.
71. Black, Robert E; Allen, Lindsay H; Bhutta, Zulfiqar A; Caulfield, Laura E; De Onis, Mercedes; Ezzati, Majid; Mathers, Colin; Rivera, Juan (2008). "Maternal and child undernutrition: Global and regional exposures and health consequences". The Lancet. 371 (9608): 243–60. doi:10.1016/S0140-6736(07)61690-0. PMID 18207566.
72. http://www.mypyramid.gov/downloads/sample_menu.pdf
73. UN (2011b). The Millennium Development Goals report 2011. New York, United Nations. http://www.un.org/en/development/desa/news/statistics/mdg-report-2011.html.
74. "World Health Organization, European Health Report 2005: Public health action for healthier children and populations"(PDF). WHO Regional Office for Europe, Copenhagen. 2005.
75. "WHO (2013b)" (PDF). World health statistics. Geneva, World Health Organization.
76. Van de Poel E; et al. (2008). "Socioeconomic inequality in malnutrition in developing countries". Bulletin of the World Health Organization. 86 (4): 241–320.
77. Polhamus B, et al. (2004). "Pediatric Nutrition Surveillance 2003 Report, Table 18D" (PDF). U.S. Department of Health and Human Services, Centers for Disease Control and Prevention, Atlanta.
78. Lassi, Zohra S; Haider, Batool A; Bhutta, Zulfiqar A (2010). Bhutta, Zulfiqar A, ed. "Community-based intervention packages for reducing maternal and neonatal morbidity and mortality and improving neonatal outcomes". Cochrane Database of Systematic Reviews (11): CD007754. doi:10.1002/14651858.CD007754.pub2. PMID 21069697.
79. Nabarro, David (2013). "Global child and maternal nutrition—the SUN rises".

The Lancet. 382 (9893): 666–7. doi:10.1016/S0140-6736(13)61086-7. PMID 23746773.

80. Commission on Life Sciences. (1985). Nutrition Education in US Medical Schools, p. 4. National Academies Press.
81. Adams KM, Lindell KC, Kohlmeier M, Zeisel SH; Lindell; Kohlmeier; Zeisel (April 2006). "Status of nutrition education in medical schools,". Am. J. Clin. Nutr. 83 (4): 941S–4S. PMC 2430660 . PMID 16600952.
82. Rosenbloom, Christine (2011). "Sports Nutrition: Updating the Practitioner". Nutrition Today. Lippincott Williams & Wilkins, Inc. 46 (4): 197–202. doi:10.1097/NT.0b013e3182261dc0. Retrieved March 31, 2015.
83. Di Pasquale, Mauro G. (2008). "Utilization of Proteins in Energy Metabolism". In Ira Wolinsky, Judy A. Driskell. Sports Nutrition: Energy metabolism and exercise. CRC Press. p. 73. ISBN 978-0-8493-7950-5.
84. "Iowa State University: Extension and outreach".
85. William D. McArdle; Frank I. Katch; Victor L. Katch (2006). Exercise Physiology: Energy, Nutrition, and Human Performance. Lippincott Williams & Wilkins. ISBN 0-8121-0682-2.
86. "Nutrition – Healthy eating: Bread, cereals and other starchy foods". BBC. July 2008. Archived from the original on September 13, 2008. Retrieved 2008-11-09.
87. Naeye, R. L.; Blanc, W.; Paul, C. (1973). "Effects of Maternal Nutrition on the Human Foetus". Pediatrics (Evanston). 52 (4): 494–503. Retrieved 2015-04-17.
88. Wahlqvist, M. L. (2011). Food and Nutrition: Food and Health Systems in Australia and New Zealand (3rd ed.). NSW, Australia: Allen & Unwin. pp. 429–441. ISBN 978 1 74175 897 9.
89. Chwang, Leh-Chii (September 2012). "Nutrition and dietics in aged care". Nutrition and Dietics. 69 (3): 203–207. doi:10.1111/j.1747-0080.2012.01617.x. Retrieved March 31, 2015.
90. Villa, P.; Bouville, C.; Courtin, J.; Helmer, D.; Mahieu, E.; Shipman, P.; Belluomini, G.; Branca, M. (1986). "Cannibalism in the Neolithic". Science. 233 (4762): 431–7. Bibcode:1986Sci...233..431V. doi:10.1126/science.233.4762.431. PMID 17794567.
91. Guzzardi, Sergio (2014-07-27). Buona Pasta. Sergio Guzzardi (published 2014). p. 19. ISBN 9786050314915. Retrieved 2014-11-21. The first two certain dates in the history of pasta in Italy are: 1154, when in a sort of tour guide ahead of its [time] Arab geographer Al-Idrin mentions 'a food of flour in the form of wires,' called Triyah [...], which is packaged in Palermo and was exported in barrels throughout the peninsula [...]; [...] and 1279 [...].
92. "Lesson 1: Nutrition, history of nutrition and scope of nutrition" (PDF). Unit 1: Introduction to nutrition. Rai University. p. 1. Archived from the original (PDF) on August 24, 2006. Retrieved 2014-11-21. Food was sacred – hence eating was a sacred act. Food is magical – cooking was a magico-religous [sic]

activity, every gesture, act and dish perhaps, endowed with ritual significance.

93. History of the Study of Nutrition in Western Culture(Rai University lecture notes for General Nutrition course, 2004)
94. Daniel 1:5-16 (alternative translation)
95. Smith, R. (2004). "'Let food be thy medicine...'". BMJ. 328 (7433): 0–g. doi:10.1136/bmj.328.7433.0-g. PMC 318470 .
96. Heinemann 2e Biology Activity Manual by Judith Brotherton and Kate Mundie[page needed]
97. "Unraveling the Enigma of Vitamin D". United States National Academy of Sciences. April 29, 2002. Archived from the original on February 4, 2013.
98. Whigham, Leah D.; Israel, Barbara A.; Atkinson, Richard L. (2006). "Adipogenic potential of multiple human adenoviruses in vivo and in vitro in animals". American Journal of Physiology. 290 (1): R190–4. doi:10.1152/ajpregu.00479.2005. PMID 16166204. Lay summary – The American Physiological Society (January 30, 2006).

CHAPTER 7

Human Disease and their Classification

Chapter Contents

- **Introduction**

Human disease is an impairment of the normal state of a human being that interrupts or modifies its vital functions. human disease can be discussed, the meanings of the terms health, physical fitness, illness, and disease must be considered. Health could be defined theoretically in terms of certain measured values; for example, a person having normal body temperature, pulse and breathing rates, blood pressure, height, weight, acuity of vision, sensitivity of hearing, and other normal measurable characteristics might be termed healthy. But what does normal mean, and how is it established? It is well known that if the temperatures are taken of a large number of active, presumably healthy, individuals the temperatures will all come close to 98.6 °F (37 °C). The great preponderance of these values will fall between 98.4 °F (36.9 °C) and 98.8 °F (37.1 °C).

INTRODUCTION

A disease is a particular abnormal condition, a disorder of a structure or function that affects part or all of an organism. The study of disease is called pathology which includes the causal study of etiology. Disease is often construed as a medical condition associated with specific symptoms and signs. It may be caused by external factors such as pathogens, or it may be caused by internal dysfunctions particularly of the immune system such as an immunodeficiency, or a hypersensitivity including allergies and autoimmunity.

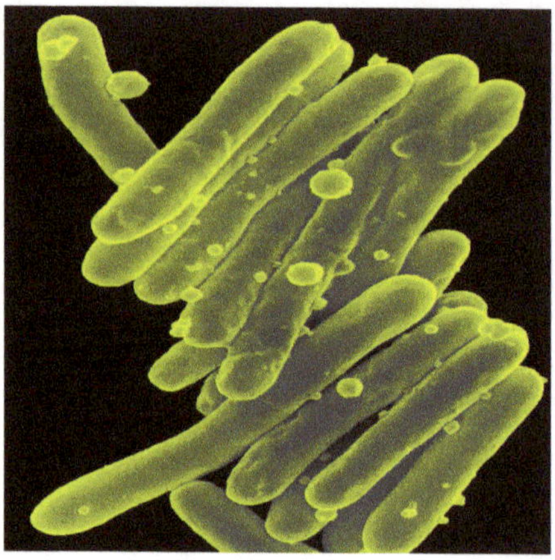

Scanning electron micrograph of *Mycobacterium tuberculosis*, a bacterium that causes tuberculosis.

In humans, *disease* is often used more broadly to refer to any condition that causes pain, dysfunction, distress, social problems, or death to the person afflicted, or similar problems for those in contact with the person. In this broader sense, it sometimes includes injuries, disabilities, disorders, syndromes, infections, isolated symptoms, deviant behaviors, and atypical variations of structure and function, while in other contexts and for other purposes these may be considered distinguishable categories. Diseases can affect people not only physically, but also emotionally, as contracting and living with a disease can alter the affected person's perspective on life.

Death due to disease is called death by natural causes. There are four main types of disease: infectious diseases, deficiency diseases, genetic diseases (both hereditary and non-hereditary), and physiological diseases. Diseases can also be classified as communicable and non-communicable. The deadliest diseases in humans are coronary artery disease (blood flow obstruction), followed by cerebrovascular disease and lower respiratory infections.

Terminology

Concepts

In many cases, terms such as *disease*, *disorder*, *morbidity* and *illness* are used interchangeably. There are situations, however, when specific terms are considered preferable.

Disease

The term *disease* broadly refers to any condition that impairs the normal functioning of the body. For this reason, diseases are associated with dysfunctioning of the body›s normal homeostatic processes. The term *disease* has both a count sense (*a disease, two diseases, many diseases*) and a noncount sense (*not much disease, less disease, a lot of disease*). Commonly, the term is used to refer specifically to infectious diseases, which are clinically evident diseases that result from the presence of pathogenic microbial agents, including viruses, bacteria, fungi, protozoa, multicellular organisms, and aberrant proteins known as prions. An infection that does not and will not produce clinically evident impairment of normal functioning, such as the presence of the normal bacteria and yeasts in the gut, or of a passenger virus, is not considered a disease. By contrast, an infection that is asymptomatic during its incubation period, but expected to produce symptoms later, is usually considered a disease. Non-infectious diseases are all other diseases, including most forms of cancer, heart disease, and genetic disease.

- *Acquired disease* disease that began at some point during one's lifetime, as opposed to disease that was already present at birth, which is congenital disease. "Acquired" sounds like it could mean "caught via contagion", but it simply means acquired sometime after birth. It also sounds like it could imply secondary disease, but acquired disease can be primary disease.
- *Acute disease* disease of a short-term nature (acute); the term sometimes also connotes a fulminant nature
- *Chronic disease* disease that is a long-term issue (chronic)
- *Congenital disease* disease that is present at birth. It is often, genetic and can be inherited. It can also be the result of a vertically transmitted infection from the mother such as HIV/AIDS.
- *Genetic disease* disease that is caused by genetic mutation. It is often inherited, but some mutations are random and de novo.
- *Hereditary or Inherited Disease* a type of genetic disease caused by mutation that is hereditary (and can run in families)
- *Idiopathic Disease* disease whose cause is unknown. As medical science has advanced, many diseases whose causes were formerly complete mysteries have been somewhat explained (for example, when it was realized that autoimmunity is the cause of some forms of diabetes mellitus type 1, even if we do not yet understand every molecular detail involved) or even extensively explained (for example, when it was realized that gastric ulcers are often associated with Helicobacter pylori infection).
- *Incurable Disease* disease that cannot be cured
- *Primary Disease* disease that came about as a root cause of illness, as opposed to secondary disease, which is a sequela of another disease
- *Secondary Disease* disease that is a sequela or complication of some other disease or underlying cause (root cause). Bacterial infections can be either primary (healthy but then bacteria arrived) or secondary to a viral infection or burn, which predisposed by creating an open wound or weakened immunity (bacteria would not have gotten established otherwise).

- *Terminal Disease* disease with death as an inevitable result

Illness

Illness is generally used as a synonym for *disease*. However, this term is occasionally used to refer specifically to the patient's personal experience of his or her disease. In this model, it is possible for a person to have a disease without being ill (to have an objectively definable, but asymptomatic, medical condition, such as a subclinical infection), and to be *ill* without being *diseased* (such as when a person perceives a normal experience as a medical condition, or medicalizes a non-disease situation in his or her life – for example, a person who feels unwell as a result of embarrassment, and who interprets those feelings as sickness rather than normal emotions). Symptoms of illness are often not directly the result of infection, but a collection of evolved responses—sickness behavior by the body—that helps clear infection. Such aspects of illness can include lethargy, depression, loss of appetite, sleepiness, hyperalgesia, and inability to concentrate.

Disorder

In medicine, a disorder is a functional abnormality or disturbance. Medical disorders can be categorized into mental disorders, physical disorders, genetic disorders, emotional and behavioral disorders, and functional disorders. The term *disorder* is often considered more value-neutral and less stigmatizing than the terms *disease* or *illness*, and therefore is a preferred terminology in some circumstances. In mental health, the term *mental disorder* is used as a way of acknowledging the complex interaction of biological, social, and psychological factors in psychiatric conditions. However, the term *disorder* is also used in many other areas of medicine, primarily to identify physical disorders that are not caused by infectious organisms, such as metabolic disorders.

Medical condition

A medical condition is a broad term that includes all diseases, lesions, disorders, or nonpathologic condition that normally receives medical treatment, such as pregnancy or childbirth. While the term *medical condition* generally includes mental illnesses, in some contexts the term is used specifically to denote any illness, injury, or disease except for mental illnesses. The Diagnostic and Statistical Manual of Mental Disorders (DSM), the widely used psychiatric manual that defines all mental disorders, uses the term *general medical condition* to refer to all diseases, illnesses, and injuries except for mental disorders. This usage is also commonly seen in the psychiatric literature. Some health insurance policies also define a *medical condition* as any illness, injury, or disease except for psychiatric illnesses.

As it is more value-neutral than terms like *disease*, the term *medical condition* is sometimes preferred by people with health issues that they do not consider deleterious. On the other hand, by emphasizing the medical nature of the condition, this term is sometimes rejected, such as by proponents of the autism rights movement.

The term *medical condition* is also a synonym for *medical state*, in which case it describes an individual patient's current state from a medical standpoint. This usage appears in statements that describe a patient as being *in critical condition*, for example.

Morbidity

Morbidity (from Latin *morbidus*, meaning "sick, unhealthy") is a diseased state, disability, or poor health due to any cause. The term may be used to refer to the existence of any form of disease, or to the degree that the health condition affects the patient. Among severely ill patients, the level of morbidity is often measured by ICU scoring systems. Comorbidity is the simultaneous presence of two or more medical conditions, such as schizophrenia and substance abuse.

In epidemiology and actuarial science, the term "morbidity rate" can refer to either the incidence rate, or the prevalence of a disease or medical condition. This measure of sickness is contrasted with the mortality rate of a condition, which is the proportion of people dying during a given time interval.

Syndrome

A syndrome is the association of several medical signs, symptoms, and or other characteristics that often occur together. Some syndromes, such as Down syndrome, have only one cause; for these, the names "syndrome" and "disease" can be synonymous. For example, Charcot–Marie–Tooth syndrome is also called Charcot–Marie–Tooth disease. Others, such as Parkinsonian syndrome, have multiple possible causes. For example, acute coronary syndrome is not a disease but rather the manifestation of any of several diseases, such as myocardial infarction secondary to coronary artery disease. In yet other syndromes, the cause is unknown. A familiar syndrome name often remains in use even after an underlying cause has been found, or when there are a number of different possible primary causes.

Predisease

Predisease is a subclinical or prodromal vanguard of a disease state. Prediabetes and prehypertension are common examples. The nosology or epistemology of predisease is contentious, though, because there is seldom a bright line differentiating a legitimate concern for subclinical/prodromal/premonitory status (on one hand) and conflict of interest–driven disease mongering or medicalization (on the other hand). Identifying legitimate predisease can result in useful preventive measures, such as motivating the person to get a healthy amount of physical exercise, but labeling a healthy person with an unfounded notion of predisease can result in overtreatment, such as taking drugs that only help people with severe disease or paying for drug prescription instances whose benefit–cost ratio is minuscule (placing it in the waste category of CMS› «waste, fraud, and abuse» classification).

Types by Body System

Mental

Mental illness is a broad, generic label for a category of illnesses that may include affective or emotional instability, behavioral dysregulation, cognitive dysfunction or impairment. Specific illnesses known as mental illnesses include major depression, generalized anxiety disorders, schizophrenia, and attention deficit hyperactivity disorder, to name a few. Mental illness can be of biological (e.g., anatomical, chemical, or genetic) or psychological (e.g., trauma or conflict) origin. It can impair the affected person's ability to work or study and can harm interpersonal relationships. The term insanity is used technically as a legal term.

Organic

An organic disease is one caused by a physical or physiological change to some tissue or organ of the body. The term sometimes excludes infections. It is commonly used in contrast with mental disorders. It includes emotional and behavioral disorders if they are due to changes to the physical structures or functioning of the body, such as after a stroke or a traumatic brain injury, but not if they are due to psychosocial issues.

Stages

"Flareup" redirects here. For the Transformers character, see Flareup (Transformers).

In an infectious disease, the incubation period is the time between infection and the appearance of symptoms. The latency period is the time between infection and the ability of the disease to spread to another person, which may precede, follow, or be simultaneous with the appearance of symptoms. Some viruses also exhibit a dormant phase, called viral latency, in which the virus hides in the body in an inactive state. For example, varicella zoster virus causes chickenpox in the acute phase; after recovery from chickenpox, the virus may remain dormant in nerve cells for many years, and later cause herpes zoster (shingles).

Acute Disease

An acute disease is a short-lived disease, like the common cold.

Chronic Disease

A chronic disease is one that lasts for a long time, usually at least six months. During that time, it may be constantly present, or it may go into remission and periodically relapse. A chronic disease may be stable (does not get any worse) or it may be progressive (gets worse over time). Some chronic diseases can be permanently cured. Most chronic diseases can be beneficially treated, even if they cannot be permanently cured.

Flare-Up

A flare-up can refer to either the recurrence of symptoms or an onset of more severe symptoms.

Refractory Disease

A refractory disease is a disease that resists treatment, especially an individual case that resists treatment more than is normal for the specific disease in question.

Progressive Disease

Progressive disease is a disease whose typical natural course is the worsening of the disease until death, serious debility, or organ failure occurs. Slowly progressive diseases are also chronic diseases; many are also degenerative diseases. The opposite of progressive disease is *stable disease* or *static disease*: a medical condition that exists, but does not get better or worse.

Cure

A cure is the end of a medical condition or a treatment that is very likely to end it, while remission refers to the disappearance, possibly temporarily, of symptoms. Complete remission is the best possible outcome for incurable diseases.

Clinical Disease

One that has clinical consequences, i.e., the stage of the disease that produces the characteristic signs and symptoms of that disease. AIDS is the clinical disease stage of HIV infection.

Subclinical Disease

Also called silent disease, silent stage, or asymptomatic disease. This is a stage in some diseases before the symptoms are first noted.

Terminal Phase

If a person will die soon from a disease, regardless of whether that disease typically causes death, then the stage between the earlier disease process and active dying is the terminal phase.

Extent

Localized Disease

A localized disease is one that affects only one part of the body, such as athlete's foot or an eye infection.

Disseminated Disease

A disseminated disease has spread to other parts; with cancer, this is usually called metastatic disease.

Systemic Disease

A systemic disease is a disease that affects the entire body, such as influenza or high blood pressure.

Classifications

Diseases may be classified by etiology (cause), pathogenesis (mechanism by which the disease is caused), or by symptom(s). Alternatively, diseases may be classified according to the organ system involved, though this is often complicated since many diseases affect more than one organ.

A chief difficulty in nosology is that diseases often cannot be defined and classified clearly, especially when etiology or pathogenesis are unknown. Thus diagnostic terms often only reflect a symptom or set of symptoms (syndrome).

Classical classification of human disease derives from observational correlation between pathological analysis and clinical syndromes. Today it is preferred to classify them by their etiology if it is known.

The most known and used classification of diseases is the World Health Organization's ICD. This is periodically updated. Currently the last publication is the ICD-10

Causes

Only some diseases such as influenza are contagious and commonly believed infectious. The micro-organisms that cause these diseases are known as pathogens and include varieties of bacteria, viruses, protozoa and fungi. Infectious diseases can be transmitted, e.g. by hand-to-mouth contact with infectious material on surfaces, by bites of insects or other carriers of the disease, and from contaminated water or food (often via fecal contamination), etc. In addition, there are sexually transmitted diseases. In some cases, microorganisms that are not readily spread from person to person play a role, while other diseases can be prevented or ameliorated with appropriate nutrition or other lifestyle changes.

Some diseases, such as most (but not all) forms of cancer, heart disease, and mental disorders, are non-infectious diseases. Many non-infectious diseases have a partly or completely genetic basis (see genetic disorder) and may thus be transmitted from one generation to another.

Social determinants of health are the social conditions in which people live that determine their health. Illnesses are generally related to social, economic, political, and environmental circumstances. Social determinants of health have been recognized by several health organizations such as the Public Health Agency of Canada and the World Health Organization to greatly influence collective and personal well-being. The World

Health Organization's Social Determinants Council also recognizes Social determinants of health in poverty.

When the cause of a disease is poorly understood, societies tend to mythologize the disease or use it as a metaphor or symbol of whatever that culture considers evil. For example, until the bacterial cause of tuberculosis was discovered in 1882, experts variously ascribed the disease to heredity, a sedentary lifestyle, depressed mood, and overindulgence in sex, rich food, or alcohol—all the social ills of the time.

Types of Causes

Airborne

An airborne disease is any disease that is caused by pathogens and transmitted through the air.

Infectious

Infectious diseases, also known as transmissible diseases or communicable diseases, comprise clinically evident illness (i.e., characteristic medical signs and/or symptoms of disease) resulting from the infection, presence and growth of pathogenic biological agents in an individual host organism. Included in this category are *contagious diseases* – an infection, such as influenza or the common cold, that commonly spreads from one person to another – and *communicable diseases* – an disease that can spread from one person to another, but does not necessarily spread through everyday contact.

Non-Communicable

A non-communicable disease is a medical condition or disease that is non-transmissible. Non-communicable diseases cannot be spread directly from one person to another. Heart disease and cancer are examples of non-communicable diseases in humans.

Foodborne

Foodborne illness or food poisoning is any illness resulting from the consumption of food contaminated with pathogenic bacteria, toxins, viruses, prions or parasites.

Lifestyle

A lifestyle disease is any disease that appears to increase in frequency as countries become more industrialized and people live longer, especially if the risk factors include behavioral choices like a sedentary lifestyle or a diet high in unhealthful foods such as refined carbohydrates, trans fats, or alcoholic beverages.

Prevention

Many diseases and disorders can be prevented through a variety of means. These include sanitation, proper nutrition, adequate exercise, vaccinations and other self-care and public health measures.

Treatments

Medical therapies or treatments are efforts to cure or improve a disease or other health problem. In the medical field, therapy is synonymous with the word *treatment*. Among psychologists, the term may refer specifically to psychotherapy or "talk therapy". Common treatments include medications, surgery, medical devices, and self-care. Treatments may be provided by an organized health care system, or informally, by the patient or family members.

Preventive healthcare is a way to avoid an injury, sickness, or disease in the first place. A treatment or cure is applied after a medical problem has already started. A treatment attempts to improve or remove a problem, but treatments may not produce permanent cures, especially in chronic diseases. Cures are a subset of treatments that reverse diseases completely or end medical problems permanently. Many diseases that cannot be completely cured are still treatable. Pain management (also called pain medicine) is that branch of medicine employing an interdisciplinary approach to the relief of pain and improvement in the quality of life of those living with pain.

Treatment for medical emergencies must be provided promptly, often through an emergency department or, in less critical situations, through an urgent care facility.

Epidemiology

Epidemiology is the study of the factors that cause or encourage diseases. Some diseases are more common in certain geographic areas, among people with certain genetic or socioeconomic characteristics, or at different times of the year.

Epidemiology is considered a cornerstone methodology of public health research, and is highly regarded in evidence-based medicine for identifying risk factors for disease. In the study of communicable and non-communicable diseases, the work of epidemiologists ranges from outbreak investigation to study design, data collection and analysis including the development of statistical models to test hypotheses and the documentation of results for submission to peer-reviewed journals. Epidemiologists also study the interaction of diseases in a population, a condition known as a syndemic. Epidemiologists rely on a number of other scientific disciplines such as biology (to better understand disease processes), biostatistics (the current raw information available), Geographic Information Science (to store data and map disease patterns) and social science disciplines (to better understand proximate and distal risk factors). Epidemiology can help identify causes as well as guide prevention efforts.

In studying diseases, epidemiology faces the challenge of defining them. Especially for poorly understood diseases, different groups might use significantly different definitions. Without an agreed-on definition, different researchers may report different numbers of cases and characteristics of the disease.

Some morbidity databases are compiled with data supplied by states and territories health authorities, at national level (National hospital morbidity database (NHMD), for example), or at European scale (European Hospital Morbidity Database or HMDB) but not yet at world scale.

Burdens of Disease

Disease burden is the impact of a health problem in an area measured by financial cost, mortality, morbidity, or other indicators.

There are several measures used to quantify the burden imposed by diseases on people. The years of potential life lost (YPLL) is a simple estimate of the number of years that a person's life was shortened due to a disease. For example, if a person dies at the age of 65 from a disease, and would probably have lived until age 80 without that disease, then that disease has caused a loss of 15 years of potential life. YPLL measurements do not account for how disabled a person is before dying, so the measurement treats a person who dies suddenly and a person who died at the same age after decades of illness as equivalent. In 2004, the World Health Organization calculated that 932 million years of potential life were lost to premature death.

The quality-adjusted life year (QALY) and disability-adjusted life year (DALY) metrics are similar, but take into account whether the person was healthy after diagnosis. In addition to the number of years lost due to premature death, these measurements add part of the years lost to being sick. Unlike YPLL, these measurements show the burden imposed on people who are very sick, but who live a normal lifespan. A disease that has high morbidity, but low mortality, has a high DALY and a low YPLL. In 2004, the World Health Organization calculated that 1.5 billion disability-adjusted life years were lost to disease and injury. In the developed world, heart disease and stroke cause the most loss of life, but neuropsychiatric conditions like major depressive disorder cause the most years lost to being sick.

Disease category	Percent of all YPLLs lost, worldwide	Percent of all DALYs lost, worldwide	Percent of all YPLLs lost, Europe	Percent of all DALYs lost, Europe	Percent of all YPLLs lost, US and Canada	Percent of all DALYs lost, US and Canada
Cancer	8%	5%	19%	11%	25%	13%
Cardiovascular diseases, principally heart attacks and stroke	14%	10%	35%	23%	26%	14%
Infectious and parasitic diseases, especially lower respiratory tract infections, diarrhea, AIDS, tuberculosis, and malaria	37%	26%	9%	6%	5%	3%
Injuries, especially motor vehicle accidents	14%	12%	18%	13%	18%	10%
Neuropsychiatric conditions, e.g. depression	2%	13%	3%	19%	5%	28%
Premature birth and other perinatal deaths	11%	8%	4%	2%	3%	2%

Society and Culture

Obesity was a status symbol in Renaissance culture: "The Tuscan General Alessandro del Borro", attributed to Andrea Sacchi, 1645. It is now generally regarded as a disease.

How a society responds to diseases is the subject of medical sociology.

A condition may be considered a disease in some cultures or eras but not in others. For example, obesity can represent wealth and abundance, and is a status symbol in famine-prone areas and some places hard-hit by HIV/AIDS. Epilepsy is considered a sign of spiritual gifts among the Hmong people.

Sickness confers the social legitimization of certain benefits, such as illness benefits, work avoidance, and being looked after by others. The person who is sick takes on a social role called the sick role. A person who responds to a dreaded disease, such as cancer, in a culturally acceptable fashion may be publicly and privately honored with higher social status. In return for these benefits, the sick person is obligated to seek treatment and work to become well once more. As a comparison, consider pregnancy, which is not interpreted as a disease or sickness, even if the mother and baby may both benefit from medical care.

Most religions grant exceptions from religious duties to people who are sick. For example, one whose life would be endangered by fasting on Yom Kippur or during Ramadan is exempted from the requirement, or even forbidden from participating. People who are sick are also exempted from social duties. For example, ill health is the only socially acceptable reason for an American to refuse an invitation to the White House.

The identification of a condition as a disease, rather than as simply a variation of human structure or function, can have significant social or economic implications. The controversial recognitions as diseases of repetitive stress injury (RSI) and post-traumatic stress disorder (also known as "Soldier's heart", "shell shock", and "combat fatigue") has had a number of positive and negative effects on the financial and other responsibilities of governments, corporations and institutions towards individuals, as well as on the individuals themselves. The social implication of viewing aging as a disease could be profound, though this classification is not yet widespread.

Lepers were people who were historically shunned because they had an infectious disease, and the term "leper" still evokes social stigma. Fear of disease can still be a widespread social phenomenon, though not all diseases evoke extreme social stigma.

Social standing and economic status affect health. Diseases of poverty are diseases that are associated with poverty and low social status; diseases of affluence are diseases that are associated with high social and economic status. Which diseases are associated with which states varies according to time, place, and technology. Some diseases, such as diabetes mellitus, may be associated with both poverty (poor food choices) and affluence (long lifespans and sedentary lifestyles), through different mechanisms. The term lifestyle diseases describes diseases associated with longevity and that are more common among older people. For example, cancer is far more common in societies in which most members live until they reach the age of 80 than in societies in which most members die before they reach the age of 50.

Language of Disease

An illness narrative is a way of organizing a medical experience into a coherent story that illustrates the sick individual's personal experience.

People use metaphors to make sense of their experiences with disease. The metaphors move disease from an objective thing that exists to an affective experience. The most popular metaphors draw on military concepts: Disease is an enemy that must be feared, fought, battled, and routed. The patient or the healthcare provider is a warrior, rather than a passive victim or bystander. The agents of communicable diseases are invaders; non-communicable diseases constitute internal insurrection or civil war. Because the threat is urgent, perhaps a matter of life and death, unthinkably radical, even oppressive, measures are society's and the patient's moral duty as they courageously mobilize to struggle against destruction. The War on Cancer is an example of this metaphorical use of language. This language is empowering to some patients, but leaves others feeling like they are failures.

Another class of metaphors describes the experience of illness as a journey: The person travels to or from a place of disease, and changes himself, discovers new information, or increases his experience along the way. He may travel "on the road to recovery" or make changes to "get on the right track" or choose "pathways". Some are explicitly immigration-themed: the patient has been exiled from the home territory of health to the land of the ill, changing identity and relationships in the process. This language is more common among British healthcare professionals than the language of physical aggression.

Some metaphors are disease-specific. Slavery is a common metaphor for addictions: The alcoholic is enslaved by drink, and the smoker is captive to nicotine. Some cancer patients treat the loss of their hair from chemotherapy as a metonymy or metaphor for all the losses caused by the disease.

Some diseases are used as metaphors for social ills: "Cancer" is a common description for anything that is endemic and destructive in society, such as poverty, injustice, or racism. AIDS was seen as a divine judgment for moral decadence, and only by purging itself from the "pollution" of the "invader" could society become healthy again. More recently, when AIDS seemed less threatening, this type of emotive language was applied to avian flu and type 2 diabetes mellitus. Authors in the 19th century commonly used tuberculosis as a symbol and a metaphor for transcendence. Victims of the disease were portrayed in literature as having risen above daily life to become ephemeral objects of spiritual or artistic achievement. In the 20th century, after its cause was better understood, the same disease became the emblem of poverty, squalor, and other social problems.

REFERENCES

1. "Disease" at Dorland's Medical Dictionary
2. "What is the deadliest disease in the world?". WHO. May 16, 2012. Retrieved December 7, 2014.
3. "Mental Illness—Glossary". US National Institute of Mental Health. Retrieved 2010-04-18.
4. "Regents Prep: Living Environment: Homeostasis". Oswego City School District Regents Exam Prep Center. Retrieved 2012-11-12.
5. "Illness" at Dorland's Medical Dictionary
6. Emson HE (April 1987). "Health, disease and illness: matters for definition". CMAJ. 136 (8): 811–3. PMC 1492114 . PMID 3567788.
7. McWhinney IR (April 1987). "Health and disease: problems of definition". CMAJ. 136 (8): 815. PMC 1492121 . PMID 3567791.
8. Hart BL (1988). "Biological basis of the behavior of sick animals". Neurosci Biobehav Rev. 12 (2): 123–137. doi:10.1016/S0149-7634(88)80004-6. PMID 3050629.
9. Johnson R (2002). "The concept of sickness behavior: a brief chronological account of four key discoveries". Veterinary Immunology and Immunopathology. 87 (3–4): 443–450. doi:10.1016/S0165-2427(02)00069-7. PMID 12072271.
10. Kelley KW, Bluthe RM, Dantzer R, Zhou JH, Shen WH, Johnson RW, Broussard SR (2003). "Cytokine-induced sickness behavior". Brain Behav Immun. 17 (Suppl 1): S112–118. doi:10.1016/S0889-1591(02)00077-6. PMID 12615196.
11. Sefton, Phil. "Condition, Disease, Disorder". AMA Style Insider. American Medical Association. Retrieved 22 July 2016.
12. American Psychiatric Association. Task Force on DSM-IV (2000). Diagnostic and statistical manual of mental disorders (4th ed.). Washington, DC: American Psychiatric Association. ISBN 978-0-89042-025-6.

13. "Expat Insurance Glossary by The Insurance Page". Archived from the original on October 27, 2008. Retrieved 2008-11-20.
14. Dorland's Medical Dictionary: morbidity, Dorland's Medical Dictionary, MerckSource
15. Lenzer, Jeanne (14 August 2012). "Blood pressure drugs for mild hypertension: Not proven to prevent heart attacks, strokes, or early death - Slate Magazine". Slate. Retrieved 16 August 2012.
16. The Free Dictionary, Clinical Disease "a stage in the history of a pathological condition that begins with anatomical or physiological changes that are sufficient to produce recognizable signs and symptoms of a disease".
17.
18. Loscalzo J1, Kohane I, Barabasi AL. Human disease classification in the postgenomic era: a complex systems approach to human pathobiology. Mol Syst Biol. 2007;3:124. Epub 2007 Jul 10.
19. Alexander van Geen, et al. "Impact of population and latrines on fecal contamination of ponds in rural Bangladesh." Science Of The Total Environment 409, no. 17 (August 2011): 3174-3182.
20. Olson, James Stuart (2002). Bathsheba's breast: women, cancer & history. Baltimore: The Johns Hopkins University Press. pp. 168–170. ISBN 0-8018-6936-6.
21. Hardy, Paul A.; Hardy, Paul A. J. (1997). Chronic Pain Management: The Essentials. Cambridge University Press. p. 10. ISBN 978-1-900151-85-6. OCLC 36881282.
22. Tuller, David (4 March 2011). "Defining an illness is fodder for debate". The New York Times.
23. National hospital morbidity database retrieved 2013-07-11
24. "Hospital Morbidity Database (HMDB)". statcan.gc.ca. Retrieved 21 September 2015.
25. European Hospital Morbidity Database which contains hospital discharge data by detailed diagnosis, age and sex, which were submitted by European countries to the WHO Regional Office for Europe.; World Health Organization Regional Office for Europe ; updated oct 2012, retrieved 2013-07-11
26. "Disease and injury regional estimates for 2004". World Health Organization
27. Standard DALYs (3% discounting, age weights).
28. DALY spreadsheet
29. YLL spreadsheet.
30. Carol Gerten-Jackson. "The Tuscan General Alessandro del Borro".
31. Haslam DW, James WP (2005). "Obesity". Lancet. 366 (9492): 1197–209. doi:10.1016/S0140-6736(05)67483-1. PMID 16198769.

32. Fadiman, Anne (1997). The spirit catches you and you fall down: a Hmong child, her American doctors, and the collision of two cultures. New York: Farrar, Straus, and Giroux. ISBN 0-374-52564-1.
33. Sulik, Gayle (2010). Pink Ribbon Blues: How Breast Cancer Culture Undermines Women's Health. New York: Oxford University Press. ISBN 0-19-974045-3.
34. Martin, Judith (2005). Miss Manners' Guide to Excruciatingly Correct Behavior. New York: W.W. Norton & Co. p. 703. ISBN 978-0-393-05874-1. OCLC 57549405.
35. Gwyn, Richard (1999). "10". In Cameron, Lynne; Low, Graham. Researching and applying metaphor. Cambridge, England: Cambridge University Press. ISBN 978-0-521-64964-3. OCLC 40881885.
36. Span, Paula (22 April 2014). "Fighting Words Are Rarer Among British Doctors". The New York Times.
37. Diedrich, Lisa (2007). Treatments: language, politics, and the culture of illness. Minneapolis: University of Minnesota Press. pp. 8, 29. ISBN 978-0-8166-4697-5. OCLC 601862594.
38. Hanne M, Hawken SJ (December 2007). "Metaphors for illness in contemporary media". Med Humanit. 33 (2): 93–9. doi:10.1136/jmh.2006.000253. PMID 23674429.

CHAPTER 8

Plant Anatomy

Chapter Contents

- **Introduction**
- **History**

Anatomy is the area of biology that deals with the structure of plants and animals, usually on the scale of whole organisms and their major systems. Major divisions falling under the umbrella of anatomy include the visible structures and organs, the comparative study of related organisms, and growth and development. Bodily systems and their functions and anatomical disease states are also a part of anatomy. Sometimes included is histology, which is the study of the different types of tissues and their special cells. In animals, the field is also concerned with major systems such as the circulatory, digestive, and nervous systems.

INTRODUCTION

Plant anatomy or phytotomy is the general term for the study of the internal structure of plants. While originally it included plant morphology, which is the description of the physical form and external structure of plants, since the mid-20th century the investigations of plant anatomy are considered a separate, distinct field, and plant anatomy refers to just the internal plant structures. Plant anatomy is now frequently investigated at the cellular level, and often involves the sectioning of tissues and microscopy.

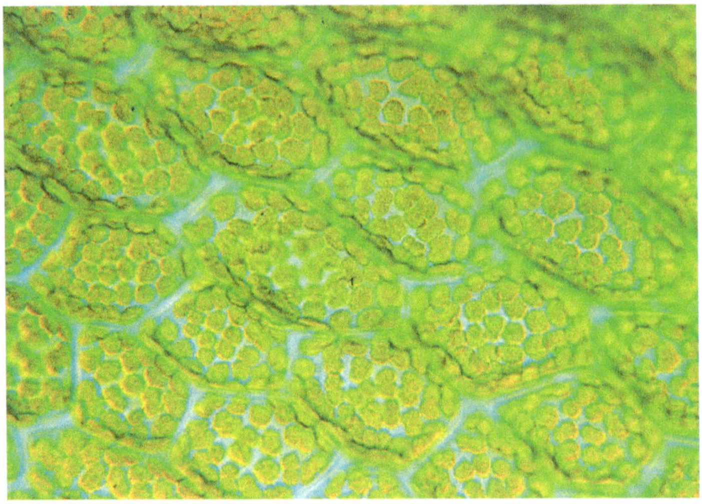

Chloroplasts in leaf cells of the moss Mnium stellare

Structural Divisions

Plant anatomy is often divided into the following categories:

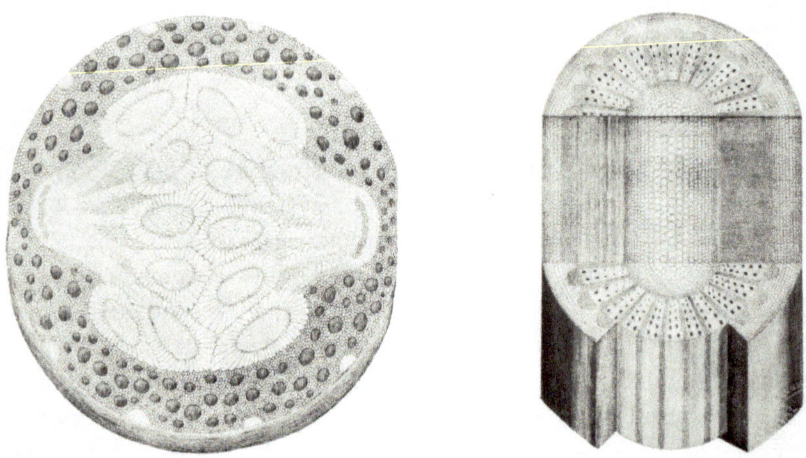

Vascular tissue of a gooseberry vine from Grew's*Anatomy of Plants*

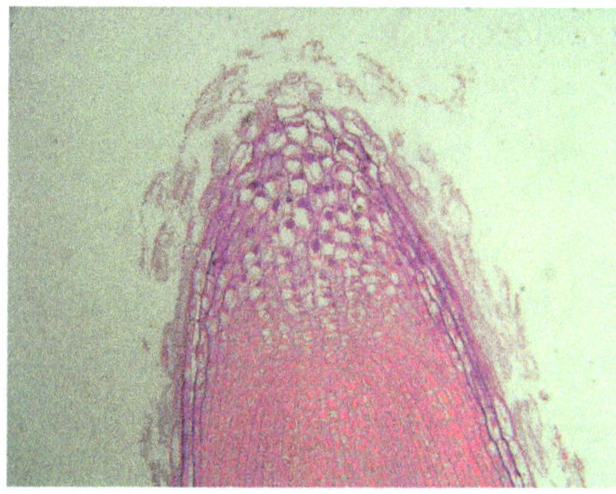

Root tip

Flower anatomy

- Calyx
- Corolla
- Androecium
- Gynoecium
- Leaf anatomy

Stem anatomy

- Stem structure

Fruit/Seed anatomy

- Ovule
- Seed structure
- Pericarp
- Accessory fruit

Wood anatomy

- Bark
- Cork
- Phloem
- Vascular cambium
- Heartwood and sapwood
- branch collar

Root anatomy

- Root structure

HISTORY

About 300 BC Theophrastus wrote a number of plant treatises, only two of which survive, *Enquiry into Plants* (Περὶ φυτῶν ἱστορία), and *On the Causes of Plants*

(Περὶ φυτῶν αἰτιῶν). He developed concepts of plant morphology and classification, which did not withstand the scientific scrutiny of the Renaissance.

A Swiss physician and botanist, Gaspard Bauhin, introduced binomial nomenclature into plant taxonomy. He published *Pinax theatri botanici* in 1596, which was the first to use this convention for naming of species. His criteria for classification included natural relationships, or 'affinities', which in many cases were structural.

It was in the late 1600s that plant anatomy became refined into a modern science. Italian doctor and microscopist, Marcello Malpighi, was one of the two founders of plant anatomy. In 1671 he published his *Anatomia Plantarum*, the first major advance in plant physiogamy since Aristotle. The other founder was the British doctor Nehemiah Grew. He published *An Idea of a Philosophical History of Plants* in 1672 and *The Anatomy of Plants* in 1682. Grew is credited with the recognition of plant cells, although he called them 'vesicles' and 'bladders'. He correctly identified and described the sexual organs of plants (flowers) and their parts.

In the Eighteenth Century, Carl Linnaeus established taxonomy based on structure, and his early work was with plant anatomy. While the exact structural level which is to be considered to be scientifically valid for comparison and differentiation has changed with the growth of knowledge, the basic principles were established by Linnaeus. He published his master work, *Species Plantarum* in 1753.

In 1802, French botanist Charles-François Brisseau de Mirbel, published *Traité d'anatomie et de physiologie végétale* (*Treatise on Plant Anatomy and Physiology*) establishing the beginnings of the science of plant cytology.

In 1812, Johann Jacob Paul Moldenhawer published *Beyträge zur Anatomie der Pflanzen*, describing microscopic studies of plant tissues.

In 1813 a Swiss botanist, Augustin Pyrame de Candolle, published *Théorie élémentaire de la botanique*, in which he argued that plant anatomy, not physiology, ought to be the sole basis for plant classification. Using a scientific basis, he established structural criteria for defining and separating plant genera.

In 1830, Franz Meyen published *Phytotomie*, the first comprehensive review of plant anatomy.

In 1838 German botanist Matthias Jakob Schleiden, published *Contributions to Phytogenesis*, stating, "the lower plants all consist of one cell, while the higher plants are composed of (many) individual cells" thus confirming and continuing Mirbel's work.

A German-Polish botanist, Eduard Strasburger, described the mitotic process in plant cells and further demonstrated that new cell nuclei can only arise from the division of other pre-existing nuclei. His *Studien über Protoplasma* was published in 1876.

Gottlieb Haberlandt, a German botanist, studied plant physiology and classified plant tissue based upon function. On this basis, in 1884 he published *Physiologische Pflanzenanatomie* (*Physiological Plant Anatomy*) in which he described twelve types of tissue systems (absorptive, mechanical, photosynthetic, etc.).

British paleobotanists Dunkinfield Henry Scott and William Crawford Williamson described the structures of fossilized plants at the end of the Nineteenth Century. Scott's *Studies in Fossil Botany* was published in 1900.

Following Charles Darwin's *Origin of Species* a Canadian botanist, Edward Charles Jeffrey, who was studying the comparative anatomy and phylogeny of different vascular plant groups, applied the theory to plants using the form and structure of plants to establish a number of evolutionary lines. He published his *The Anatomy of Woody Plants* in 1917.

The growth of comparative plant anatomy was spearheaded by British botanist Agnes Arber. She published *Water Plants: A Study of Aquatic Angiosperms* in 1920, *Monocotyledons: A Morphological Study* in 1925, and *The Gramineae: A Study of Cereal, Bamboo and Grass* in 1934.

Following World War II, Katherine Esau published, *Plant Anatomy* (1953), which became the definitive textbook on plant structure in North American universities and elsewhere, it was still in print as of 2006. She followed up with her *Anatomy of seed plants* in 1960.

REFERENCES

1. Raven, P. H.; Evert, R. F. and Eichhorn, S. E. (2005) Biology of Plants (7th edition) W. H. Freeman, New York, page 9, ISBN 0-7167-1007-2
2. Hagemann, Wolfgang (1992). "The Relationship of Anatomy to Morphology in Plants: A New Theoretical Perspective". International Journal of Plant Sciences. 153 (3(2)): S38–S48. doi:10.1086/297062. JSTOR 2995526.
3. Evert, Ray Franklin and Esau, Katherine (2006) Esau's Plant anatomy: meristems, cells, and tissues of the plant body - their structure, function and development Wiley, Hoboken, New Jersey, page xv, ISBN 0-471-73843-3
4. See e.g. Craig, Richard & Vassilyev, Andrey. "Plant Anatomy". McGraw-Hill. Archived from the original on 24 July 2010.
5. Bolam, J. (1973). "The botanical works of Nehemiah Grew, FRS (1641-1712)". Notes and Records of the Royal Society of London. 27 (2): 219–231. doi:10.1098/rsnr.1973.0017. JSTOR 530999.
6. Thomas, Hanshaw H. (1960). "Agnes Arber, 1879–1960". Biographical Memoirs of Fellows of the Royal Society. 6: 1–11. doi:10.1098/rsbm.1960.0021. JSTOR 769330.
7. Chaffey, N. (2006). "(Book Review) Esau's Plant Anatomy, Meristems, Cells, and Tissues of the Plant Body: their Structure, Function, and Development. 3rd edn.". Annals of Botany. 99 (4): 785–786. doi:10.1093/aob/mcm015.

CHAPTER 9

Classification System in Biology

Chapter Contents

- **Introduction**
- **Summary**

Biology is, quite simply, the scientific exploration and study of life. At the highest level, it includes categories based on the type of organism studied: zoology, botany, and microbiology. Each field has contributed to humanity in numerous ways such as improvements in agriculture, greater understanding of livestock and ecological systems, and the study of diseases. Modern biological studies largely center on the concepts of cell theory, evolution, gene theory, and homeostasis.

INTRODUCTION

In biology, kingdom (latin: regnum, pl. regna) is the second highest taxonomic rank below domain. Kingdoms are divided into smaller groups called phyla. Traditionally, textbooks from the United States used a system of six kingdoms (Animalia, Plantae, Fungi, Protista, Archaea/Archaeabacteria, and Bacteria/Eubacteria) while textbooks in Great Britain, India, Australia, Latin America and other countries used five kingdoms (Animalia, Plantae, Fungi, Protista and Monera). Some recent classifications based on modern cladistics have explicitly abandoned the term "kingdom", noting that the traditional kingdoms are not monophyletic, i.e., do not consist of all the descendants of a common ancestor.

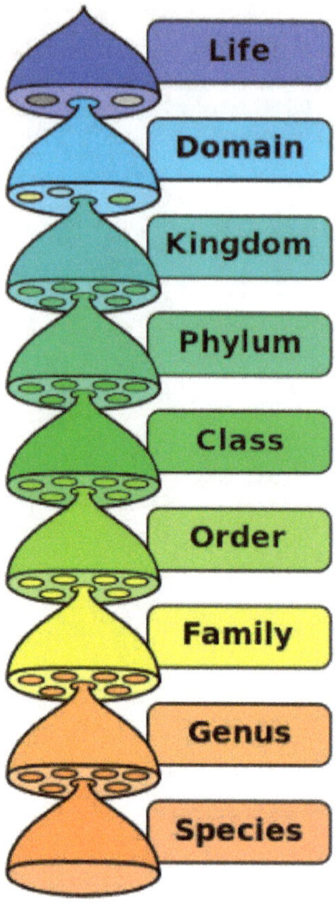

The hierarchy of biological classification's eight major taxonomic ranks. A domain contains one or more kingdoms. Intermediate minor rankings are not shown.

Definition and Associated Terms

When Carl Linnaeus introduced the rank-based system of nomenclature into biology in 1735, the highest rank was given the name "kingdom" and was followed by four other main or principal ranks: the class, order, genus and species. Later two further main ranks were introduced, making the sequence kingdom, phylum or division, class, order, family, genus and species. In the 1960s a rank was introduced above kingdom, namely domain (or empire), so that kingdom is no longer the highest rank.

Prefixes can be added so subkingdom and infrakingdom are the two ranks immediately below kingdom. Superkingdom may be considered as an equivalent of domain or empire or as an independent rank between kingdom and domain or subdomain. In some classification systems the additional rank branch (Latin: *ramus*) can be inserted between subkingdom and infrakingdom (e.g. Protostomia and Deuterostomia in the classification of Cavalier-Smith).

Modern View

The Three Domains of Life

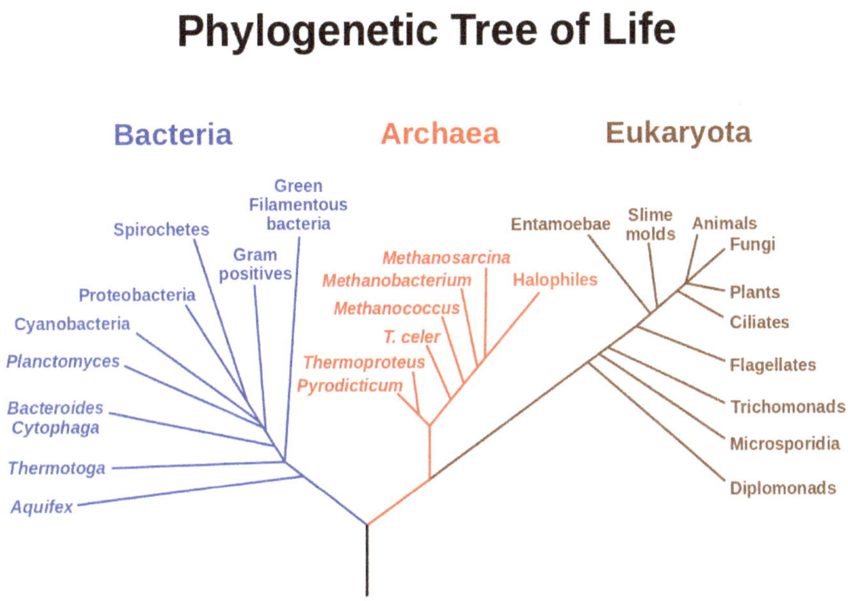

A phylogenetic tree based on rRNA data showing Woese's three-domain system. All smaller branches can be considered kingdoms.

From around the mid-1970s onwards, there was an increasing emphasis on comparisons of genes on the molecular level (initially ribosomal Ribonucleic acid genes) as the primary factor in classification; genetic similarity was stressed over outward appearances and behavior. Taxonomic ranks, including kingdoms, were to be groups of organisms with a common ancestor, whether monophyletic (*all* descendants of a common ancestor) or paraphyletic (*only some* descendants of a common ancestor). Based on such RNA studies, Carl Woese thought life could be divided into three large divisions and referred to them as the "three primary kingdom" model or "urkingdom model. In 1990, the name domain was proposed for the highest rank.This term represents a synonym for the category of dominion (lat. dominium), introduced by Moore in 1974.Unlike Moore, Whoese et al. (1990) did not suggest a Latin term for this category, which represents a further argument supporting the accurately introduced term dominion. Woese divided the prokaryotes (previously classified as the Kingdom Monera) into two groups, called Eubacteria and Archaebacteria, stressing that there was as much genetic difference between these two groups as between either of them and all eukaryotes.

According to genetic data, although eukaryote groups such as plants, fungi, and animals may look different, they are more closely related to each other than they are to either the Eubacteria or Archaea. It was also found that the eukaryotes are more closely related to the Archaea than they are to the Eubacteria. Although the primacy of the Eubacteria-Archaea divide has been questioned, it has been upheld by subsequent research.There is no consensus on how many kingdoms exist in the classification scheme proposed by Woese.

Kingdoms of the Eukaryota

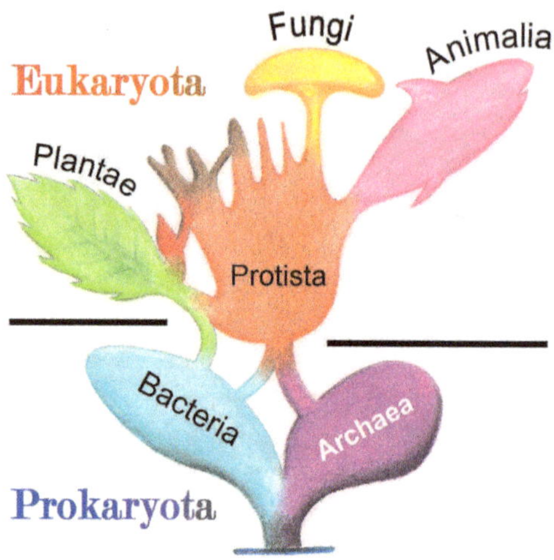

Phylogenetic and symbiogenetic tree of living organisms, showing the origins of eukaryotes & prokaryotes

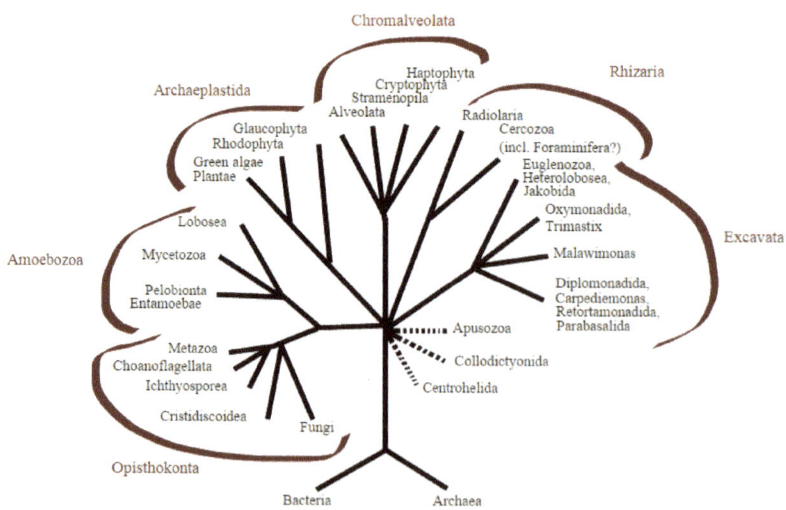

One hypothesis of eukaryotic relationships, modified from Simpson and Roger (2004).

In 2004, a review article by Simpson and Roger noted that the Protista were "a grab-bag for all eukaryotes that are not animals, plants or fungi". They held that only monophyletic groups should be accepted as formal ranks in a classification and that - while this approach had been impractical previously (necessitating "literally dozens of eukaryotic 'kingdoms'") - it had now become possible to divide the eukaryotes into "just a few major groups that are probably all monophyletic".

On this basis, the diagram opposite (redrawn from their article) showed the real ‹kingdoms' (their quotation marks) of the eukaryotes. A classification which followed this approach was produced in 2005 for the International Society of Protistologists, by a committee which worked in collaboration with specialists from many societies. It divided the eukaryotes into the same six supergroups. The published classification deliberately did not use formal taxonomic ranks, including that of kingdom.

In this system the multicellular animals (Metazoa) are descended from the same ancestor as both the unicellular choanoflagellates and the fungi which form the Opisthokonta. Plants are thought to be more distantly related to animals and fungi.

However, in the same year as the International Society of Protistologists' classification was published (2005), doubts were being expressed as to whether some of these supergroups were monophyletic, particularly the Chromalveolata, and a review in 2006 noted the lack of evidence for several of the supposed six supergroups.

As of 2010, there is widespread agreement that the Rhizaria belong with the Stramenopiles and the Alveolata, in a clade dubbed the SAR supergroup, so that Rhizaria is not one of the main eukaryote groups. Beyond this, there does not appear to be a consensus. Rogozin *et al.* in 2009 noted that "The deep phylogeny of eukaryotes is an extremely difficult and controversial problem." As of December 2010, there appears to be a consensus that the 2005 six supergroup model does not reflect the true phylogeny of the eukaryotes and hence how they should be classified, although there is no agreement as to the model which should replace it.

Historical Development

The classification of living things into animals and plants is an ancient one. Aristotle (384–322 BC) classified animal species in his *History of Animals*, while his pupil Theophrastus (c. 371–c. 287 BC) wrote a parallel work, the *Historia Plantarum*, on plants.

Carl Linnaeus (1707–1778) laid the foundations for modern biological nomenclature, now regulated by the Nomenclature Codes, in 1735. He distinguished two kingdoms of living things: *Regnum Animale* ('animal kingdom') and *Regnum Vegetabile* ('vegetable kingdom', for plants). Linnaeus also included minerals in his classification system, placing them in a third kingdom, *Regnum Lapideum*.

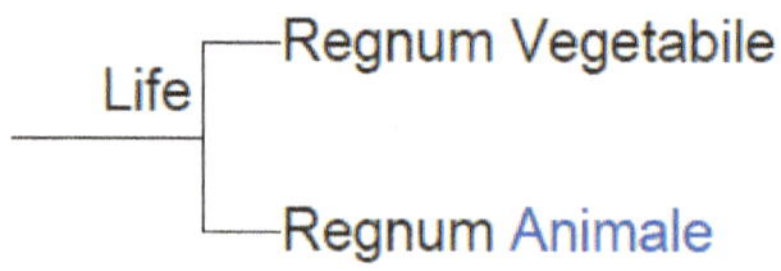

In 1674, Antonie van Leeuwenhoek, often called the "father of microscopy", sent the Royal Society of London a copy of his first observations of microscopic single-celled organisms. Until then, the existence of such microscopic organisms was entirely unknown. Despite this, Linnaeus did not include any microscopic creatures in his original taxonomy.

At first, microscopic organisms were classified within the animal and plant kingdoms. However, by the mid-19th century, it had become clear to many that "the existing dichotomy of the plant and animal kingdoms rapidly blurred at its boundaries and outmoded". In 1866, Ernst Haeckel proposed a third kingdom of life, the Protista, for "neutral organisms" which were neither animal nor plant. Haeckel revised the content of this kingdom a number of times before settling on a division based on whether organisms were unicellular (Protista) or multicellular (animals and plants).

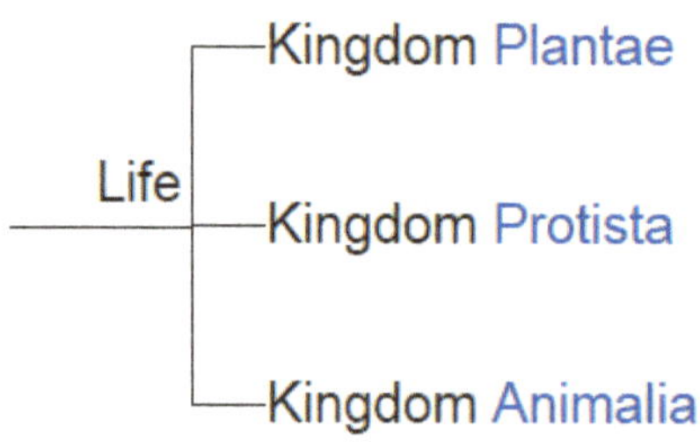

The development of the electron microscope revealed important distinctions between those unicellular organisms whose cells do not have a distinct nucleus (prokaryotes) and those unicellular and multicellular organisms whose cells do have a distinct nucleus (eukaryotes). In 1938, Herbert F. Copeland proposed a four-kingdom classification, elevating the protist classes of bacteria (Monera) and blue-green algae (Phycochromacea) to phyla in the novel Kingdom Monera.

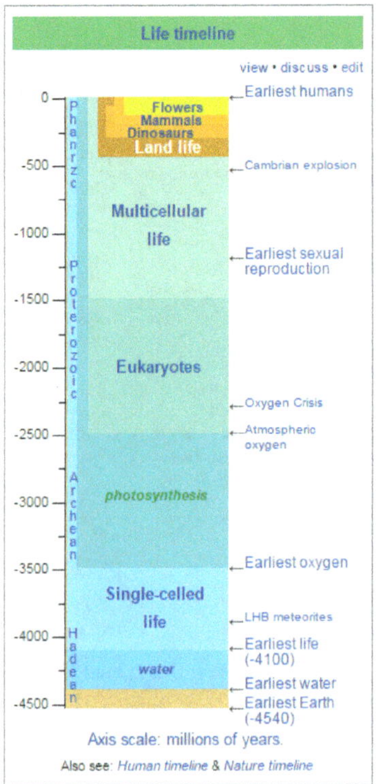

The importance of the distinction between prokaryotes and eukaryotes gradually became apparent. In the 1960s, Stanier and van Niel popularised Édouard Chatton's much earlier proposal to recognise this division in a formal classification. This required the creation, for the first time, of a rank above kingdom, a *superkingdom* or *empire*, later called a *domain*.

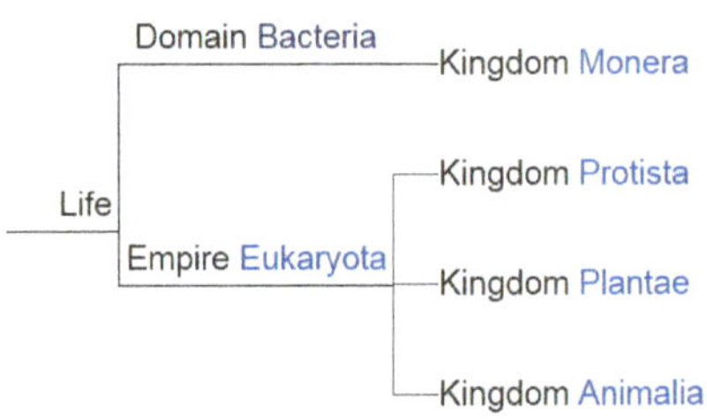

The differences between fungi and other organisms regarded as plants had long been recognised by some; Haeckel had moved the fungi out of Plantae into Protista after his original classification, but was largely ignored in this separation by scientists of his time. Robert Whittaker recognized an additional kingdom for the Fungi. The resulting five-kingdom system, proposed in 1969 by Whittaker, has become a popular standard and with some refinement is still used in many works and forms the basis for new multi-kingdom systems. It is based mainly upon differences in nutrition; his Plantae were mostly multicellular autotrophs, his Animalia multicellular heterotrophs, and his Fungi multicellular saprotrophs. The remaining two kingdoms, Protista and Monera, included unicellular and simple cellular colonies. The five kingdom system may be combined with the two empire system:

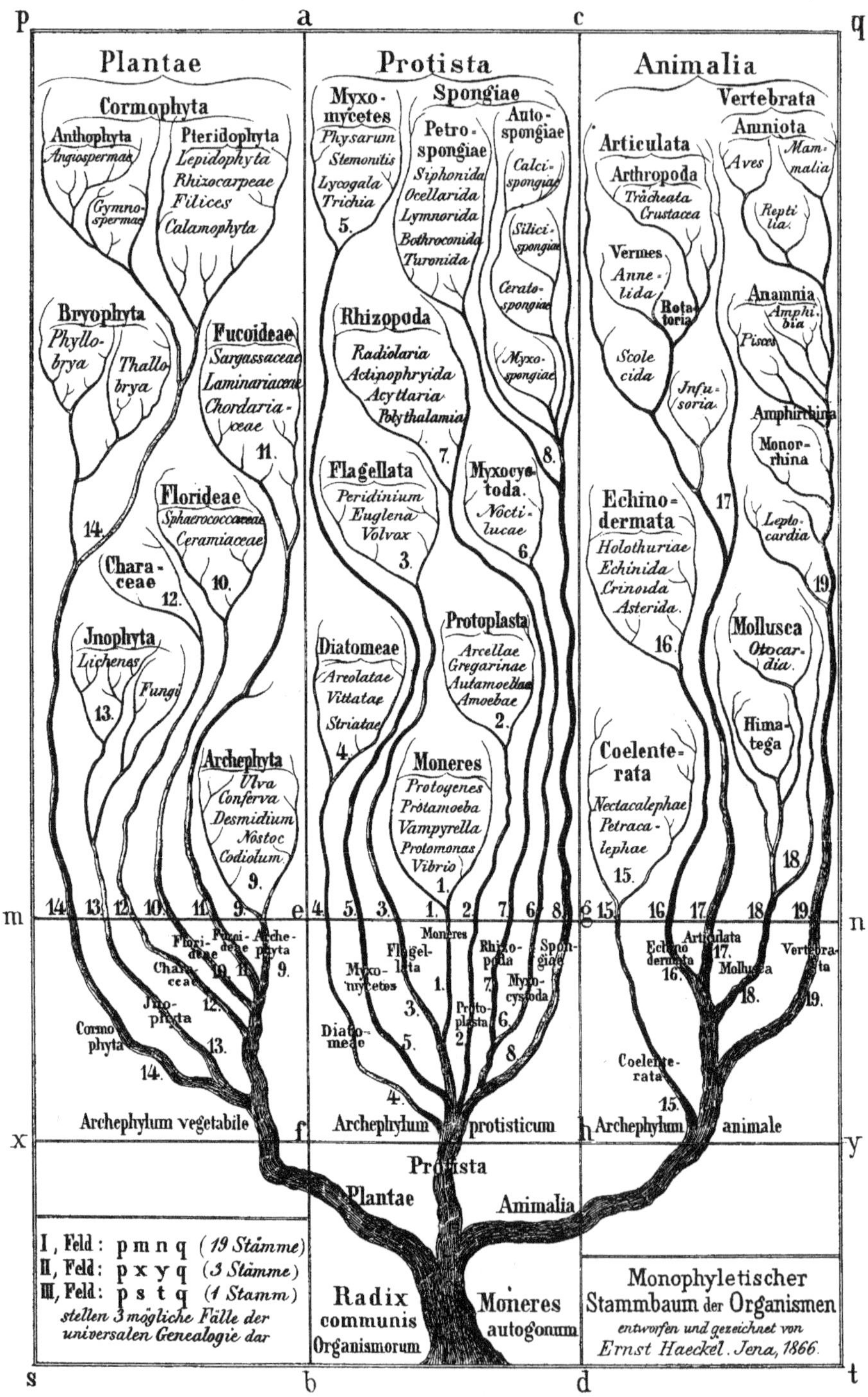

Haeckel's original (1866) conception of the three kingdoms of life, including the new kingdom Protista. Notice the inclusion of the cyanobacterium Nostoc with plants.

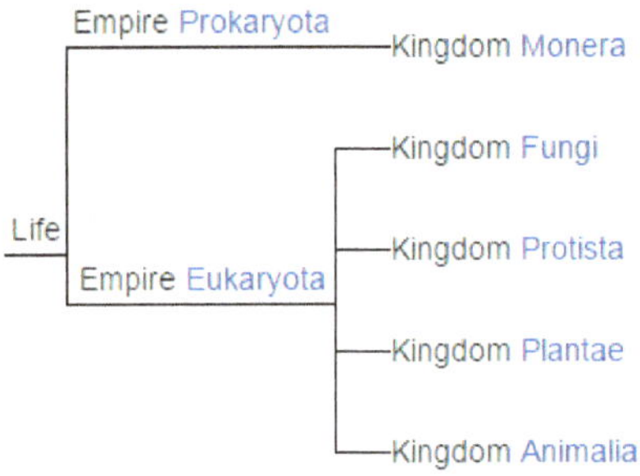

In the Whittaker system, Plantae included some algae. In other systems (e.g., Margulis system), Plantae included just the land plants (Embryophyta).

Despite the development from two kingdoms to five among most scientists, some authors as late as 1975 continued to employ a traditional two-kingdom system of animals and plants, dividing the plant kingdom into Subkingdoms Prokaryota (bacteria and cyanophytes), Mycota (fungi and supposed relatives), and Chlorota (algae and land plants).

Cavalier-Smith's Systems

Eight Kingdoms

Thomas Cavalier-Smith thought at first, as it was nearly consensually admitted at that time, that the difference between *eubacteria* and *archaebacteria* was so great (particularly considering the genetic distance of ribosomal genes) that they needed to be separated into two different kingdoms, hence splitting the empire Bacteria into two kingdoms. He then divided Eubacteria into two subkingdoms: Negibacteria (Gram negative bacteria) and Posibacteria (Gram positive bacteria). Technological advances in electron microscopy allowed the separation of the Chromista from the Plantae kingdom. Indeed, the chloroplast of the chromists is located in the lumen of the endoplasmic reticulum instead of in the cytosol. Moreover, only chromists contain chlorophyll c. Since then, many non-photosynthetic phyla of protists, thought to have secondarily lost their chloroplasts, were integrated into the kingdom Chromista.

Finally, some protists lacking mitochondria were discovered. As mitochondria were known to be the result of the endosymbiosis of a proteobacterium, it was thought that these amitochondriate eukaryotes were primitively so, marking an important step in eukaryogenesis. As a result, these amitochondriate protists were separated from the protist kingdom, giving rise to the, at the same time, superkingdom and kingdom Archezoa. This was known as the Archezoa hypothesis. This superkingdom was opposed to the Metakaryota superkingdom, grouping together the five other eukaryotic kingdoms (Animalia, Protozoa, Fungi, Plantae and Chromista).

Six Kingdoms

In 1998, Cavalier-Smith published a six-kingdom model, which has been revised in subsequent papers. The version published in 2009 is shown below.(Compared to the version he published in 2004, the alveolates and the rhizarians have been moved from Kingdom Protozoa to Kingdom Chromista.) Cavalier-Smith no longer accepts the importance of the fundamental eubacteria–archaebacteria divide put forward by Woese and others and supported by recent research. His Kingdom Bacteria includes Archaebacteria as a phylum of the subkingdom Unibacteria which comprises only one other phylum: the Posibacteria. The two subkingdoms Unibacteria and Negibacteria of kingdom Bacteria (sole kingdom of empire Prokaryota) are opposed according to their membrane topologies. The bimembranous-unimembranous transition is thought to be far more fundamental than the long branch of genetic distance of Archaebacteria, viewed as having no particular biological significance. Cavalier-Smith does not accept the requirement for taxa to be monophyletic (holophyletic in his terminology) to be valid. He defines Prokaryota, Bacteria, Negibacteria, Unibacteria and Posibacteria as valid paraphyletic (therefore monophyletic in the sense he uses this term) taxa, marking important innovations of biological significance (in regard of the concept of biological niche).

In the same way, his paraphyletic kingdom Protozoa includes the ancestors of Animalia, Fungi, Plantae and Chromista. The advances of phylogenetic studies allowed Cavalier-Smith to realize that all the phyla thought to be archezoans (i.e. primitively amitochondriate eukaryotes) had in fact secondarily lost their mitochondria, most of the time by transforming them into new organelles: hydrogenosomes. This means that all living eukaryotes are in fact metakaryotes, according to the significance of the term given by Cavalier-Smith. Some of the members of the defunct kingdom Archezoa, like the phylum Microsporidia, were reclassified into kingdom Fungi. Others were reclassified in kingdom Protozoa like Metamonada which is now part of infrakingdom Excavata.

The diagram below does not represent an evolutionary tree.

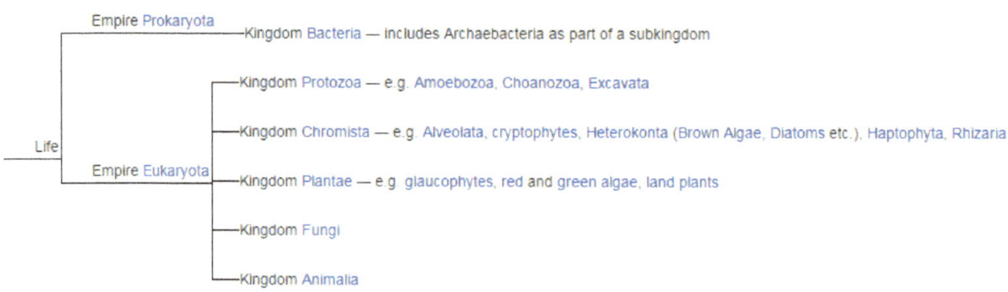

Seven Kingdoms

Cavalier-Smith and his collaborators revised the classification in 2015, and published it in PLOS ONE. In this scheme they reintroduced the division of prokaryotes into two kingdoms, Bacteria (=Eubacteria) and Archaea (=Archaebacteria). This is based on the consensus in the Taxonomic Outline of Bacteria and Archaea (TOBA) and the Catalogue of Life.

SUMMARY

A summary of the different kinds of proposed classification schemes presented in this article is summarized in the table below.

Linnaeus 1735	Haeckel 1866	Chatton 1925	Copeland 1938	Whittaker 1969	Woese *et al.* 1977	Woese *et al.* 1990	Cavalier-Smith 1993	Cavalier-Smith 1998	Ruggiero *et al.* 2015
2 kingdoms	3 kingdoms	2 empires	4 kingdoms	5 kingdoms	6 kingdoms	3 domains	8 kingdoms	6 kingdoms	7 kingdoms
(not treated)	Protista	Prokaryota	Monera	Monera	Eubacteria	Bacteria	Eubacteria	Bacteria	Bacteria
					Archaebacteria	Archaea	Archaebacteria		Archaea
		Eukaryota	Protista	Protista	Protista	Eucarya	Archezoa	Protozoa	Protozoa
							Protozoa		
							Chromista	Chromista	Chromista
Vegetabilia	Plantae		Plantae	Plantae	Plantae		Plantae	Plantae	Plantae
				Fungi	Fungi		Fungi	Fungi	Fungi
Animalia	Animalia		Animalia	Animalia	Animalia		Animalia	Animalia	Animalia

The kingdom-level classification of life is still widely employed as a useful way of grouping organisms, notwithstanding some problems with this approach:

- Kingdoms such as Bacteria represent grades rather than clades, and so are rejected by phylogenetic classification systems.
- The most recent research does not support the classification of the eukaryotes into any of the standard systems. As of April 2010, no set of kingdoms is sufficiently supported by research to attain widespread acceptance. In 2009, Andrew Roger and Alastair Simpson emphasized the need for diligence in analyzing new discoveries: "With the current pace of change in our understanding of the eukaryote tree of life, we should proceed with caution."

Viruses

There is ongoing debate as to whether viruses, obligate intracellular parasites that lack metabolism and are not capable of replication outside of a host, can be included in the tree of life. A principal reason for inclusion comes from the discovery of unusually large and complex viruses, such as Mimivirus, that possess typical cellular genes.

REFERENCES

1. C. Linnaeus (1735). "Systemae Naturae, sive regna tria naturae, systematics proposita per classes, ordines, genera & species".
2. McNeill, J.; et al., eds. (2006), International Code of Botanical Nomenclature (Vienna Code) Adopted by the Seventeenth International Botanical Congress, Vienna, Austria, July 2005 (electronic ed.), Vienna: International Association for Plant Taxonomy, archived from the original on 6 October 2012, retrieved 2011-02-20, article 3.1
3. Cavalier-Smith, T. (1998), "A revised six-kingdom system of life", Biological Reviews, 73 (3): 203–66, doi:10.1111/j.1469-185X.1998.tb00030.x, PMID 9809012

4. Balch, W.E.; Magrum, L.J.; Fox, G.E.; Wolfe, C.R. & Woese, C.R. (August 1977), "An ancient divergence among the bacteria", J. Mol. Evol., 9 (4): 305–11, doi:10.1007/BF01796092, PMID 408502
5. Woese, C.R.; Kandler, O.; Wheelis, M.L. (1990). "Towards a natural system of organisms: proposal for the domains Archaea, Bacteria, and Eucarya". Proceedings of the National Academy of Sciences of the United States of America. 87 (12): 4576–9. Bibcode:1990PNAS...87.4576W. doi:10.1073/pnas.87.12.4576. PMC 54159 . PMID 2112744.
6. Moore R.T. (1974). "Proposal for the recognition of super ranks" (PDF). Taxon. 23 (4): 650–652.
7. Luketa S. (2012). "New views on the megaclassification of life" (PDF). Protistology. 7 (4): 218–237.
8. Dagan, T.; Roettger, M.; Bryant & Martin, W. (2010), "Genome Networks Root the Tree of Life between Prokaryotic Domains", Genome Biology and Evolution, 2:: 379–92, doi:10.1093/gbe/evq025
9. Simpson, Alastair G.B. & Roger, Andrew J. (2004), "The real 'kingdoms' of eukaryotes", Current Biology, 14 (17): R693–6, doi:10.1016/j.cub.2004.08.038, PMID 15341755
10. Adl, Sina M.; et al. (2005), "The New Higher Level Classification of Eukaryotes with Emphasis on the Taxonomy of Protists", Journal of Eukaryotic Microbiology, 52 (5): 399–451, doi:10.1111/j.1550-7408.2005.00053.x, PMID 16248873
11. Harper, J.T.; Waanders, E. & Keeling, P. J. (2005), "On the monophyly of chromalveolates using a six-protein phylogeny of eukaryotes" (PDF), Nt. J. System. Evol. Microbiol., 55 (Pt 1): 487–496, doi:10.1099/ijs.0.63216-0, PMID 15653923
12. Parfrey, Laura W.; Barbero, Erika; Lasser, Elyse; Dunthorn, Micah; Bhattacharya, Debashish; Patterson, David J. & Katz, Laura A. (2006), "Evaluating Support for the Current Classification of Eukaryotic Diversity", PLoS Genet., 2 (12): e220, doi:10.1371/journal.pgen.0020220, PMC 1713255 , PMID 17194223
13. Burki et al. 2007, p. 4
14. Cavalier-Smith, Thomas (2009), "Kingdoms Protozoa and Chromista and the eozoan root of the eukaryotic tree", Biology Letters, 6 (3): 342–5, doi:10.1098/rsbl.2009.0948, PMC 2880060 , PMID 20031978
15. Burki, Fabien; Shalchian-Tabrizi, Kamran; Minge, Marianne; Skjæveland, Åsmund; Nikolaev, Sergey I.; Jakobsen, Kjetill S. & Pawlowski, Jan (2007), Butler, Geraldine, ed., Phylogenomics Reshuffles the Eukaryotic Supergroups, PLoS ONE, 2 (8): e790, Bibcode:2007PLoSO...2..790B, doi:10.1371/journal.pone.0000790, PMC 1949142 , PMID 17726520
16. Burki, Fabien; Shalchian-Tabrizi, Kamran & Pawlowski, Jan (2008), Phylogenomics reveals a new ‹megagroup' including most photosynthetic eukaryotes, Biology Letters, 4 (4): 366–369, doi:10.1098/rsbl.2008.0224, PMC 2610160 , PMID 18522922.

17. Burki, F.; Inagaki, Y.; Brate, J.; Archibald, J. M.; Keeling, P. J.; Cavalier-Smith, T.; Sakaguchi, M.; Hashimoto, T.; et al. (2009), Large-Scale Phylogenomic Analyses Reveal That Two Enigmatic Protist Lineages, Telonemia and Centroheliozoa, Are Related to Photosynthetic Chromalveolates, Genome Biology and Evolution, 1: 231–8, doi:10.1093/gbe/evp022, PMC 2817417 , PMID 20333193

18. Hackett, J.D.; Yoon, H.S.; Li, S.; Reyes-Prieto, A.; Rummele, S.E. & Bhattacharya, D. (2007), Phylogenomic analysis supports the monophyly of cryptophytes and haptophytes and the association of Rhizaria with chromalveolates, Mol. Biol. Evol., 24 (8): 1702–13, doi:10.1093/molbev/msm089, PMID 17488740

19. Rogozin, I.B.; Basu, M.K.; Csürös, M. & Koonin, E.V. (2009), Analysis of Rare Genomic Changes Does Not Support the Unikont–Bikont Phylogeny and Suggests Cyanobacterial Symbiosis as the Point of Primary Radiation of Eukaryotes, Genome Biology and Evolution, 1: 99–113, doi:10.1093/gbe/evp011, PMC 2817406 , PMID 20333181

20. Kim, E.; Graham, L.E. & Redfield, Rosemary Jeanne (2008), Redfield, Rosemary Jeanne, ed., EEF2 analysis challenges the monophyly of Archaeplastida and Chromalveolata, PLoS ONE, 3 (7): e2621, Bibcode:2008PLoSO...3.2621K, doi:10.1371/journal.pone.0002621, PMC 2440802 , PMID 18612431

21. Singer, Charles J. (1931), A short history of biology, a general introduction to the study of living things, Oxford: Clarendon Press, OCLC 1197036

22. Scamardella, Joseph M. (1999), Not plants or animals: a brief history of the origin of Kingdoms Protozoa, Protista and Protoctista, International Microbiology, 2 (4): 207–16, PMID 10943416

23. Stanier, R.Y. & Van Neil, C.B. (1962), The concept of a bacterium, Archiv für Mikrobiologie, 42 (1): 17–35, doi:10.1007/BF00425185, PMID 13916221

24. Whittaker, R.H. (January 1969), New concepts of kingdoms or organisms. Evolutionary relations are better represented by new classifications than by the traditional two kingdoms, Science, 163 (3863): 150–60, Bibcode:1969Sci...163..150W, doi:10.1126/science.163.3863.150, PMID 5762760

25. Palmer, E. Laurence; Fowler, Seymour H (January 1975), Fieldbook of Natural History: Second Edition, McGraw-Hill, ISBN 978-0070484252

26. Cavalier-Smith, Thomas (March 26, 1987). Eucaryotes with no mitochondria. Nature. 326 (6111): 332–333. doi:10.1038/326332a0.

27. Cavalier-Smith, T. (2004), Only six kingdoms of life (PDF), Proc. R. Soc. Lond. B, 271 (1545): 1251–62, doi:10.1098/rspb.2004.2705, PMC 1691724 , PMID 15306349, retrieved 2010-04-29

28. Ruggiero, Michael A.; Gordon, Dennis P.; Orrell, Thomas M.; Bailly, Nicolas; Bourgoin, Thierry; Brusca, Richard C.; Cavalier-Smith, Thomas; Guiry, Michael D.; Kirk, Paul M.; Thuesen, Erik V. (2015). A higher level classification of all living organisms. PLOS ONE. 10 (4): e0119248. doi:10.1371/journal.pone.0119248. PMC 4418965 . PMID 25923521.

29. Linnaeus, C. (1735). Systemae Naturae, sive regna tria naturae, systematics proposita per classes, ordines, genera & species.

30. Haeckel, E. (1866). Generelle Morphologie der Organismen. Reimer, Berlin.
31. Chatton, É. (1925). Pansporella perplexa. Réflexions sur la biologie et la phylogénie des protozoaires. Annales des Sciences Naturelles - Zoologie et Biologie Animale. 10-VII: 1–84.
32. Chatton, É. (1937). Titres et Travaux Scientifiques (1906–1937). Sette, Sottano, Italy.
33. Copeland, H. (1938). The kingdoms of organisms. Quarterly Review of Biology. 13: 383–420. doi:10.1086/394568.
34. Copeland, H. F. (1956). The Classification of Lower Organisms. Palo Alto: Pacific Books. p. 6. doi:10.5962/bhl.title.4474.
35. Whittaker, R. H. (January 1969). New concepts of kingdoms of organisms. Science. 163 (3863): 150–60. Bibcode:1969Sci...163..150W. doi:10.1126/science.163.3863.150. PMID 5762760.
36. Woese, C. R.; Balch, W. E.; Magrum, L. J.; Fox, G. E.; Wolfe, R. S. (August 1977). An ancient divergence among the bacteria. Journal of Molecular Evolution. 9 (4): 305–311. doi:10.1007/BF01796092. PMID 408502.
37. Woese, C. R.; Fox, G. E. (November 1977). Phylogenetic structure of the prokaryotic domain: the primary kingdoms. Proceedings of the National Academy of Sciences of the United States of America. 74 (11): 5088–90. Bibcode:1977PNAS...74.5088W. doi:10.1073/pnas.74.11.5088. PMC 432104 . PMID 270744.
38. Woese, C.; Kandler, O.; Wheelis, M. (1990). Towards a natural system of organisms: proposal for the domains Archaea, Bacteria, and Eucarya.. Proceedings of the National Academy of Sciences of the United States of America. 87 (12): 4576–9. Bibcode:1990PNAS...87.4576W. doi:10.1073/pnas.87.12.4576. PMC 54159 . PMID 2112744.
39. Cavalier-Smith, T. (1981). Eukaryote kingdoms: seven or nine?. Bio Systems. 14 (3–4): 461–481. doi:10.1016/0303-2647(81)90050-2. PMID 7337818.
40. Cavalier-Smith, T. (1992). Origins of secondary metabolism. Ciba Foundation symposium. 171: 64–80; discussion 80–7. PMID 1302186.
41. Cavalier-Smith, T. (1993). Kingdom protozoa and its 18 phyla. Microbiological reviews. 57 (4): 953–994. PMC 372943 . PMID 8302218.
42. Cavalier-Smith, T. (1998), A revised six-kingdom system of life, Biological Reviews, 73 (03): 203–66, doi:10.1111/j.1469-185X.1998.tb00030.x, PMID 9809012
43. Cavalier-Smith, T. (2004), Only six kingdoms of life (PDF), Proceedings of the Royal Society B: Biological Sciences, 271: 1251–62, doi:10.1098/rspb.2004.2705, PMC 1691724 , PMID 15306349, retrieved 2010-04-29
44. Cavalier-Smith T (June 2010). Kingdoms Protozoa and Chromista and the eozoan root of the eukaryotic tree. Biol. Lett. 6 (3): 342–5. doi:10.1098/rsbl.2009.0948. PMC 2880060 . PMID 20031978.

45. Ruggiero, Michael A.; Gordon, Dennis P.; Orrell, Thomas M.; Bailly, Nicolas; Bourgoin, Thierry; Brusca, Richard C.; Cavalier-Smith, Thomas; Guiry, Michael D.; Kirk, Paul M.; Thuesen, Erik V. (2015). A higher level classification of all living organisms. PLOS ONE. 10 (4): e0119248. doi:10.1371/journal.pone.0119248. PMC 4418965 . PMID 25923521.

46. Roger, A.J. & Simpson, A.G.B. (2009), Evolution: Revisiting the Root of the Eukaryote Tree, Current Biology, 19 (4): R165–7, doi:10.1016/j.cub.2008.12.032, PMID 19243692

47. Moreira, David; Purificación López-García (2009). Ten reasons to exclude viruses from the tree of life. Nature Reviews Microbiology. 7 (4): 306–311. doi:10.1038/nrmicro2108. PMID 19270719.

48. Hegde, Nagendra; Maddur, Mohan S.; Kaveri, Srini V. & Bayry, Jagadeesh (2009). Reasons to include viruses in the tree of life. Nature Reviews Microbiology. 7 (8): 615. doi:10.1038/nrmicro2108-c1.

49. Luketa, Stefan (2012). New views on the megaclassification of life (PDF). Protistology. 7 (4): 218–237.

50. Raoult, Didier; Audic, Stéphane; Robert, Catherine; Abergel, Chantal; Renesto, Patricia; Ogata, Hiroyuki; La Scola, Bernard; Suzan, Marie; Claverie, Jean-Michel (2004). The 1.2-Megabase Genome Sequence of Mimivirus. Science. 306 (5700): 1344–1350. Bibcode:2004Sci...306.1344R. doi:10.1126/science.1101485. PMID 15486256.

CHAPTER 10

Synthetic Biology

Chapter Contents

- **Introduction**
- **Summary**

Synthetic biology is form of biological study that involves creating a holistic understanding of a concept by combining several areas of research. More recently, the term synthetic biology has come to represent a form of research that combines science with engineering. In this way, new biological systems and functions can be designed and built.

INTRODUCTION

Synthetic biology is an interdisciplinary branch of biology and engineering. The subject combines various disciplines from within these domains, such as biotechnology, evolutionary biology, genetic engineering, molecular biology, molecular engineering, systems biology, biophysics, and computer engineering.

Descriptions of synthetic biology depend on how the user approaches it, as a biologist or as an engineer. Originally seen as a subset of biology, in recent years the role of electrical and chemical engineering has become more important. For example, one description designates synthetic biology as "an emerging discipline that uses engineering principles to design and assemble biological components". Another description, by Jan Staman Director of the Rathenau Institute in The Hague in 2006, portrayed it as "a new emerging scientific field where ICT, biotechnology and nanotechnology meet and strengthen each other".

The definition of synthetic biology is debated, not only among natural scientists and engineers but also in the human sciences, arts and politics. One popular definition is "designing and constructing biological modules, biological systems, and biological machines for useful purposes." However, the functional aspects of this definition are rooted in molecular biology and biotechnology.

As usage of the term has expanded to many interdisciplinary fields, synthetic biology has been recently defined as the artificial design and engineering of biological systems and living organisms for purposes of improving applications for industry or biological research.

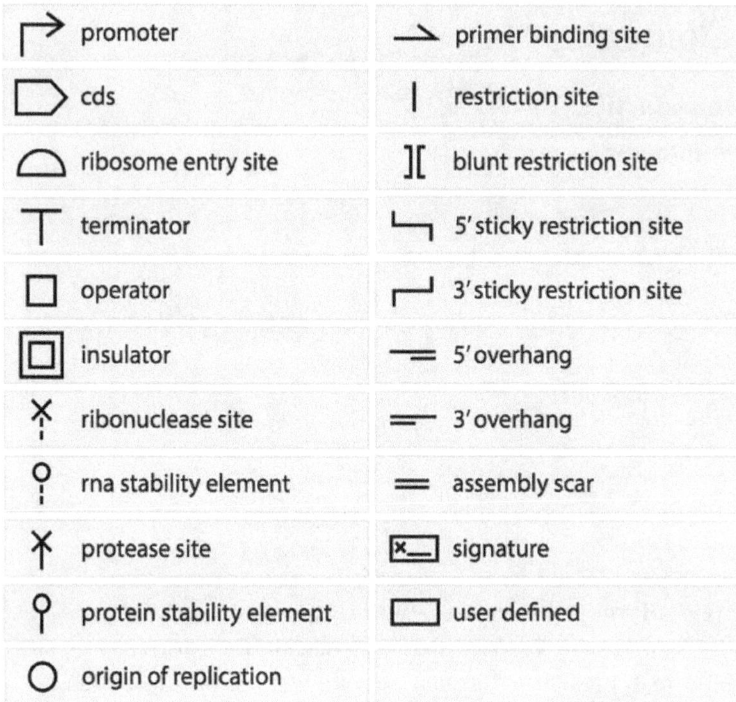

Synthetic Biology Open Language (SBOL) standard visual symbols for use with BioBricks Standard

History

The first identifiable use of the term "synthetic biology" was in Stéphane Leduc's publication of *Théorie physico-chimique de la vie et générations spontanées*(1910) and his *La Biologie Synthétique* (1912).

Sixty-four years later, in 1974, the term gained its more modern usage when Polish geneticist Wacław Szybalski used the term "synthetic biology", writing:

Let me now comment on the question what next. Up to now we are working on the descriptive phase of molecular biology. ... But the real challenge will start when we enter the synthetic phase of research in our field. We will then devise new control elements and add these new modules to the existing genomes or build up wholly new genomes. This would be a field with an unlimited expansion potential and hardly any limitations to building new better control circuits or ..... finally other synthetic organisms, like a new better mouse. ... I am not concerned that we will run out of exciting and novel ideas, ... in the synthetic biology, in general.

When in 1978 Arber, Nathans and Smith won the Nobel Prize in Physiology or Medicine for the discovery of restriction enzymes, Wacław Szybalski wrote in an editorial comment in the journal *Gene*:

The work on restriction nucleases not only permits us easily to construct recombinant DNA molecules and to analyze individual genes, but also has led us into the new era of synthetic biology where not only existing genes are described and analyzed but also new gene arrangements can be constructed and evaluated.

A notable advance in synthetic biology occurred in 2000, when two articles in Nature by Michael B. Elowitz and Stanislas Leibler discussed the creation of synthetic biological circuit devices of a genetic toggle switch and a biological clock by combining genes within E. coli cells.

Perspectives

Engineering

Engineers view biology as a *technology* – the *systems biotechnology* or *systems biological engineering*. Synthetic biology includes the broad redefinition and expansion of biotechnology, with the ultimate goals of being able to design and build engineered biological systems that process information, manipulate chemicals, fabricate materials and structures, produce energy, provide food, and maintain and enhance human health and our environment.

Studies in synthetic biology can be subdivided into broad classifications according to the approach they take to the problem at hand: standardization of biological parts, biomolecular engineering, genome engineering. Biomolecular engineering includes approaches which aim to create a toolkit of functional units that can be introduced to present new technological functions in living cells. Genetic engineering includes approaches to construct synthetic chromosomes for whole or minimal organisms. Biomolecular design refers to the general idea of de novo design and additive combination of biomolecular components. Each of these approaches share a similar task: to develop a more synthetic entity at a higher level of complexity by inventively manipulating a simpler part at the preceding level.

Re-Writing

Re-writers are synthetic biologists interested in testing the irreducibility of biological systems. Due to the complexity of natural biological systems, it would be simpler to re-build the natural systems of interest from the ground up; In order to provide engineered surrogates that are easier to comprehend, control and manipulate.Re-writers draw inspiration from refactoring, a process sometimes used to improve computer software.

Key Enabling Technologies

Several key enabling technologies are critical to the growth of synthetic biology. The key concepts include standardization of biological parts and hierarchical abstraction to permit using those parts in increasingly complex synthetic systems. Achieving this is greatly aided by basic technologies of reading and writing of DNA (sequencing and fabrication), which are improving in price/performance exponentially (Kurzweil 2001). Measurements under a variety of conditions are needed for accurate modeling and computer-aided-design (CAD).

Standardized DNA Parts

The most used[22–23] standardized DNA parts are BioBrick plasmids invented by Tom Knight in 2003. Biobricks are stored at the Registry of Standard Biological Parts in Cambridge, Massachusetts and the BioBrick standard has been used by thousands of students worldwide in the international Genetically Engineered Machine (iGEM) competition.[22–23]

DNA Synthesis

In 2007 it was reported that several companies were offering the synthesis of genetic sequences up to 2000 bp long, for a price of about $1 per base pair and a turnaround time of less than two weeks. Oligonucleotides harvested from a photolithographic or inkjet manufactured DNA chip combined with DNA mismatch error-correction allows inexpensive large-scale changes of codons in genetic systems to improve gene expression or incorporate novel amino-acids (see George M. Church's and Anthony Forster's synthetic cell projects.) This favors a synthesis-from-scratch approach.

Additionally, the CRISPR/Cas system has emerged as a promising technique for gene editing. It was hailed by The Washington Post as "the most important innovation in the synthetic biology space in nearly 30 years." While other methods take months or years to edit gene sequences, CRISPR speeds that time up to weeks. However, due to its ease of use and accessibility, it has raised a number of ethical concerns, especially surrounding its use in the biohacking space.

DNA Sequencing

DNA sequencing is determining the order of the nucleotide bases in a molecule of DNA. Synthetic biologists make use of DNA sequencing in their work in several ways. First, large-scale genome sequencing efforts continue to provide a wealth of information on naturally occurring organisms. This information provides a rich substrate from which synthetic biologists can construct parts and devices. Second, synthetic biologists use sequencing to verify that they fabricated their engineered system as intended. Third, fast, cheap, and reliable sequencing can also facilitate rapid detection and identification of synthetic systems and organisms.

Modular Protein Assembly

While DNA is most important for information storage, a large fraction of the cell's activities are carried out by proteins. Therefore, it is important to have tools to send proteins to specific regions of the cell and to link different proteins together, as desired. Ideally the interaction strength between protein partners should be tunable between a lifetime of seconds (desirable for dynamic signaling events) up to an irreversible interaction (desirable when building devices stable over days or resilient to harsh conditions). Interactions such as coiled coils, SH3 domain-peptide binding or SpyTag/SpyCatcher have helped to give such control. In addition it is important to be able to regulate protein-protein interactions in cells, such as with light (using Light-oxygen-voltage-sensing domains) or cell-permeable small molecules by Chemically induced dimerization.

Modeling

Models inform the design of engineered biological systems by allowing synthetic biologists to better predict system behavior prior to fabrication. Synthetic biology will benefit from better models of how biological molecules bind substrates and catalyze reactions, how DNA encodes the information needed to specify the cell and how multi-component integrated systems behave. Recently, multiscale models of gene regulatory networks have been developed that focus on synthetic biology applications. Simulations have been used that model all biomolecular interactions in transcription, translation, regulation, and induction of gene regulatory networks, guiding the design of synthetic systems.

Research Examples

Synthetic DNA

Driven by dramatic decreases in costs of making oligonucleotides ("oligos"), the sizes of DNA constructions from oligos have increased to the genomic level. For example, in 2000, researchers at Washington University reported synthesis of the 9.6 kbp (kilo base pair) Hepatitis C virus genome from chemically synthesized 60 to 80-mers. In 2002 researchers at SUNY Stony Brook succeeded in synthesizing the 7741 base poliovirus genome from its published sequence, producing the second synthetic genome. This took about two years of work. In 2003 the 5386 bp genome of the bacteriophage Phi X 174 was assembled in about two weeks. In 2006, the same team, at the J. Craig Venter Institute, had constructed and patented a synthetic genome of a novel minimal bacterium, *Mycoplasma laboratorium* and were working on getting it functioning in a living cell.

Synthetic Transcription Factors

Studies have also been performed on the components of the DNA translation mechanism. One desire of scientists creating synthetic biological circuits is to be able to control the translation of synthetic DNA in prokaryotes and eukaryotes. One study tested the adjustability of synthetic transcription factors (sTFs) in areas of transcription output and cooperative ability among multiple transcription factor complexes. Researchers were able to mutate zinc fingers, the DNA specific component of sTFs, to decrease their affinity for DNA, and thus decreasing the amount of translation. They were also able to use the zinc fingers as components of complex forming sTFs, which are the eukaryotic translation mechanisms.

Applications

Synthetic Life

One important topic in synthetic biology is *synthetic life*, that is, artificial life created *in vitro* from biomolecules and their component materials. Synthetic life experiments attempt to either probe the origins of life, study some of the properties of life, or more ambitiously to recreate life from non-living (abiotic) components. Synthetic biology attempts to create new biological molecules and even novel living species capable of carrying out a range of important medical and industrial functions, from manufacturing pharmaceuticals to detoxifying polluted land and water. In medicine, it offers prospects of using designer biological parts as a starting point for an entirely new class of therapies and diagnostic tools.

In the area of synthetic biology, a living artificial cell has been defined as a completely synthetically-made cell that can capture energy, maintain ion gradients, contain macromolecules as well as store information and have the ability to mutate. Nobody has been able to create such an artificial cell.

The first living organism with ‹artificial› DNA was produced by scientists at the Scripps Research Institute as *E. coli* was engineered to replicate an expanded genetic alphabet.

A completely synthetic genome was produced by Craig Venter, and his team introduced it to genomically emptied bacterial host cells, and allowed the host cells to grow and replicate.

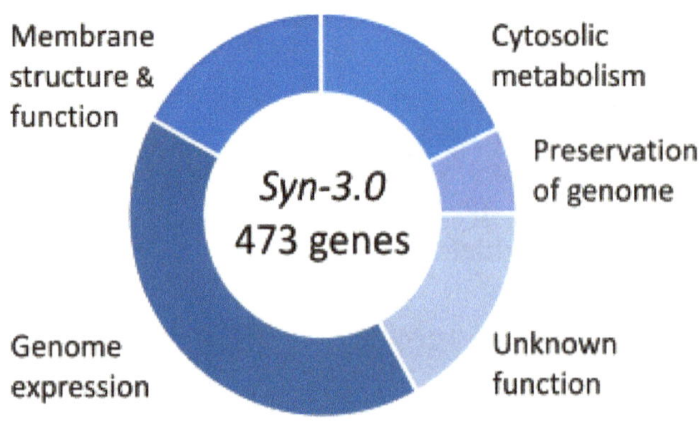

Gene functions in the minimal genome of the synthetic organism, Syn 3.

Cell Transformation

Currently, entire organisms are not being created from scratch, but instead living cells are being transformed with inserts of new DNA. There are several ways of constructing synthetic DNA components and even entire synthetic genomes, but once the desired genetic code is obtained, it is integrated into a living cell that is expected to manifest the desired new capabilities or phenotypes while growing and thriving. Cell transformation is used to create biological circuits, which can be manipulated to yield desired outputs.

Information Storage

Scientists can encode vast amounts of digital information onto a single strand of synthetic DNA. In 2012, George M. Church encoded one of his books about synthetic biology in DNA. The 5.3 Mb of data from the book is more than 1000 times greater than the previous largest amount of information to be stored in synthesized DNA. A similar project had encoded the complete sonnets of William Shakespeare in DNA.

Synthetic Genetic Pathways

Traditional metabolic engineering has been bolstered by the introduction of combinations of foreign genes and optimization by directed evolution. Perhaps the best known application of synthetic biology to date is engineering *E. coli* and yeast for commercial production of a precursor of the antimalarial drug, Artemisinin, by the laboratory of Jay Keasling

Unnatural Nucleotides

Many technologies have been developed for incorporating unnatural nucleotides and amino acids into nucleic acids and proteins, both in vitro and in vivo. For example, in May 2014, researchers announced that they had successfully introduced two new artificial nucleotides into bacterial DNA. By including individual artificial nucleotides in the culture media, were able to exchange the bacteria 24 times; they did not generate mRNA or proteins able to use the artificial nucleotides.

Unnatural Amino Acids

Another common topic of investigation is expansion of the normal repertoire of 20 amino acids. Excluding stop codons, there are 61 codons, but only 20 amino acids are coded generally in all organisms. Certain codons are engineered to code for alternative amino acids including: nonstandard amino acids such as O-methyl tyrosine; or exogenous amino acids such as 4-fluorophenylalanine. Typically, these projects make use of re-coded nonsense suppressor tRNA-Aminoacyl tRNA synthetase pairs from other organisms, though in most cases substantial engineering is still required.

Reduced Amino-Acid Libraries

Instead of expanding the genetic code, other researchers have investigated the structure and function of proteins by reducing the normal set of 20 amino acids. Limited protein sequence libraries are made by generating proteins where certain groups of amino acids may be substituted with a single amino acid. For instance, several non-polar amino acids within a protein can all be replaced with a single non-polar amino acid. One project demonstrated that an engineered version of Chorismate mutase still had catalytic activity when only 9 amino acids were used.

Designed Proteins

The Top7 protein was one of the first proteins designed for a fold that had never been seen before in nature

While there are methods to engineer natural proteins such as by directed evolution, there are also projects to design novel protein structures that match or improve on the functionality of existing proteins. One group generated a helix bundle that was capable of binding oxygen with similar properties as hemoglobin, yet did not bind carbon monoxide. A similar protein structure was generated to support a variety of oxidoreductase activities. Another group generated a family of G-protein coupled receptors which could be activated by the inert small molecule clozapine-N-oxide but insensitive to the native ligand, acetylcholine.

Biosensors

A biosensor refers to an engineered organism, usually a bacterium, which is capable of reporting some ambient phenomenon such as the presence of heavy metals or toxins. In this capability, a very widely used system is the Lux operon of *Aliivibrio fischeri*. The Lux operon codes for an enzyme which is the source bacterial bioluminescence, and can be placed after a respondent promoter to express the luminescence genes in response to a specific environmental stimulus. One such sensor created in Oak Ridge National Laboratory, and named "critter on a chip", consisted of a bioluminescent bacterial coating on a photosensitive computer chip to detect certain petroleum pollutants. When the bacteria sense the pollutant, they begin to luminesce.

Materials Production

By integrating synthetic biology approaches with materials sciences, it would be possible to envision cells as microscopic molecular foundries to produce materials with properties that can be genetically encoded. Recent advances towards this include the re-engineering of curli fibers, the amyloid component of extracellular material of biofilms, as a platform for programmable nanomaterial. These nanofibers have been genetically constructed for specific functions, including: adhesion to substrates; nanoparticle templating; and protein immobilization.

Industrial Enzymes

Researchers and companies utilizing synthetic biology aim to synthesize enzymes with high activity, to produce products with optimal yields and effectiveness. These synthesized enzymes aim to improve products such as detergents and lactose-free dairy products, as well as make them more cost effective.

The improvements of metabolic engineering by synthetic biology is an example of a biotechnological technique utilized in industry to discover pharmaceuticals and fermentative chemicals. Synthetic biology may investigate modular pathway systems in biochemical production and increase yields of metabolic production. Artificial enzymatic activity and subsequent effects on metabolic reaction rates and yields may develop "efficient new strategies for improving cellular properties . . . for industrially important biochemical production."

Space Exploration

Synthetic biology raised NASA's interest as it could help to produce resources for astronauts from a restricted portfolio of compounds sent from Earth. On Mars, in particular, synthetic biology could also lead to production processes based on local resources, making it a powerful tool in the development of manned outposts with minimal dependence on Earth.

REFERENCES

1. IEEE Xplore Abstract - Intellectual Property and the Commons in Synthetic Biology: Strategies to Facilitate an Emerging Tec...
2. W97 binnenwerk-8 - Rathenau Constructing Life 2006.pdf
3. "Synthetic biology: promises and perils of modern biotechnology". Marsilius Academy Heidelberg – Summer school. Heidelberg University. Retrieved 2014-09-11.
4. Nakano, Tadashi (2013). Molecular Communication. Cambridge. ISBN 978-1-107-02308-6.
5. "Registry of Standard Biological Parts". Retrieved 2014-09-11.
6. "Synthetic-biology firms shift focus". Nature. 505 (7485): 598. doi:10.1038/505598a. Retrieved 2014-09-11.

7. Osbourn, Anne E.; O'Maille, Paul E.; Rosser, Susan J.; Lindsey, Keith (2012-11-01). "Synthetic biology". New Phytologist. 196 (3): 671–677. doi:10.1111/j.1469-8137.2012.04374.x. ISSN 1469-8137.
8. Théorie physico-chimique de la vie et générations spontanées, S. Leduc, 1910
9. Leduc, Stéphane (1912). Poinat, A., ed. La biologie synthétique, étude de biophysique.
10. Wacław Szybalski, In Vivo and in Vitro Initiation of Transcription, Page 405. In: A. Kohn and A. Shatkay (Eds.), Control of Gene Expression, pp. 23–4, and Discussion pp. 404–5 (Szybalski›s concept of Synthetic Biology), 411–2, 415–7. New York: Plenum Press, 1974
11. Szybalski, W; Skalka, A (November 1978). Nobel prizes and restriction enzymes. Gene. 4 (3): 181–2. doi:10.1016/0378-1119(78)90016-1. PMID 744485.
12. Elowitz, Michael B.; Leibler, Stanislas (January 2000). A synthetic oscillatory network of transcriptional regulators. Nature. 403 (6767): 335–338. doi:10.1038/35002125. PMID 10659856.
13. Collins, James J.; Gardner, Timothy S.; Cantor, Charles R. (January 2000). Construction of a genetic toggle switch in Escherichia coli. Nature. 403 (6767): 339–342. doi:10.1038/35002131. PMID 10659857.
14. Zeng, Jie (Bangzhe). On the concept of systems bio-engineering. Coomunication on Transgenic Animals, June 1994, CAS, PRC. 6.
15. Chopra, Paras; Akhil Kamma. Engineering life through Synthetic Biology. In Silico Biology. 6. Retrieved 2008-06-09.
16. Channon, Kevin; Bromley, Elizabeth HC; Woolfson, Derek N (August 2008). Synthetic Biology through Biomolecular Design and Engineering. Current Opinion in Structural Biology. 18 (4): 491–8. doi:10.1016/j.sbi.2008.06.006. PMID 18644449.
17. Stone, M (2006). Life Redesigned to Suit the Engineering Crowd (PDF). Microbe. 1 (12): 566–570. Archived from the original (PDF) on 2014-08-29.
18. Group, Bio FAB; Baker D; Church G; Collins J; Endy D; Jacobson J; Keasling J; Modrich P; Smolke C; Weiss R (June 2006). Engineering life: building a fab for biology. Scientific American. 294 (6): 44–51. doi:10.1038/scientificamerican0606-44. PMID 16711359.
19. Synthetic Biology – A Primer. World Scientific. 2012. ISBN 978-1-84816-863-3.
20. Tom Knight (2003). Idempotent Vector Design for Standard Assembly of Biobricks. Retrieved 2014-09-26.
21. Pollack, Andrew (2007-09-12). How Do You Like Your Genes? Biofabs Take Orders. The New York Times. ISSN 0362-4331. Retrieved 2007-12-28.
22. Forster, AC; Church GM (2006-08-22). Towards synthesis of a minimal cell. Mol Syst Biol. 2 (1): 45. doi:10.1038/msb4100090. PMC 1681520 . PMID 16924266.
23. Basulto, Dominic (November 4, 2015). Everything you need to know about why CRISPR is such a hot technology. Washington Post. Retrieved 5 December 2015.

24. Kahn, Jennifer (November 9, 2015). The Crispr Quandary. New York Times. Retrieved 5 December 2015.

25. Ledford, Heidi (June 3, 2015). CRISPR, the disruptor. Nature. Nature News. Retrieved 5 December 2015.

26. Higginbotham, Stacey (4 December 2015). Top VC Says Gene Editing Is Riskier Than Artificial Intelligence. Fortune. Retrieved 5 December 2015.

27. Rollie; et al. (2012). Designing biological systems: Systems Engineering meets Synthetic Biology. Chemical Engineering Science. 69 (1): 1–29. doi:10.1016/j.ces.2011.10.068.

28. Woolfson DN (2012). New currency for old rope: from coiled-coil assemblies to α-helical barrels. Curr Op Struct Biol. 22: 432. doi:10.1016/j.sbi.2012.03.002. PMID 22445228.

29. Dueber JE (2009). Synthetic protein scaffolds provide modular control over metabolic flux. Nat Biotech. 27: 753. doi:10.1038/nbt.1557. PMID 19648908.

30. Reddington SC (2015). Secrets of a covalent interaction for biomaterials and biotechnology: SpyTag and SpyCatcher. Curr Op Chem Biol. 29: 94. doi:10.1016/j.cbpa.2015.10.002. PMID 26517567.

31. Bayle JH (2006). Rapamycin analogs with differential binding specificity permit orthogonal control of protein activity. Chem Biol. 13: 99. doi:10.1016/j.chembiol.2005.10.017. PMID 16426976.

32. Kaznessis YN (2007). Models for synthetic biology. BMC Systems Biology. 1 (1): 47. doi:10.1186/1752-0509-1-47. PMC 2194732 . PMID 17986347.

33. Kosuri, S. & Church, G.M. (2014). Large-scale de novo DNA synthesis: technologies and applications. Nature Methods. 11 (5): 499–507. doi:10.1038/nmeth.2918.

34. Blight KJ; Kolykhalov AA; Rice CM (2000-12-08). Efficient initiation of HCV RNA replication in cell culture. Science. 290 (5498): 1972–4. doi:10.1126/science.290.5498.1972. PMID 11110665.

35. Couzin J (2002). Virology. Active poliovirus baked from scratch. Science. 297 (5579): 174–5. doi:10.1126/science.297.5579.174b. PMID 12114601.

36. Smith, Hamilton O.; Clyde A. Hutchison; Cynthia Pfannkoch; J. Craig Venter (2003-12-23). Generating a synthetic genome by whole genome assembly: {phi}X174 bacteriophage from synthetic oligonucleotides. Proc. Natl. Acad. Sci. U.S.A. 100 (26): 15440–5. doi:10.1073/pnas.2237126100. PMC 307586 . PMID 14657399.

37. Wade, Nicholas (2007-06-29). Scientists Transplant Genome of Bacteria. The New York Times. ISSN 0362-4331. Retrieved 2007-12-28.

38. Gibson, DG; Benders GA; Andrews-Pfannkoch C; Denisova EA; Baden-Tillson H; Zaveri J; Stockwell TB; Brownley A; Thomas DW; Algire MA; Merryman C; Young L; Noskov VN; Glass JI; Venter JC; Hutchison CA 3rd; Smith HO (2008-01-24). Complete chemical synthesis, assembly, and cloning of a Mycoplasma genitalium genome. Science. 319 (5867): 1215–20. doi:10.1126/

science.1151721. PMID 18218864.

39. Khalil, Ahmad S.; Lu, Timothy K.; Bashor, Caleb J.; Ramirez, Cherie L.; Pyenson, Nora C.; Joung, J. Keith; Collins, James J. (2012-03-08). A Synthetic Biology Framework for Programming Eukaryotic Transcription Functions. Cell. 150 (3): 647–658. doi:10.1016/j.cell.2012.05.045. ISSN 0092-8674. PMC 3653585 . PMID 22863014.
40. Khalil, Ahmad S.; Lu, Timothy K.; Bashor, Caleb J.; Ramirez, Cherie L.; Pyenson, Nora C.; Joung, J. Keith; Collins, James J. (2012-03-08). A Synthetic Biology Framework for Programming Eukaryotic Transcription Functions. Cell. 150 (3): 647–658. doi:10.1016/j.cell.2012.05.045. ISSN 0092-8674. PMC 3653585 . PMID 22863014.
41. Hutchison, Clyde A.; Chuang, Ray-Yuan; Noskov, Vladimir N.; Assad-Garcia, Nacyra; Deerinck, Thomas J.; Ellisman, Mark H.; Gill, John; Kannan, Krishna; Karas, Bogumil J. (2016-03-25). Design and synthesis of a minimal bacterial genome. Science. 351 (6280): aad6253. doi:10.1126/science.aad6253. ISSN 0036-8075. PMID 27013737.
42. Connor, Steve (1 December 2014). Major synthetic life breakthrough as scientists make the first artificial enzymes. The Independent. London. Retrieved 2015-08-06.
43. Deamer, D (July 2005). A giant step towards artificial life?. Trends in Biotechnology. 23 (7): 336–8. doi:10.1016/j.tibtech.2005.05.008. PMID 15935500.
44. Malyshev, Denis A.; Dhami, Kirandeep; Lavergne, Thomas; Chen, Tingjian; Dai, Nan; Foster, Jeremy M.; Corrêa, Ivan R.; Romesberg, Floyd E. A semi-synthetic organism with an expanded genetic alphabet. Nature. 509 (7500): 385–388. doi:10.1038/nature13314. PMC 4058825 . PMID 24805238.
45. Gibson, D. G.; Glass, J. I.; Lartigue, C.; Noskov, V. N.; Chuang, R.-Y.; Algire, M. A.; Benders, G. A.; Montague, M. G.; Ma, L.; Moodie, M. M.; Merryman, C.; Vashee, S.; Krishnakumar, R.; Assad-Garcia, N.; Andrews-Pfannkoch, C.; Denisova, E. A.; Young, L.; Qi, Z.-Q.; Segall-Shapiro, T. H.; Calvey, C. H.; Parmar, P. P.; Hutchison, C. A.; Smith, H. O.; Venter, J. C. (20 May 2010). Creation of a Bacterial Cell Controlled by a Chemically Synthesized Genome. Science. 329 (5987): 52–56. doi:10.1126/science.1190719. PMID 20488990.
46. Scientists Reach Milestone On Way To Artificial Life. 2010-05-20. Retrieved 2010-06-09.
47. Connor, Steve (28 March 2014). Eureka! Scientists unveil giant leap towards synthetic life. The Independent. Retrieved 2015-08-06.
48. Church, G.M.; et al. (2012). Next-Generation Digital Information Storage in DNA. Science. 337 (6102): 1628. doi:10.1126/science.1226355. PMID 22903519.
49. Huge amounts of data can be stored in DNA. Sky News. 23 January 2013. Retrieved 24 January 2013.
50. Westfall, P. J., Pitera, D. J., Lenihan, J. R., Eng, D., Woolard, F. X., Regentin, R., ... & Fickes, S. (2012). Production of amorphadiene in yeast, and its conversion to dihydroartemisinic acid, precursor to the antimalarial agent artemisinin. Proceedings of the National Academy of Sciences, 109(3), E111-E118.

51. Pollack, Andrew (May 7, 2014). Researchers Report Breakthrough in Creating Artificial Genetic Code. New York Times. Retrieved May 7, 2014.

52. Callaway, Ewen (May 7, 2014). First life with ‹alien› DNA. Nature. doi:10.1038/nature.2014.15179. Retrieved May 7, 2014.

53. Malyshev, Denis A.; Dhami, Kirandeep; Lavergne, Thomas; Chen, Tingjian; Dai, Nan; Foster, Jeremy M.; Corrêa, Ivan R.; Romesberg, Floyd E. (May 7, 2014). A semi-synthetic organism with an expanded genetic alphabet. Nature. 509 (7500): 385–388. doi:10.1038/nature13314. PMC 4058825 . PMID 24805238. Retrieved May 7, 2014.

54. Wang, Qian; Parrish, Angela R; Wang, Lei (2009). Expanding the genetic code for biological studies. Chemistry and Biology. 16 (3): 323–36. doi:10.1016/j.chembiol.2009.03.001. PMC 2696486 . PMID 19318213.

55. Davidson, AR; Lumb, KJ; Sauer, RT (1995). Cooperatively folded proteins in random sequence libraries. Nature Structural Biology. 2 (10): 856–864. doi:10.1038/nsb1095-856.

56. Kamtekar, S.; Schiffer, J.; Xiong, H.; Babik, J.; Hecht, M. (1993). Protein design by binary patterning of polar and nonpolar amino acids. Science. 262 (5140): 1680–1685. doi:10.1126/science.8259512. PMID 8259512.

57. Walter, K.U.; Vamvaca, K.; Hilvert, D. (2005). An active enzyme constructed from a 9-amino acid alphabet. The Journal of Biological Chemistry. 280 (45): 37742–6. doi:10.1074/jbc.M507210200. PMID 16144843.

58. Kuhlman, B; Dantas, G; Ireton, GC; Varani, G; Stoddard, BL; Baker, D (November 21, 2003). Design of a novel globular protein fold with atomic-level accuracy.. Science. 302 (5649): 1364–8. Bibcode:2003Sci...302.1364K. doi:10.1126/science.1089427. PMID 14631033.

59. Koder, RL; et al. (2009). Design and engineering of an O(2) transport protein. Nature. 458 (7236): 305–9. doi:10.1038/nature07841. PMC 3539743 . PMID 19295603.

60. Farid, TA; Kodali, G.; Solomon, LA; et al. (2013). Elementary tetrahelical protein design for diverse oxidoreductase functions. Nature Chemical Biology. 9 (12): 826–33. doi:10.1038/nchembio.1362. PMID 24121554.

61. Armbruster BN; Li X; Pausch MH; Herlitze S; Roth BL (2007). Evolving the lock to fit the key to create a family of G protein-coupled receptors potently activated by an inert ligand. PNAS USA. 104 (12): 5163–8. doi:10.1073/pnas.0700293104. PMC 1829280 . PMID 17360345.

62. Gibbs, W. Wayt (1997). Critters on a Chip. Scientific American. Retrieved 2 Mar 2009.

63. Nguyen, Peter; Botyanszki, Zsofia; Tay, Pei-Kun; Joshi, Neel (Sep 17, 2014). Programmable biofilm-based materials from engineered curli nanofibres. Nature Communications. 5: 4945. doi:10.1038/ncomms5945. PMID 25229329.

64. Synthetic Biology Applications. www.thermofisher.com. Retrieved 2015-11-12.

65. Liu, Yanfeng; Shin, Hyun-dong; Li, Jianghua; Liu, Long (2014-12-31). Toward

metabolic engineering in the context of system biology and synthetic biology: advances and prospects. Applied Microbiology and Biotechnology. 99 (3): 1109–1118. doi:10.1007/s00253-014-6298-y. ISSN 0175-7598.

66. Verseux, C.; Paulino-Lima, I.; Baque, M.; Billi, D.; Rothschild, L. (2016). Synthetic Biology for Space Exploration: Promises and Societal Implications. Ambivalences of Creating Life. Societal and Philosophical Dimensions of Synthetic Biology, Publisher: Springer-Verlag: 73–100. doi:10.1007/978-3-319-21088-9_4.
67. Menezes, A; Cumbers, J; Hogan, J; Arkin, A (2014). Towards synthetic biological approaches to resource utilization on space missions. Journal of the Royal Society, Interface. 12.
68. Montague, M; et al. (2012). The Role of Synthetic Biology for In Situ Resource Utilization (ISRU).. Astrobiology. 12 (12): 1135–1142. doi:10.1089/ast.2012.0829.
69. Bügl, H.; et al. (2007). DNA synthesis and biological security. Nature Biotechnology. 25 (6): 627–629. doi:10.1038/nbt0607-627.
70. Ethical Issues in Synthetic Biology: An Overview of the Debates (PDF).
71. Presidential Commission for the study of Bioethical Issues, December 2010 NEW DIRECTIONS The Ethics of Synthetic Biology and Emerging TechnologiesRetrieved 2012-04-14.
72. SYNBIOSAFE official site
73. Schmidt M; Ganguli-Mitra A; Torgersen H; Kelle A; Deplazes A; Biller-Andorno N (2009). A priority paper for the societal and ethical aspects of synthetic biology(PDF). Systems and Synthetic Biology. 3 (1–4): 3–7. doi:10.1007/s11693-009-9034-7. PMC 2759426 . PMID 19816794.
74. Schmidt M. Kelle A. Ganguli A, de Vriend H. (Eds.) 2009. Synthetic Biology. The Technoscience and its Societal Consequences. Springer Academic Publishing.
75. Kelle A (2009). Ensuring the security of synthetic biology—towards a 5P governance strategy (PDF). Systems and Synthetic Biology. 3 (1–4): 85–90. doi:10.1007/s11693-009-9041-8. PMC 2759433 . PMID 19816803.
76. Schmidt M (2008). Diffusion of synthetic biology: a challenge to biosafety(PDF). Systems and Synthetic Biology. 2 (1–2): 1–6. doi:10.1007/s11693-008-9018-z. PMC 2671588 . PMID 19003431.
77. COSY: Communicating Synthetic Biology
78. Kronberger, N; Holtz, P; Kerbe, W; Strasser, E; Wagner, W (2009). Communicating Synthetic Biology: from the lab via the media to the broader public (PDF). Systems and Synthetic Biology. 3 (1–4): 19–26. doi:10.1007/s11693-009-9031-x. PMC 2759424 . PMID 19816796.
79. Cserer A; Seiringer A (2009). Pictures of Synthetic Biology: A reflective discussion of the representation of Synthetic Biology (SB) in the German-language media and by SB experts (PDF). Systems and Synthetic Biology. 3 (1–4): 27–35. doi:10.1007/s11693-009-9038-3. PMC 2759430 . PMID 19816797.
80. COSY/SYNBIOSAFE Documentary

81. Report of IASB Technical solutions for biosecurity in synthetic biologyArchived July 19, 2011, at the Wayback Machine., Munich, 2008
82. Parens E., Johnston J., Moses J. Ethical Issues in Synthetic Biology. 2009.
83. NAS Symposium official site
84. Presidential Commission for the study of Bioethical Issues, December 2010 FAQ
85. Synthetic Biology F.A.Q.›s | Presidential Commission for the Study of Bioethical Issues
86. Erickson, Brent; Singh, Rina; Winters, Paul (2011-09-02). Synthetic Biology: Regulating Industry Uses of New Biotechnologies. Science. 333 (6047): 1254–1256. doi:10.1126/science.1211066. ISSN 0036-8075. PMID 21885775.
87. Katherine Xue for Harvard Magazine. September–October 2014 Synthetic Biology's New Menagerie
88. Yojana Sharma for Scidev.net March 15, 2012. NGOs call for international regulation of synthetic biology
89. The New Synthetic Biology: Who Gains? (2014-05-08), Richard C. Lewontin, New York Review of Books
90. Savulescu, Julian; Pugh, Jonathan; Douglas, Thomas; Gyngell, Christopher (2015-06-26). The moral imperative to continue gene editing research on human embryos. Protein & Cell. 6 (7): 476–479. doi:10.1007/s13238-015-0184-y. ISSN 1674-800X. PMC 4491050 . PMID 26113289.

Index

A

Arabidopsis thaliana 49, 62, 157

B

Biological scholarship 7
Biological Species Concept (BSC) 48
Biological studies 241

C

Cell theory 2, 13
Cognitive ecology 104, 128
Community ecology 97

D

Demographic history 178
deoxyribonucleic acid (DNA) 147
Dietary fiber 190
dietitians are registered (RD) 200
disability-adjusted life year (DALY) 229
Drosophila melanogaster 35, 155, 156, 157, 173

E

Ecological dynamics 91, 127
Environmental science 88
Evolutionary biology 257
Evolutionary processes 27
Evolutionary synthesis 167, 168, 169, 170, 179
Evolutionary theory 11, 16
Evolutionary tree 155

F

Fatty acids 186, 187, 189, 190, 192, 195, 209
Flower anatomy 237

G

Gene flow 37, 42, 48, 174, 175
Genetic code 263, 269
Genetic drift 41, 68, 173, 181
Genetic information 33, 42
Geneticists examine 139
Genetic processes 140
Geographical Distribution 12
Geographical isolation 49
Geographic space 109

H

Highly energetic chemistry 52
Human Genome 143, 159, 162, 166
Human Nutrition 185, 210
Human science 88, 108

I

intelligent design (ID) 57
International Society 245

L

last universal common ancestor (LUCA) 27
Leaf anatomy 237
licensed (LD) 200
Life processes 88
Living organisms 87, 95
Long Term Ecological Network (LTER) 101

M

Mathematical models 28
Mendelian-chromosome theory 15, 16, 18
Mendelian inheritance 168
Modern ecology 90, 118
Modern genetics 141
Molecular Biology 16, 17, 18, 23, 24, 25
Morphological evolution 173
Multicellular animals 245
Mycobacterium tuberculosis 220

N

National hospital morbidity database (NHMD) 229
Natural philosophy 3, 5, 9, 10
Natural selection 28, 34, 39, 40, 69, 78
Nervous systems 235
Neutral theory 41, 69
Nutritional education 194
Nutrition-learning nexus 194

O

Occupied Palestinian Territory (OPT) 202
Origin-fixation 170, 177

P

Philosophical History 238
Physical disorders 222
Plant anatomy 235, 236, 239
polymerase chain reaction (PCR) 21, 159
polymerase chain reaction (PCR). 107
Population genetics 168, 171, 172, 177
Predator-prey interactions 103
Punnett square 144, 145

Q

quality-adjusted life year (QALY) 229

R

Recombinant techniques 20
Recommended Dietary Allowances (RDA) 207
Recommended Dietary Allowances (RDAs) 210
repetitive stress injury (RSI) 231
Restriction enzymes 20
Root structure 237

S

Schematic relationship 157
Scientific analysis 88
Social ecological 105
Speciation events 49
Statistical Manual of Mental Disorders (DSM) 222
Swiss physician 238
Symbiogenetic tree 244
Synthetic biology 257, 259, 261, 262, 265, 266
Synthetic Biology Open Language (SBOL) 258
Synthetic life 262, 268

T

Taxonomic Outline of Bacteria and Archaea (TOBA) 250
Taxonomic rank 241
Traditional metabolic 263

U

Unicellular organisms 246

V

Vascular tissue 236

W

Wood anatomy 237

Y

years of potential life lost (YPLL) 229